Solomon Golomb's Course
on Undergraduate Combinatorics

Solomon W. Golomb • Andy Liu

Solomon Golomb's Course on Undergraduate Combinatorics

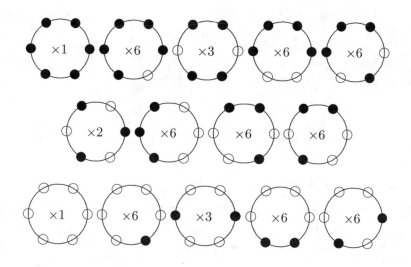

 Springer

Solomon W. Golomb (*Deceased*)
Los Angeles, CA, USA

Andy Liu
Mathematical and Statistical Sciences
University of Alberta
Edmonton, AB, Canada

ISBN 978-3-030-72230-2   ISBN 978-3-030-72228-9   (eBook)
https://doi.org/10.1007/978-3-030-72228-9

Mathematics Subject Classification (2020): 05-01

This Springer imprint is published by the registered company Springer Nature Switzerland AG
The registered company address is: Gewerbestrasse 11, 6330 Cham, Switzerland

Foreword : Golombinatorics
Beatrice A. Golomb, MD, PhD

Solomon W. Golomb

1932 − 2016

Solomon Golomb had a lifelong love affair with mathematics and numbers, and joyfully shared this love with those around him. I hope through this book, a piece of that joy can be passed to you.

For fun, Sol knew π to 1000 decimal places. (He quipped that this coarse approximation usually sufficed for his needs.) When the date of an event decades previously was mentioned, he would stun by relating the day of the week (promptly and correctly). His in-his-head calculations were the stuff of legend — and the lore began early.

His high school Biology teacher opined that two options for each of 24 chromosomes[1] meant that the number of combinations afforded by each parent was a staggering 2^{24}, "and you all know what that is," she quipped. Sol promptly responded, "Yes, 16,777,216," having figured the answer on the spot. She said, "Ha ha, Sol. The real answer is," and was forced, on looking down, to read the selfsame number. But for those who knew him well, such feats were an everyday matter.

A genius by any metric, Solomon Golomb contributed seminally to many areas of mathematics and beyond. He started college young, and graduated as a sophomore. He substantially developed the field of Shift Register Sequences, beginning in graduate school. He was part of the team at the Jet Propulsion Laboratory (JPL) that launched the first earth-orbiting US satellite, Explorer I — launched only 80 days from when they were told to start. (JPL was not yet under NASA. The launch succeeded the first time, with no test launch, and I gather their success rates much surpassed those of later programs — despite that their followers could build on their work.) While still twenty-something, he oversaw all 25 researchers in the Information Processing group at JPL. He authored (with two others) the first book referencing either "digital communications" or "space communications" in the title — developing codes that permitted communication in space — and on cell phones, and GPS. I have heard him called the Father of Modern Digital Communications (which makes me its sibling). Indeed, as physicist Stephen Wolfram noted in his obituary of Sol, he is responsible for the most used algorithm in history, with a conservative estimate of an octillion usages.

Sol liked to teach, and made USC his home, spurning other offers. He raked in the awards — National Academy of Sciences and of Engineering, the Sigma Xi Proctor Prize, the IEEE Hamming Medal, the Benjamin Franklin Medal. Sol's Shannon lecture, on receiving the Shannon Prize, was the only one Claude Shannon (the founder of Information Theory) ever attended — other than his own. And Sol received the National Medal of Science from President Obama in a lovely ceremony at the White House.

Given the heady achievements, one might think that for him, mathematics was not all fun and games. But that is just what it was. He developed the field of Polyominoes, aka Golombinoes (so I say — in that, the "b" is silent) — from which we have the popular game Tetris.

---

[1] Sol shared that at the time, it was believed that humans had 24 (pairs of) chromosomes rather than the 23 that we know today. I affirmed this via a 1930s science book, owned by a colleague of Sol, which indeed averred that humans had 24 (pairs of) chromosomes.

He had puzzle columns in three venues, the *Los Angeles Times* (for a time), the *Johns Hopkins Magazine* ("Golomb's Gambits" were so popular that they continue to be reprinted after his death); and the *IEEE Information Theory Society Newsletter*. When he received a "Magic Cube" from England (later introduced in the US as Rubik's Cube), he promptly unriddled its group theoretic properties. These were, as he published, the same as for quarks. He enjoyed inspecting a Cube someone had scrambled, then solving it behind his back. He worked with his former student Herb Taylor to train a high school student, Minh Thai, on algorithms they developed. Minh Thai went on to win the (televised) US and then World Rubik's Cube solving championships (events invested with more drama and intrigue than one might suppose).

When once asked how he justified spending his time on recreational mathematics, Sol replied that to him, all mathematics was recreational. Conversely, he opined that all mathematics was useful. Give me any beautiful mathematical result, he said, and he could find an application. Indeed, he felt that his career was defined by taking the abstract mathematics that his Professors at Harvard (where he got his PhD) had proudly asserted had no possible application, and applying them.

Sol had a gift for producing examples that made concepts self-evident. The unusual was so usual with him that I didn't retain these — but I recall one, since shared shortly before his death. I'd mentioned a case in which a quantity was different approaching the limit than at the limit. Sol said, "Oh, that's quite common" and gave (as usual) a lovely example. In the square of side s below, with steps from the upper left to bottom right, as the "steps" become successively smaller, the sum of the vertical and horizontal step segments remains the same: $2s$. (Attaching all the vertical segments end-to-end equals one side, all the horizontal segments another, however small the steps get.) But at the limit, it is the diagonal $\sqrt{2}s$.

He loved all mathematics, but discrete mathematics was Sol's métier. I hope you can come to have the same joy with combinatorics — Golombinatorics? — that he had.

(For a longer Biography, see Sol's A Career in Engineering, IEEE Transactions on Information Theory Volume 64(4) April 2018, pages 2825–2836, with Prologue starting on page 2805.)

Solomon W. Golomb

On December 21, 2012, President Obama announced the recipients of the National Medal of Science for 2011 . This is the country's highest distinction for contributions to scientific research. According to a news release from the Office of Science and Technology Policy, "The National Medal of Science honors individuals for pioneering scientific research in a range of fields, including physical, biological, mathematical, social, behavioral, and engineering sciences, that enhances our understanding of the world and leads to innovations and technologies that give the United States its global economic edge." The National Science Foundation administers the award, which was established by Congress in 1959. **Golomb** was among the twelve honorees. The medal was for his "pioneering work in shift register sequences that changed the course of communications from analog to digital and for numerous innovations in reliable and secure space, radar, cellular, wireless and spread-spectrum communications."

The *Notices* asked Alfred W. Hales to comment on the work of Golomb. Hales is a professor emeritus at the University of California Los Angeles and is currently with the Center for Communications Research of the Institute for Defense Analyses. Hales responded: "For the past fifty-plus years, Solomon Golomb has been a world leader in the development and application of mathematics for communications and coding theory, especially digital and space communications. In his remarkable career, first at the Jet Propulsion Laboratory and then at the University of Southern California (in electrical engineering and mathematics), his research contributions have ranged over a wide spectrum of science and technology. Perhaps he is best known for his mathematical analysis of shift register sequences in his classic book with that title and in numerous journal publications — such sequences are ubiquitous in radar, space communications, cryptography, cell phones, etc. Other noteworthy contributions in this direction include Golomb (entropy) coding, Golomb rulers, and the Golomb construction for Costas arrays. For all this work the IEEE had honored him with both its Hamming Medal and the Shannon Award in Information Theory. Golomb's Harvard Ph.D. was in analytic number theory, and he has extensive (and seminal) publications in number theory, combinatorics, algebra, and various other fields (including even molecular genetics).

Acknowledgement.

This article is excerpted and modified from **Elaine Kehoe**, "Mazur and Golomb Awarded National Medal of Science", which appeared in the *Notices of the American Mathematical Society*, Volume 60, June/July 2013, pages 758–759. ©2013 American Mathematical Society www.ams.org/notices/201306/rnoti-p758.pdf

"He is also a noted expert in mathematical game theory (polyominoes were his invention), and he continues to publish a number of puzzle columns in various journals. In addition to his many research contributions, Golomb has had a great influence on future generations through his generous and insightful mentoring of young people: his students, postdocs, and many others."

Solomon Golomb was born in Baltimore, Maryland. He received his B.A. from Johns Hopkins University and his Ph.D. in mathematics from Harvard University in 1957 under the direction of David Widder. He is currently professor of electrical engineering at the University of Southern California and is best known to the general public and fans of mathematical games as the inventor of polyominoes, the inspiration for the computer game Tetris. While he was working toward his Ph.D. he held a Fulbright Fellowship at the University of Oslo. He worked at the Jet Propulsion Laboratory at the California Institute of Technology from 1956 to 1963, where he was a senior research mathematician and later supervisor of the research group and assistant chief of the Telecommunications Research Section. In the latter position he played a key role in formulating the design of deep-space communications for the subsequent lunar and planetary explorations. He joined the University of Southern California in 1963.

In 1992 he received the medal of the U.S. National Security Agency for his research. He has also been the recipient of the Lomonosov Medal of the Russian Academy of Science and the Kapitsa Medal of the Russian Academy of Natural Sciences. His honors from the IEEE include the Shannon Award (1985) and the Richard W. Hamming Medal (2000) for his exceptional contributions to information sciences and systems. He was awarded the USC Presidential Medallion in 1985. He has been a major contributor in coding and information theory for more than forty years and is recognized for his ability to apply advanced mathematics to problems in digital communications. His books include *Shift Register Sequences* (Holden-Day, 1967) , *Polyominoes* (Princeton University Press, 2nd ed., 1996), and *Signal Design for Good Correlation* (coauthored with Guang Gong, Cambridge University Press, 2005). He is a member of the National Academy of Engineering, the Institute of Electrical and Electronics Engineers (IEEE), the American Association for the Advancement of Science(AAAS) , and a fellow of the American Mathematical Society (AMS) . He became a foreign member of the Russian Academy of Natural Science in 1994.

Preface

I am a devoted follower of the Polyomino cult, and **Solomon Golomb** was the acknowledged cult leader. I first met him at the Strens Conference in 1986, which celebrated the seventieth birthday of **Richard Guy**. It was only later that I learned of Sol's status as a world-class scientist. His passing in 2016 was the end to an era in mathematics, both serious and recreational.

Although I claimed no formal academic lineage with him, he had been most influential on my career. This book came into being after a chance conversation we had, about the teaching of combinatorics courses at the senior undergraduate level. Neither of us was completely satisfied with any of the texts we had experimented with over the years. This was not surprising as all people had their own pet peeves, and we were no exceptions. Thus we had both been using our own lecture notes to supplement whatever textbooks we were using at the time.

It occurred to Sol that it might not be a bad idea if we pooled our resources and came up with a package that would please at least the two of us. I was delighted to have an opportunity to collaborate with him, and the die was cast.

We agreed from the outset that books written by two different people do not work well. As there was no question who the senior author was, I took on the grunt work under Sol's guidance, based on the classroom notes of his and mine, including assignments and test papers. He carefully proofread what I wrote, and made insightful comments and suggestions. Often, he supplied me with very clever illustrative examples which are not found elsewhere, which add greatly to the value of this book.

We had a serious argument about the style of the book. I pointed out from my own teaching experience that many students were confounded by mathematics long before they got to any content that should confound them. Thus they were confounded, not by the content of mathematics, but by its language. While the Theorem-Proof style of writing was bread and butter for research mathematicians, it was not the most conducive to learning.

Sol asked for a specific example. I cited Pólya' Enumeration Theorem. It was a compact statement full of symbols. Typical problems involve counting the number of ways of painting various features of geometric objects. In my notes, I broke the problem down to three stages. First, there is the symmetry group of the geometric object. We then specify which features of the geometric object are to be painted. This brings in the cycle index. Finally, we specify the desired color patterns. This brings in the inventory function. It is quite long drawn out, but my own feeling is that it is well worth it.

So Chapter Five was the first to be finished. After reading it carefully, Sol finally came around to my proposal of using informal language. He made sure that this did not inherit any loss of rigor. He also liked the last section on combinatorial symmetry, which is not treated in most textbooks.

Distribution problems have been used as a framework around which combinatorics books are based. The eight cases cover whether objects are identical or distinct, whether containers are identical or distinct, and whether containers may not empty or otherwise. I raised the possibility of whether objects or containers are neither *all* identical nor *all* distinct. This case is handled in that last section on combinatorial symmetry.

Sol felt strongly that the Principle of Inclusion-Exclusion should be introduced as early as possible. He said that there are quite a lot of problems for which there is simply no better way to solve them than to invoke the Principle. In most textbooks, it is often treated as a special topic, about halfway into the book. We put it in our very first section. Some calculus texts are called *ET* for early treatment of transcendental functions. I pointed out to Sol that our book could be called *EIEIO*, for early treatment of inclusion-exclusion, induction and order.

We introduced the Pigeonhole Principle in a non-traditional way. We started with the Extremal Value Principle, which stated that in a finite non-empty set of real numbers, there exists a maximum and a minimum. It is the discrete version of the Maximality Principle in advanced mathematics. An immediate corollary is the Mean Value Principle, which states that in a finite non-empty set of real numbers, there exists one not below average and one not above average. Of course, the maximum cannot be below average and the minimum cannot be above average. The Pigeonhole Principle is now just a special case of the Mean Value Principle. Most textbooks move in the other direction, and called the Mean Value Principle the Generalized Pigeonhole Principle.

There are also two other cases for the Pigeonhole Principle. If there are more holes than pigeons, then at least one hole will be empty. What happens if the number of holes is equal to the number of pigeons? Not much can be said, but in a special case, we have what is known as a one-to-one correspondence, a most important concept in mathematics in general and in combinatorics in particular.

The last section of the long Chapter Zero deals with critical measures. This is not usually covered in other combinatorics textbooks. In problem-solving books, there is often a topic called *invariance*. In our terminology, an invariant is simply a critical measure which does not change, but there are other non-constant critical measures which nevertheless provide valuable reference frameworks amidst continual changes.

The material on permutation is usually scattered all over the place in most combinatorics textbooks. It appears early to provide the factorial formula for the binomial coefficient, but then only reappears much later, as applications of the Principle of Inclusion-Exclusion. We feel that permutation problems are in general more difficult than combination problems, just as non-Abelian algebra is more difficult than Abelian algebra. So we defer permutation to Chapter Three, and provide a unified treatment of the subject.

We have also taken exponential generating functions from their traditional spot and placed them in this chapter. Their similarity to ordinary generating functions is only superficial. The main technique for the latter is the resolution of partial fractions, whereas the main technique for the former is the shifting of indices in summations. Also, permutations with repetitions provide a clear motivation for putting $n!$ under x^n.

Our book contains a good collection of problems, many of which are drawn from the International Mathematics Tournament of the Towns, a famed contest organized in Moscow by the great **Nikolai Konstantinov**.

Sol did not feel that we need a bibliography section. It is almost impossible to keep up to date, and much information can be obtained from the internet. However, this jarring break from tradition may not be well-received. So I have included a list of other combinatorics textbooks and resources, mostly obtained from the internet. I would also mention that combinatorics books are filed under the call numbers QA 164 to 166 in the libraries.

I am deeply indebted to my friend and colleague **Sean Graves** for proof-reading and field-testing several drafts of the manuscript. I am grateful to all staff members of Springer Nature, in particular **Loretta Bartolini, Anne Comment, Jan Holland** and **Kathleen Moriaty** for their editorial guidance. I am also grateful to Project Manager **Savita Rockey Samuel**, Production Editor **Shobha Karappiah** and their Straive team from India for technical support.

Andy Liu,
Edmonton, 2021.

Introduction

The target audience of this book are students in undergraduate programs with mathematics as a major. Combinatorics courses used to be offered at the senior level, but nowadays some are offered at the introductory level. Our treatment is sufficiently deep, but at the same time accessible to all, and is particularly suited for self-study.

In this book, the student will learn about combinatorial structures. A combinatorial structure is one which has combinatorial properties. Combinatorial properties are those possessed by combinatorial structures. So formal definitions are not getting us anywhere.

We shall leave combinatorial structure as an undefined term. We learn about combinatorial structures by studying examples and their properties. Two important examples are sets and sequences.

A *set*, which is another undefined term, is loosely speaking a collection of objects, called its *elements*. For instance, "the bunch of you" is a very impolite way of saying "the set of readers of this book". Examples more closely related to mathematics are the set of the positive integers, the set of the positive real numbers, and so on.

By convention, if an element is listed more than once in a set, the repetition is ignored. However, if we want to keep track of the number of copies of the same object in a collection, we use the term *multi-set*. The number of copies of an element is called its *multiplicity*. A set is by definition a multi-set in which every element has multiplicity 1. On the other hand, a multi-set with at least one element which has multiplicity at least 2 is not a set.

A *sequence* is a listing of the elements of a multi-set. For example, we have the sequence $\{1,2,3,4,5,6,\ldots,\}$ of the positive integers. Note that this is not the same thing as the set of the positive integers, because $\{2,1,4,3,6,5,\ldots\}$ is another sequence based on the same set. The main difference is that in a set, order does not matter. In a sequence, order does matter.

Of particular interest to us are the *binary* sequences in which every term is 0 or 1. Similarly, a *ternary* sequence is one in which every term is 0, 1 or 2, a *quaternary* sequences is one in which every term is 0, 1, 2 or 3, and so on.

The first question about a combinatorial structure is whether it even exists. If it does not, we are not talking about anything. Once existence is established, it is natural to ask how many such structures are there. If the enumeration process turns out many instances, they may be compared with one another to see which is the best under specific criteria.

Thus the three fundamental considerations in combinatorics are **exis-tence problems**, **enumeration problems** and **extremal problems**. Of the three, enumeration problems have been developed most systematically, and their study will form the main part of this book. Existence and extremal problems are treated in Chapters Zero and Six.

The book divides itself naturally into three parts. The first part consists of the rather long Chapter Zero. It can stand on its own as a textbook for a problem-solving course. Thus it is important not to devote too much time in a combinatorics course to it. We recommend that after covering Sections 1 and 5 of Chapter Zero, the course should proceed directly to Chapter One, returning to Chapter Zero from time to time when the need arises.

The second part of the book comprises Chapters One, Two and Three. These are the basic topics in combinatorics. In each chapter, the last section may be considered as an extra topic which instructors may skip over at their discretion without interrupting the flow of the text. The third part of the book comprises Chapters Four, Five and Six. These are additional topics which instructors may choose either to cover or omit.

We offer many fun and challenging exercises in every section. However, some of them could be a little too challenging. To balance things, we introduce Practice Questions for each of Chapters Zero, One, Two and Three. Answers are provided.

The text has been field-tested several times before publication. The universal request from the readers is to have more exercises and examples, in greater varieties both in scope and in level of difficulty. Twelve exercises to each chapter have been added under Appendix A and six Examples to each chapter have been added under Appendix B. Solutions to all odd-numbered exercises, whether original or additional, are given in Appendix C. A separate manual contains solutions to even-numbered exercises.

Table of Contents

Chapter Zero: Basic Techniques

Section 0.1. Set Theoretic Counting

Let us begin with a review of Set Theory. The undefined terms are *sets* and *elements*. Loosely speaking, the elements are objects and the sets are collections of objects. There is also an undefined relation to indicate whether a certain object belongs to a particular collection. An element x is either *in* a set A or not. These two alternatives are denoted respectively by $x \in A$ and $x \notin A$.

The *union* of two sets A and B is the set consisting of those elements which belong to A or B or both. It is denoted by

$$A \cup B = \{x : x \in A \text{ or } x \in B\}.$$

The *intersection* of two sets A and B is the set consisting of those elements which belong to both A and B. It is denoted by

$$A \cap B = \{x : x \in A \text{ and } x \in B\}.$$

The operations $\cup$ and $\cap$ have the following properties.

Idempotent Laws: $A \cup A = A$, $A \cap A = A$.

Commutative Laws: $A \cup B = B \cup A$, $A \cap B = B \cap A$.

Associative Laws: $(A \cup B) \cup C = A \cup (B \cup C)$, $(A \cap B) \cap C = A \cap (B \cap C)$.

Distributive Laws:
$A \cup (B \cap C) = (A \cup B) \cap (A \cup C)$, $A \cap (B \cup C) = (A \cap B) \cup (A \cap C)$.

There is a set with no elements. It is called the *empty set* and its standard symbol is $\emptyset$. We may also consider the set consisting of all elements under discussion. This set is called the *universal set* of the discussion. The usual symbol for it is S, but this is by no means standard.

Universal Laws: $A \cup \emptyset = A$, $A \cap \emptyset = \emptyset$, $A \cup S = S$, $A \cap S = A$.

A set A is said to be a *subset* of a set B if every element of A is also an element of B. We write $A \subseteq B$. Note that $A \cap B \subseteq A \subseteq A \cup B$ and $A \cap B \subseteq B \subseteq A \cup B$ for any A and B. In fact, both $A \cap B = A$ and $A \cup B = B$ are equivalent to $A \subseteq B$. We also have $\emptyset \subseteq A \subseteq S$ for all A.

The *complement* of a subset A of S is the set consisting of all elements in S that are not in A. It is denoted by

$$\overline{A} = \{x : x \in S \text{ and } x \notin A\}.$$

Complement Laws: $\overline{\emptyset} = S$, $\overline{S} = \emptyset$, $\overline{\overline{A}} = A$, $A \cup \overline{A} = S$, $A \cap \overline{A} = \emptyset$.

De Morgan's Laws: $\overline{A \cup B} = \overline{A} \cap \overline{B}$, $\overline{A \cap B} = \overline{A} \cup \overline{B}$.

© The Author(s), under exclusive license to Springer Nature Switzerland AG 2021
S. W. Golomb, A. Liu, *Solomon Golomb's Course on Undergraduate Combinatorics*,
https://doi.org/10.1007/978-3-030-72228-9_1

The *difference* of a set A from another set B is the set consisting of those elements which belong to A but not to B. It is denoted by

$$A - B = \{x : x \in A \text{ and } x \notin B\}.$$

Note that $\overline{A}$ is the same as $S - A$.

The *Cartesian product* of two sets A and B is the set consisting of all ordered pairs of elements, the first element in a pair belonging to A and the second element in a pair belonging to B. It is denoted by

$$A \times B = \{(x, y) : x \in A \text{ and } y \in B\}.$$

The number of elements in a finite set A is called its *cardinality* and is denoted by $|A|$. In set theoretic counting, there are two basic results.

Multiplication Principle.
$|A_1 \times A_2 \times \cdots \times A_n| = |A_1| \cdot |A_2| \cdots |A_n|.$

Addition Principle.
$|A_1 \cup A_2 \cup \cdots \cup A_n| = |A_1| + |A_2| + \cdots + |A_n|$ if $A_i \cap A_j = \emptyset$ for all $i \neq j$.

Suppose there are three parties in a certain parliament. The Red Party has 54 members, the Blue Party has 33 and the Green Party 7. If we wish to choose a Chairman from among all members, the Addition Principle tells us that there are 54+33+7=94 different choices. If instead we wish to choose a Steering Committee with one member from each party, the Multiplication Principle tells us that there are $54 \times 33 \times 7 = 12474$ different choices.

The Distribution Problem.
Suppose n objects are to be distributed into k boxes. We consider the following eight cases:
(1) distinct boxes, distinct objects, empty boxes allowed;
(2) distinct boxes, distinct objects, empty boxes not allowed;
(3) distinct boxes, identical objects, empty boxes allowed;
(4) distinct boxes, identical objects, empty boxes not allowed;
(5) identical boxes, distinct objects, empty boxes allowed;
(6) identical boxes, distinct objects, empty boxes not allowed;
(7) identical boxes, identical objects, empty boxes allowed;
(8) identical boxes, identical objects, empty boxes not allowed.

In Case (1), each object may be put into any of the k boxes. By the Multiplication Principle, there are altogether k^n distributions. We will encounter and solve the remaining cases in later chapters of the book.

We next give an application of the Multiplication Principle to number theory. For a positive integer n, the function $\tau(n)$ denotes the number of positive divisors of n. We have $\tau(1) = 1$, $\tau(4) = 3$, $\tau(6) = 4$ and $\tau(p) = 2$ for any prime number p.

We wish to determine $\tau(720)$. Note that the prime factorization of 720 is $720 = 2^4 \times 3^2 \times 5$. The prime factorization of any positive divisor of 720 has no primes other than 2, 3 and 5. We can take from 0 to 4 copies of 2, from 0 to 2 copies of 3 and from 0 to 1 copy of 5. It follows that we have $\tau(720) = (4+1)(2+1)(1+1) = 30$.

More generally, let the prime factorization of a positive integer m be $p_1^{k_1} p_2^{k_2} \cdots p_n^{k_n}$. The prime factorization of a positive divisor of m consists of 0 to k_i copies of p_i for $1 \le i \le n$. Hence $\tau(m) = (k_1+1)(k_2+1) \cdots (k_n+1)$.

We now give an application of the Addition Principle to number theory. Suppose we wish to determine the maximum value of n such that 2^n divides 100!. From every multiple of 2 up to 100, we can extract a factor of 2 of 100!. The number of such factors is $\frac{100}{2} = 50$. From every multiple of 4 up to 100, we can extract another factor of 2 of 100!. The number of such factors is $\frac{100}{4} = 25$. From every multiple of 8 up to 100, we can extract a third factor of 2 of 100!. The number of such factors is $\lfloor \frac{100}{8} \rfloor = 12$, where $\lfloor x \rfloor$ is the greatest integer not exceeding x. Continuing this argument and applying the Addition Principle, the maximum value of n is given by

$$\frac{100}{2} + \frac{100}{4} + \left\lfloor \frac{100}{8} \right\rfloor + \left\lfloor \frac{100}{16} \right\rfloor + \left\lfloor \frac{100}{32} \right\rfloor + \left\lfloor \frac{100}{64} \right\rfloor + \left\lfloor \frac{100}{128} \right\rfloor + \cdots$$

$$= 50 + 25 + 12 + 6 + 3 + 1 + 0 + \cdots = 97.$$

From the Addition Principle, we can derive the following result.

Subtraction Rule.
$|B - A| = |B| - |A \cap B|$; in particular, $|B - A| = |B| - |A|$ if $A \subseteq B$.

Proof:
Every element in $A \cap B$ belongs to A while no element in $B - A$ belongs to A. It follows that $(A \cap B) \cap (B - A) = \emptyset$. If an element in B belongs to A, it belongs to $A \cap B$. Otherwise, it belongs to $B - A$. It follows that $B = (A \cap B) \cup (B - A)$. By the Addition Principle, $|B| = |A \cap B| + |B - A|$ so that $|B - A| = |B| - |A \cap B|$. If $A \subseteq B$, we have $A \cap B = A$ so that $|B - A| = |B| - |A|$.

The simpler form of the above result, namely $|B - A| = |B| - |A|$, comes with a condition, namely $A \subseteq B$. To remove this condition, we modify the result to $|B - A| = |B| - |A \cap B|$. Another variation is the following result.

Complement Rule.
$|\overline{A}| = |S| - |A|$.

Proof:
This follows from the Subtraction Rule by taking $B = S$.

We now remove the condition in the Addition Principle, and obtain the following result.

Addition Rule.

$$|A_1 \cup A_2 \cup \cdots \cup A_n| = \sum |A_i| - \sum |A_i \cap A_j| + \cdots + (-1)^{n-1} |A_1 \cap A_2 \cap \cdots \cap A_n|.$$

Here $\sum_n |A_i|$ means $\sum_{1 \le i \le n} |A_i|$, $\sum |A_i \cap A_j|$ means $\sum_{1 \le i < j \le n} |A_i \cap A_j|$, ... The proof of the Addition Rule will be deferred to Section 0.5. A very important corollary of the Addition Rule is the following result.

Principle of Inclusion-Exclusion.

$$|\overline{A_1} \cap \overline{A_2} \cap \cdots \cap \overline{A_n}| = |S| - \sum |A_i| + \sum |A_i \cap A_j| - \cdots + (-1)^n |A_1 \cap A_2 \cap \cdots \cap A_n|.$$

Proof:

This follows from The Addition Rule, the Complement Rule and de Morgan's Law.

The Principle of Inclusion-Exclusion is a very powerful tool. There are problems for which it is really the only approach that can be used. It has a very broad application. The only drawback is that if the desired answer is a symbolic expression, it is usually rather messy. However, most problems ask for numerical answers.

A cigarette company issued the following report. Of 100000 people it surveyed, 40000 were males, 80000 were smokers and 10000 had cancer. Only 1000 were males with cancer, 2000 were smokers with cancer and 3000 were male smokers. Finally, there were only 100 male smokers with cancer. Analyse mathematically the credibility of this report.

Let S, M, K and C denote respectively the set of those surveyed, males surveyed, smokers surveyed and cancer patients surveyed. Then we have $|S| = 100000$, $|M| = 40000$, $|K| = 80000$ and $|M \cap K| = 3000$. By the Principle of Inclusion-Exclusion,

$$\begin{aligned}
|\overline{M} \cap \overline{K} \cap \overline{C}| &= |S| - |M| - |K| - |C| \\
&\quad + |M \cap K| + |K \cap C| + |C \cap M| - |M \cap K \cap C| \\
&= 100000 - 40000 - 80000 - 10000 \\
&\quad + 3000 + 2000 + 1000 - 100 \\
&= -24100.
\end{aligned}$$

Even simpler, $|\overline{M} \cap \overline{K}| = 100000 - 40000 - 80000 + 3000 = -17000$. We have a clear case of inconsistent data.

We turn now to another example from number theory. For a positive integer n, the function $\phi(n)$ denotes the number of positive integers less than or equal to n and relatively prime to n, that is, having no positive common divisor with n other than 1. Thus $\phi(1) = \phi(2) = 1$, $\phi(3) = \phi(4) = \phi(6) = 2$ and $\phi(p) = p - 1$ for any prime number p. We wish to determine $\phi(210)$. Let S be the set of positive integers up to 210. Let A, B, C and D be the subsets of S consisting of those divisible by 2, 3, 5 and 7 respectively.

By the Principle of Inclusion-Exclusion and the identity

$$
\begin{aligned}
(1-a)(1-b)(1-c)(1-d) \;=\; & 1 - (a+b+c+d) \\
& + (ab + ac + ad + bc + bd + cd) \\
& - (abc + abd + acd + bcd) + abcd,
\end{aligned}
$$

we have

$$
\begin{aligned}
\phi(210) \;=\; & |\overline{A} \cap \overline{B} \cap \overline{C} \cap \overline{D}| \\
=\; & |S| - |A| - |B| - |C| - |D| + |A \cap B| + |A \cap C| + |A \cap D| \\
& + |B \cap C| + |B \cap D| + |C \cap D| - |A \cap B \cap C| - |A \cap B \cap D| \\
& - |A \cap C \cap D| - |B \cap C \cap D| + |A \cap B \cap C \cap D| \\
=\; & 210 - \frac{210}{2} - \frac{210}{3} - \frac{210}{5} - \frac{210}{7} + \frac{210}{2 \times 3} + \frac{210}{2 \times 5} + \frac{210}{2 \times 7} \\
& + \frac{210}{3 \times 5} + \frac{210}{3 \times 7} + \frac{210}{5 \times 7} - \frac{210}{2 \times 3 \times 5} - \frac{210}{2 \times 3 \times 7} \\
& - \frac{210}{2 \times 5 \times 7} - \frac{210}{3 \times 5 \times 7} + \frac{210}{2 \times 3 \times 5 \times 7} \\
=\; & 210 \left(1 - \frac{1}{2}\right)\left(1 - \frac{1}{3}\right)\left(1 - \frac{1}{5}\right)\left(1 - \frac{1}{7}\right) \\
=\; & 48.
\end{aligned}
$$

More generally, let the prime factorization of a positive integer m be $p_1^{k_1} p_2^{k_2} \cdots p_n^{k_n}$. Let S be the set of positive integers up to m. For $1 \le i \le n$, let A_i be the subset of S consisting of those divisible by the prime p_i. By the Principle of Inclusion-Exclusion,

$$
\begin{aligned}
\phi(m) \;=\; & |\overline{A_1} \cap \overline{A_2} \cap \cdots \cap \overline{A_n}| \\
=\; & |S| - \sum_n |A_i| + \sum_n |A_i \cap A_j| - \cdots + (-1)^n |A_1 \cap A_2 \cap \cdots \cap A_n| \\
=\; & m - \sum_n \frac{m}{p_i} + \sum_n \frac{m}{p_i p_j} - \cdots + (-1)^n \frac{m}{p_1 p_2 \cdots p_n} \\
=\; & m \left(1 - \frac{1}{p_1}\right)\left(1 - \frac{1}{p_2}\right) \cdots \left(1 - \frac{1}{p_n}\right).
\end{aligned}
$$

Examples

Example 0.1.1.
In how many ways can the positive integers from 1 to 1000 inclusive be arranged in a row such that every number except the last is followed, not necessarily immediately, by a number differing from it by 1?

Solution:
More generally, consider the set $\{a, a+1, \ldots, b-1, b\}$. We claim that the first number in the arrangement must be a or b. Suppose to the contrary it is k where $a < k < b$. The removal of k partitions the set into two subsets $A = \{a, a+1, \ldots, k-1\}$ and $B = \{k+1, k+2, \ldots, b-1, b\}$ of consecutive integers. By symmetry, we may assume that the last number in the arrangement comes from B. Then the integer which appears last among those from A is not the last number and yet cannot be followed by an integer differing from it by 1. This contradiction justifies our claim. Hence we have two choices for the first integer of the arrangement, namely, the first from the set and the last from the set. We are then left with another set of consecutive integers, and the same argument can be applied. Thus we have two choices in every step except the last when only one number is left. Hence the total number of ways is 2^{b-a} by the Multiplication Principle. When $a = 1$ and $b = 1000$, the answer is 2^{999}.

Example 0.1.2.
How many binary sequences of length 5 are there with no isolated 1s?

Solution:
Let S be the set of all binary sequences of length 5. For $1 \leq i \leq 5$, let A_i be the subset of S with an isolated 1 in the i-th position. Then $|S| = 2^5 = 32$. We have $|A_1| = |A_5| = 2^3 = 8$ and $|A_2| = |A_3| = |A_4| = 2^2 = 4$. Of the pairwise intersections, $|A_1 \cap A_3| = |A_1 \cap A_5| = |A_3 \cap A_5| = 2$ and $|A_1 \cap A_4| = |A_2 \cap A_4| = |A_2 \cap A_5| = 1$. The only other non-empty intersection is $|A_1 \cap A_3 \cap A_5| = 1$. Hence

$$|\overline{A_1} \cap \overline{A_2} \cap \overline{A_3} \cap \overline{A_4} \cap \overline{A_5}| = 32 - 2 \times 8 - 3 \times 4 + 3 \times 2 + 3 - 1 = 12.$$

Example 0.1.3.
How many integers up to 1000000 are neither squares nor cubes?

Solution:
The number of squares up to 1000000 is 1000. The number of cubes up to 1000000 is 100. The number of sixth-powers up to 1000000 is 10. By the Principle of Inclusion-Exclusion, the number of integers up to 1000000 which are neither squares nor cubes is 1000000-1000-100+10=998910.

Exercises

1. A positive integer is said to be odd-looking if each of its digits is odd. How many odd-looking numbers with at most four digits are there?

2. Among twenty objects, ten are identical while the other ten are distinct. In how many different ways can ten of them be chosen?

3. In a class of 12 students, 8 passed calculus, 6 passed linear algebra and 5 passed abstract algebra. There were 5 students who passed both calculus and linear algebra, 3 who passed both linear algebra and abstract algebra, and 4 who passed both abstract algebra and calculus. The number of students who passed all three subjects was 3. How many students did not pass any of these three subjects?

4. In a class of 25 students, 17 passed calculus, 13 passed linear algebra and 8 passed abstract algebra. No one passed all three, but at least 6 failed all three. Exactly how many failed all three?

5. Determine the number of positive integers less than or equal to 210 which are divisible by both 2 and 7, but by neither 3 nor 5.

6. Determine the number of positive integers less than or equal to 210 which are divisible by both 3 and 5, but by neither 2 nor 7.

Section 0.2. Extremal Value Principle

Most combinatorics problems deal with finite sets. The finiteness of a set gives rise to many of its basic properties, of which the following is of fundamental importance.

Extremal Value Principle

Every non-empty finite set of real numbers has a maximum and a minimum.

Note that the maximal and minimal elements need not be unique. Moreover, we could get by with postulating the existence of just a maximum or just a minimum. The existence of the other can be deduced simply by switching the signs of all the numbers and then applying the previous result. However, we prefer to state this principle in the symmetric form.

We can give a "proof" of the existence of a maximum as follows. Pick any two of the numbers and throw away the smaller one, or either one if they are equal. Repeat this process. Since the set of numbers is *finite*, the process eventually terminates, and the number we still have is a maximum.

We are trying to prove the existence of an object with certain properties. The Extremal Value Principle tells us to pick an object which maximizes or minimizes some function. The resulting object is then shown to have the desired property by showing that slight perturbation (variation) would further increase or decrease the given function. If there are several optimizing objects, then it is usually immaterial which one we use.

The best way to learn to use the Extremal Value Principle is by solving problems. Suppose that at the Mathematical Sciences Society's Christmas Ball, each dance was between a boy and a girl. Every boy danced with some girl but no girl danced with all the boys. Prove that there were two boys A and B and two girls C and D such that A danced with C but not D, and B danced with D but not C.

Let A be any girl for now. Since no girl danced with all the boys, there was a boy D with whom A had not danced. Since every boy danced with at least one girl, there was a girl B with whom D had danced. Among all those who had danced with A, we need a boy C who had not danced with B. Let A be a girl who danced with the highest number of boys. If C did not exist, then B would have danced with every boy who had danced with A, as well as D. This contradicts the maximality assumption on A. Hence C must exist.

Here is another illustration of the method. There are n cars on a circular track. Together they have just enough gas for one car to complete a lap. Prove that there is a car which can complete a lap by collecting gas from the other cars on its way around.

Suppose there is a jeep with enough gas to complete a lap. Let it start anywhere on the circular track. It collects all the gas from the cars as it passes them. When it returns to its starting point, the jeep has the same amount of gas as before its start. The amount of gas it has goes steadily down until it comes upon a car. At some point A during the lap, the amount m of gas in it is lowest. This must occur just as it comes upon a car, so that there has to be a car at A. We claim that this car can complete the lap. Otherwise, it runs out of gas at some point B before it reaches the next car. This means that the amount of gas in the jeep at B is also m. However, there is no car at B, and the amount of gas in the jeep will continue to decrease until it reaches the next car. This contradicts the assumption that the minimum amount of gas is m.

The Extremal Value Principle can also be helpful in analysing games. Here is an example. The game starts with several piles of chips. Two players move alternately. A move consists in splitting every pile with more than one chip into two non-empty but otherwise arbitrary piles. The player who makes the last move wins. For what initial conditions does the first player win and what is her winning strategy?

Suppose the first player can leave behind a position in which each pile consists of a single chip. Clearly she wins. If not, the piles consisting of single chips do not matter. This suggests that the focus should be on the piles consisting of the highest number of chips. Suppose she leaves behind a position in which the highest number of chips in any pile is m. As observed above, she wins if $m = 1$. Suppose $m > 1$. The second player cannot keep m at the same value as every pile must be split. He cannot reduce m to less than half of its current value as every pile is split into two. Working backwards, the winning positions are $2^k - 1$ for any positive integer k. Hence the first player wins if and only if the initial value of m is not of the form $2^k - 1$. Suppose $2^k - 1 < m < 2^{k+1} - 1$. Then she can lower m to $2^k - 1$. Now the second player must leave behind a position in which $2^{k-1} + 1 < m < 2^k - 1$, and the first player can win by continuing with this strategy.

The next illustration highlights the difference between finite and infinite sets. Two wizards are dueling over the Dead Sea, each levitated 100 meters above water. One of them chooses the sword while the other chooses the pen. The pen-wizard prepares a scroll of spells of the form (x, y), where $0 < x < y$. The spells are available to both wizards, and may be used any number of times. The sword-wizard casts the first spell, and they take turns thereafter. When the sword-wizard casts an (x, y) spell, the pen-wizard drops y meters and then the sword-wizard drops x meters. When the pen-wizard casts an (x, y) spell, the pen-wizard drops x meters and then the sword-wizard drops y meters. Whoever touches water first loses. Can the pen-wizard win by putting the right spells into the scroll?

At first, it appears that the pen-wizard is faced with an impossible task. The sword-wizard keeps casting the (x, y) spell for which $y - x$ is maximum. The best the pen-wizard can do is catch up, but as the pen-wizard drops first no matter who is casting, the sword-wizard cannot lose.

However, this argument presupposes that the number of spells in the scroll is finite. To win the duel, the pen-wizard must put an infinite number of spells in the scroll. For instance, the scroll may consist of all spells of the form $(\frac{1}{n}, 100 - \frac{1}{n})$ where n is an arbitrary positive integer. When the sword-wizard casts such a spell, the pen-wizard counters with the spell $(\frac{1}{n+1}, 100 - \frac{1}{n+1})$. It is easy to verify that the pen-wizard wins by casting only this spell.

Closely related to the Extremal Value Principle is the **Well-Ordering Principle**, which states that every non-empty subset of the set of positive integers, finite or infinite, has a minimum. At first, this may appear as an easy corollary of the Extremal Value Principle, because if the set is non-empty, it must contain some positive integer, say 53. Then there are only finitely many candidates for the minimum, namely, 1, 2, 3, and so on, up to 53. By the Extremal Value Principle, this finite set has a minimum.

However, suppose the subset under consideration is actually empty. We cannot appeal to the Extremal Value Principle then. The Well-Ordering Principle may be applied to show that the subset is indeed empty. Perhaps an alternative formulation of the Well-Ordering Principle is as follows. If a set of positive integers does not have a minimum, then it must be empty.

As an illustration, we prove that $\sqrt{3}$ is irrational. Suppose to the contrary that $\sqrt{3} = \frac{a}{b}$ for some positive integers a and b. Let S be the subset of the set of the positive integers consisting of those elements k such that $k\sqrt{3}$ is a positive integer. S is non-empty since b is in it. By the Well-Ordering Principle, S has a smallest element m. Since $1 < \sqrt{3} < 2$, we have $0 < \sqrt{3} - 1 < 1$ so that $0 < m\sqrt{3} - m < m$. Let $n = m\sqrt{3} - m$. Since m and $m\sqrt{3}$ are both integers, so is n, which is positive. Hence n is an element of S smaller than m. This is a contradiction. It follows that S is empty. In other words, $\sqrt{3}$ is irrational.

Here is another approach in using the Well-Ordering Principle. Let S be a subset of the set of the positive integers. If we can show that for any element n in S, there exists another element m in S such that $m < n$, then S must be empty. Otherwise, we can produce smaller and smaller positive integers at will, which contradicts the Well-Ordering Principle. This approach is often called the method of **infinite descent**.

We wish to prove that the equation $x^2 + y^2 = xy$ has no solutions in positive integers. Suppose there are positive integer solutions. Then there exists one in which y is minimum. If both x and y are odd, then the left side is even while the right side is odd. If one of x and y is even and the other one is odd, then the left side is odd while the right side is even. Hence both x and y must be even, so that $x = 2r$ and $y = 2s$ for some positive integers r and s. Then $r^2 + s^2 = \frac{1}{4}(x^2 + y^2) = \frac{1}{4}xy = rs$. We have found a new solution (r, s) with $s < y$, a contradiction.

We will see in Section 0.5 that the concept known as mathematical induction is closely related to the Well-Ordering Principle and its associated method of infinite descent.

Examples

Example 0.2.1.

In a country, there are 100 towns. Some pairs of towns are joined by roads. The roads do not intersect one another except meeting at towns. It is possible to go from any town to any other town by road. Prove that it is possible to pave some of the roads so that the number of paved roads at each town is odd.

Solution:

Let m be the number of towns with an odd number of paved roads. If there are no paved roads, then $m = 0$. Each new paved road affects exactly two towns, changing the parity of their numbers of paved roads. Hence m is always even. Among all possible paving schemes, consider the one for which m is maximum. We claim that in this case, $m = 100$. Suppose $m < 100$. Then there are two towns A and B each with an even number of paved roads. Take an arbitrary path going from A to B by road. Pave all the roads along the path that are unpaved, and unpave all the roads along the path that are paved. The parity of the number of paved roads of any town along the path remains unchanged except for A and B. This would increase m by 2, a contradiction.

Example 0.2.2.

Every participant in a tournament plays with every other participant exactly once. No game is a draw. After the tournament, every player makes a list with the names of all players who were beaten by him and those who were beaten by the players beaten by him. Prove that the list of some player contains the names of all other players.

Solution:

Let A be a participant who has won the largest number of games. If A would not have the property of the problem, then there would be another player B, who has won against A and against all players who were beaten by A. So B would have won more games than A. This contradicts the choice of A.

Example 0.2.3.

Prove that $\frac{1}{1\times2} + \frac{1}{2\times3} + \cdots + \frac{1}{n(n+1)} = \frac{n}{n+1}$ for all $n \geq 1$.

Solution:

Let S be the subset of the set of positive integers consisting of those elements n for which the given expression does not hold. Suppose S is non-empty. By the Well-Ordering Principle, it has a smallest element m. Now $m \neq 1$ since both sides of the expression reduce to $\frac{1}{2}$ when $n = 1$. Hence $m - 1$ is a positive integer not in S. This means that $\frac{1}{1\times2} + \frac{1}{2\times3} + \cdots + \frac{1}{(m-1)m} = \frac{m-1}{m}$. Adding $\frac{1}{m(m+1)}$ to both sides, we have $\frac{1}{1\times2} + \frac{1}{2\times3} + \cdots + \frac{1}{m(m+1)} = \frac{m}{m+1}$. Hence the expression holds for $n = m$, and this contradicts m's membership in S. It follows that S is empty and the given expression holds for all $n \geq 1$.

Exercises

1. In the parliament of a certain country, every member has at most three enemies among the remaining members. Prove that one can partition the parliament into two committees so that every member has at most one enemy within the committee.

2. A banquet is attended by $2n$ ambassadors. Each has at most $n - 1$ enemies among the others. Prove that they can be seated at a round table so that none sits next to an enemy.

3. In the Mathematical Sciences Society, there are at least two members who are friends. Any two members with the same number of friends in the society have no common friends in the society. Prove that there is a member with exactly one friend in the society.

4. In the Mathematical Sciences Society, there are n members from the Pure Mathematics Division, n from the Applied Mathematics Division and n from the Statistics Division. Each member has altogether $n + 1$ friends in the other two divisions. Prove that one can select one member from each division so that the three selected members are all friends.

5. Prove that the equation $x^3 = 3y^3 + 9z^3$ has no solutions in positive integers.

6. Prove that the equation $w^2 + x^2 = 3(y^2 + z^2)$ has no solutions in positive integers.

Section 0.3. Mean Value Principle

An important corollary of the Extremal Value Principle in Section 0.2 is the following result.

Mean Value Principle
In every non-empty finite set of real numbers, there is at least one which is not less than the average of the set, and at least one not greater than the average of the set.

This follows from the Extremal Value Principle, since a maximum and a minimum of the set can play the roles of these two numbers, respectively. As a simple application, consider twenty-five crates of apples of three different sorts, and all the apples in each crate are of the same sort. Then among these crates there are at least nine containing the same sort of apple. This is because the average number of crates of apples of a kind is $\frac{25}{3} > 8$. The desired conclusion follows from the Mean Value Principle.

A very nice application of this principle is the following result.

Erdős-Szekeres Theorem.
Let m and n be positive integers. If the positive integers 1 to $mn + 1$ are written down in a sequence, one can either find an increasing subsequence consisting of $m + 1$ terms, or a decreasing subsequence consisting of $n + 1$ terms.

Proof:
Let the terms of the sequence be a_1, a_2, ..., a_{mn+1}. Assign the label 1 to a_1. For $k \geq 2$, assign the label $i + 1$ to a_k if i is the largest label among all integers a_j such that $1 \leq j < k$ and $a_j < a_k$. If $a_j > a_k$ for $1 \leq j < k$, assign the label 1 to a_k. Continue until a label has been assigned to a_{mn+1}. Suppose the largest label is at least $m + 1$. A term with label $m + 1$ is assigned that label because it is greater than an earlier term with label m. Backtracking yields an increasing subsequence consisting of $m + 1$ terms. Suppose the largest label is at most m. Then the average number of terms with the same label is at least $\frac{mn+1}{m} > n$. Hence there exist $n + 1$ terms with the same label, and they form a decreasing subsequence consisting of $n + 1$ terms.

Suppose the alphabet of a certain language contains 22 consonants and 11 vowels. Any string of these letters is a word in this language, so long as no two consonants are together and no letter is used twice. The alphabet is divided into 6 non-empty subsets. We claim that the letters in at least one of these subsets form a word in the language.

For $1 \leq k \leq 6$, let the k-th subset consist of c_k consonants and v_k vowels. Then $(c_1 - v_1) + (c_2 - v_2) + \cdots + (c_6 - v_6) = 22 - 11 = 11$. The total difference 11 comes from 6 different subsets. Hence the average number of contribution from each subset is $\frac{11}{6} < 2$. By the Mean Value Principle, there is at least one subset in which this difference is less than 2. By alternating the consonants and vowels in this subset, we have formed a word.

Suppose a warehouse contains 200 boots of size 41, 200 boots of size 42 and 200 boots of size 43. Of these 600 boots, there are 300 left boots and 300 right boots. Prove that one can find among these boots at least 100 pairs each consisting of one left boot and one right boot of the same size.

We may assume there are at least as many left boots as right boots of size 41. Suppose this is also the case with boots of size 42. There are 300 right boots in all, and at most 200 of them are of size 43. Hence the total number of right boots of size 41 or 42 is at least 100. Since each of these sizes contains at least as many left boots as right boots, there is a match for each right boot. Hence there are at least 100 matching pairs in the warehouse.

The same argument applies if there are also as many left boots as right boots of size 43. Suppose there are more right boots than left boots of each of size 42 and 43. The same argument can still be used with the roles of the left and right boots interchanged.

For our next illustration, we turn to combinatorial number theory. Given 8 different positive integers, none greater than 15, prove that at least three pairs of them have the same positive difference. The pairs need not be disjoint as sets.

In tackling this problem, we encounter a seemingly insuperable obstacle. There are 14 possible differences between the 8 given numbers, the values of the differences being 1 through 14. However, there are 28 pairs of the given numbers, and therefore 28 differences. The average number of differences for each value is $\frac{28}{14} = 2$, which does not yield our desired conclusion.

The key observation is that the difference 14 can be obtained in only one way with positive integers not greater than 15, namely, $14 = 15 - 1$. Hence there are at least 27 differences of value between 1 and 13 inclusive. Now the average number of differences for each value is $\frac{27}{13} > 2$, and the Mean Value Principle delivers the desired result.

Our final illustration comes from combinatorial geometry. A $20 \times 20 \times 20$ cube is built of $1 \times 2 \times 2$ bricks. Prove that one can run a needle through the cube without piercing any of the bricks.

In each of the three directions, there are 19×19 potential lines along which a needle can run without piercing any of the bricks. So a total of $3 \times 361 = 1083$ lines need to be blocked. The number of bricks to block them is $(20 \times 20 \times 20) \div (2 \times 2 \times 1) = 2000$. The average number of bricks to block each line is $\frac{2000}{1083} < 2$, so that some line ℓ is blocked by at most one brick. If it is not blocked by any brick, the needle can run along ℓ. Suppose it is blocked by exactly one brick. Cut the cube into four pieces by two planes parallel to the faces of the cube and containing ℓ. Each piece contains exactly 1 unit cube of the blocking brick, and either 0, 2 or 4 unit cubes of any other brick. Hence the total number of unit cubes in each piece is odd. However, this number must be a multiple of 20, and we have a contradiction.

Examples

Example 0.3.1.
Ten students solved a total of 35 problems in a mathematics contest. Each problem was solved by exactly one student. There is at least one student who solved exactly one problem, at least one student who solved exactly two problems, and at least one student who solved exactly three problems. Prove that there is also at least one student who has solved at least five problems.

Solution:
Take aside one student who solved exactly one problem, one student who solved exactly two problems and one student who solved exactly three problems. They had taken care of six of the 35 problems. The remaining 29 problems were solved by the other seven students. The average number of these problems solved by each of these students is $\frac{29}{7} > 4$. Hence one of them solved at least 5 problems.

Example 0.3.2.
Twenty different positive integers are all less than 70. Prove that among their pairwise differences, there are four equal numbers.

Solution:
Denote the 20 integers a_1 to a_{20}. Then $0 < a_1 < \cdots < a_{20} < 70$. We want to prove that there is a k so that $a_j - a_i = k$ has at least four solutions. Now

$$0 < (a_2 - a_1) + (a_3 - a_2) + \cdots + (a_{20} - a_{19}) = a_{20} - a_1 \leq 68.$$

We will prove that among the differences $a_2-a_1, a_3-a_2, \ldots, a_{20}-a_{19}$, there will be four equal ones. Suppose there are at most three equal differences. The minimum value of these 19 differences is $3(1+2+3+4+5+6)+7 = 70$. However, $70 \leq 68$ is a contradiction!

Example 0.3.3.
The 49 dots in a 7×7 array are painted two colors. What is the minimum number of different rectangles with vertices of the same color and with sides parallel to the sides of the array? Two rectangles are still different if they share two common vertices.

Solution:
The answer is 21. We call a pair of dots in the same row of the same color a *good* pair. Suppose there are k white and $7 - k$ black cells in some row. Then there are $\frac{k(k-1)}{2} + \frac{(7-k)(6-k)}{2} = k^2 - 7k + 21$ good pairs. This term is minimal for $k = 3$ and $k = 4$ and is equal to 9. Thus there are at least 9 good pairs in each row, and in the whole square at least 63. We label a good pair $C(i, j)$ if the dots are in color C and are in columns i and j. There are 42 different labels since we have two colors and $\frac{7 \times 6}{2} = 21$ pairs of columns.

We choose a maximum set S of good pairs with distinct labels. Now $|S| \leq 42$. Hence there are at least $63 - 42 = 21$ good pairs not in S, each of which has the same label as some pair in S. Hence we have at least 21 rectangles of the specified kind. The array in the diagram below shows that this is the best possible result.

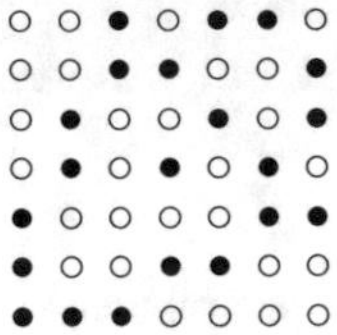

```
○ ○ ● ○ ● ● ○
○ ○ ● ● ○ ○ ●
○ ● ○ ○ ● ○ ●
○ ● ○ ● ○ ● ○
● ○ ○ ○ ○ ● ●
● ○ ○ ● ● ○ ○
● ● ● ○ ○ ○ ○
```

Exercises

1. A bag contains 111 balls, each of which is green, red, white or blue. If 100 balls are drawn at random, there will always be 4 balls of different colors among them. What is the smallest number of balls that must be drawn, at random, in order to guarantee that there will be 3 balls of different colors among them?

2. A bag contains 100 balls, each of which is red, white or blue. If 26 balls are drawn at random, there will always be 10 balls of the same color among them. What is the smallest number of balls that must be drawn, at random, in order to guarantee that there will be 30 balls of the same color among them?

3. A one-digit positive integer is placed on each square of an 8×8 board. You may select any 3×3 or 4×4 subboard and add 1 to each number on its squares. When a number reaches 10, it immediately becomes 0. Is it always possible to make every number 0?

4. The numbers from 1 to 81 are written on the squares of a 9×9 board. Prove that there exist two neighbors which differ by at least 6.

5. What is the maximum number of the squares of an 8×8 checkerboard which can be painted green so that wherever the shape shown in the diagram below is placed on the board, covering exactly three of the 64 squares, at least one of the squares covered is not green? The piece may be rotated.

6. What is the minimum number of checkers such that no matter how they are placed on an 8×8 checkerboard, there are always five of them with no two on the same row and no two on the same column?

Section 0.4. Pigeonhole Principle

The following result is a special case of the Mean Value Principle.

Pigeonhole Principle
Finitely many pigeons are put into finitely many holes. If there are more pigeons than holes, then at least one hole contains at least two pigeons. If there are more holes than pigeons, then there is at least one empty hole.

If there are more pigeons than holes, then the average number of pigeons per hole is greater than one. It is understood, especially by the pigeons, that they are not to be carved up. By the Mean Value Principle, there is at least one hole containing no less than the average number of pigeons. This means at least two pigeons are in it. Similarly, if there are more holes than pigeons, then the average number of pigeons is less than one. There is a hole which contains less than this number of pigeon, which means that it must be empty.

The Pigeonhole Principle can be applied to prove that in any group of at least two people, there are two who have an identical number of friends within the group. Here, the holes are the possible numbers of acquaintances for any person. The pigeons are the n people in the group, where n is some integer greater than 1.

Suppose somebody has $n - 1$ acquaintances. Then no person may have 0 acquaintances, so that there are only $n - 1$ holes, $1, 2, \ldots, n - 1$. On the other hand, if nobody has $n - 1$ acquaintances, then there are also only $n - 1$ holes, $0, 1, \ldots, n - 2$. It follows from the Pigeonhole Principle that two of the n pigeons must be in the same hole, meaning that two people must have the same number of acquaintances.

We turn now to combinatorial number theory. Suppose a positive integer n is relatively prime to 10. We claim that there is a multiple of n in which every digit is 1, and a power of n which ends in a 1 preceded immediately by an arbitrary number of 0s which are preceded by some other digits.

Consider the n integers $1, 11, \ldots, 11\ldots1$, where the last integer has n digits. If any of them is a multiple of n, there is nothing further to prove. Suppose each of them leave a non-zero remainder when divided by n. Since there are only $n - 1$ possible remainders, namely, $1, 2, \ldots, n - 1$, two of the remainders must be the same. Hence the difference of these two numbers is a multiple of n. However, the difference of two such numbers has the form $11\ldots100\ldots0$. Since n is relatively prime to 10, we can strike out the zeros at the end and get a number consisting only of 1s and divisible by n.

To prove the second claim, let k be an arbitrary positive integer. Consider the 10^k powers n, n^2, n^3, ..., n^{10^k}. Divide them by 10^k. Since n is relatively prime to 10, the remainder cannot be 0. Thus there are only $10^k - 1$ possible remainders, 1, 2, ..., $10^k - 1$. Hence two of the powers n^i and n^j $(i < j)$ will have the same remainder, and so their difference $n^j - n^i = n^i(n^{j-i} - 1)$ will be divisible by 10^k. Since n is relatively prime to 10, $n^{j-i} - 1$ is divisible by 10^k. It follows that $n^{j-1} = 10^k q + 1$ for some positive integer q, and this number ends in a 1 preceded by at least k 0s.

If $n + 1$ numbers are chosen from $\{1, 2, ..., 2n\}$, then one of them is divisible by another. Let the chosen numbers be a_1, a_2, ..., a_{n+1}. For $1 \le i \le n + 1$, let b_i be the largest odd divisor of a_i. These $n + 1$ odd numbers b_1, b_2, ..., b_{n+1} all lie in the interval $[1, 2n]$, which contains only n odd numbers, By the Pigeonhole Principle, there exist i and j such that $b_i = b_j$. Then one of the numbers a_i and a_j is divisible by the other.

The next illustration comes from combinatorial geometry. The shape shown in the diagram below covers three unit squares.

We wish to cover all but one of the 25 unit squares in a 5×5 chessboard with eight copies of this shape. Where can the uncovered unit square be? The diagram below shows that it can be at the central square, a corner square or the middle square along an edge.

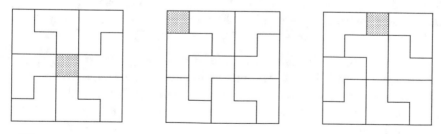

We now prove that the uncovered square must be one of the nine shaded squares in the diagram below. This is because each copy of the given shape can cover at most one of the shaded squares. By the Pigeonhole Principle, at least one shaded square must be uncovered.

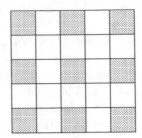

What can we say about the critical case of the Pigeonhole Principle in which the number of pigeons is equal to the number of holes? Among many possible distributions, we may very well have exactly one pigeon in each hole. In this case, we say that there is a **one-to-one correspondence** between the pigeons and the holes. Sometimes, there may be a many-to-one correspondence between the elements of two sets.

Exhibiting a one-to-one correspondence is a very useful way of showing that two sets have the same size. As an illustration of this technique, consider quaternary sequences of length n. Let a_n denote the number of those with an even number of 0s and an even number of 1s, b_n those with an odd number of 0s and an odd number of 1s, c_n those with an even number of 0s and an odd number of 1s, and d_n those with an odd number of 0s and an even number of 1s. We wish to determine a_n, b_n, c_n and d_n.

Note by symmetry that $c_n = d_n$. The total number of quaternary sequences of length n is 4^n. Apart from the sequences in which every term is 2 or 3, the others may be put into one-to-one correspondence so that the two sequences in each pair differ only in the first term which is neither 2 nor 3, with a 0 in one sequence and a 1 in the other. Then each pair consists of one sequence with an even number of 0s and one with an odd number of 0s. It follows that $a_n + c_n = \frac{4^n - 2^n}{2} + 2^n = \frac{4^n + 2^n}{2}$ and $b_n + d_n = \frac{4^n - 2^n}{2}$.

Consider the $\frac{4^n + 2^n}{2}$ sequences counted in $a_n + c_n$. Apart from those in which every term is 0 or 3, the others may be put into one-to-one correspondence so that the two sequences in each pair differ only in the first term which is neither 0 nor 3, with a 1 in one sequence and a 2 in the other. Then each pair consists of one sequence with an even number of 1s and one with an odd number of 1s. Hence $a_n = \frac{4^n - 2^{n+1}}{4} + 2^n = 4^{n-1} + 2^{n-1}$ and $c_n = 4^{n-1} = d_n$, so that $b_n = 4^{n-1} - 2^{n-1}$.

A **composition** of a positive integer is an expression of the integer as a sum of one or more positive integers. Two compositions with the same summands are different if the orders of the terms are different. Thus the number 5 has 16 compositions, namely, 5, 4+1, 1+4, 3+2, 2+3, 3+1+1, 1+3+1, 1+1+3, 2+2+1, 2+1+2, 1+2+2, 2+1+1+1, 1+2+1+1, 1+1+2+1, 1+1+1+2 and 1+1+1+1+1.

Counting the number of compositions of a positive integer this way is not illuminating. We shall represent the positive integer n by n circles. In each of the $n-1$ spaces between adjacent circles, we can decide to put a marker, or not. By the Multiplication Principle, there are 2^{n-1} possible diagrams. Since each diagram corresponds to a unique composition, and vice versa, the number of compositions is also 2^{n-1}. We illustrate with the case $n = 5$ below.

Diagrams									Compositions
0	1	0	1	0	1	0	1	0	1+1+1+1+1
0		0	1	0	1	0	1	0	2+1+1+1
0	1	0		0	1	0	1	0	1+2+1+1
0	1	0	1	0		0	1	0	1+1+2+1
0	1	0	1	0	1	0		0	1+1+1+2
0		0	1	0		0	1	0	2+2+1
0		0	1	0	1	0		0	2+1+2
0	1	0		0	1	0		0	1+2+2
0		0		0	1	0	1	0	3+1+1
0	1	0		0		0	1	0	1+3+1
0	1	0	1	0		0		0	1+1+3
0		0		0	1	0		0	3+2
0		0	1	0		0		0	2+3
0		0		0		0	1	0	4+1
0	1	0		0		0		0	1+4
0		0		0		0		0	5

A **partition** of a positive integer is also an expression of the integer as a sum of one or more positive integers. Two partitions with the same terms are considered the same, regardless of the orders of the terms. Thus the number 5 has only 7 partitions, namely, 5, 4+1, 3+2, 3+1+1, 2+2+1, 2+1+1+1 amd 1+1+1+1+1. Although their number is smaller than the number of compositions, they are a lot harder to count.

There are 5 partitions of 5 consisting of at most 3 terms, namely, 5, 4+1, 3+2, 3+1+1 and 2+2+1. There are also 5 partitions of 5 containing no terms bigger than 3, namely 3+2, 3+1+1, 2+2+1, 2+1+1+1 and 1+1+1+1+1. It turns out that every positive integer also has an equal number of these two types of partitions. However, the one-to-one correspondence is not so easy to spot this time.

Once again, we appeal to diagrams. A partition may be represented by what is called a **Ferrers graph**. Each term is represented by a column with the appropriate number of dots, and the columns are arranged in non-ascending order of height. The Ferrers graphs for the 5 partitions of 5 containing no terms bigger than 3 are shown below.

```
 :  :      :        ·        ·       · · · · ·
 :  :    · · ·    · · ·    · · · ·
 · ·              
```

The one-to-one correspondence we seek is by counting the dots horizontally instead of vertically. Then the above diagrams may be modified to become the Ferrers graphs of the partitions 2+2+1, 3+1+1, 3+2, 4+1 and 5. It is also true that the number of partitions of any positive integer into exactly 3 terms is equal to the number of partitions of the same number into terms the biggest of which is 3.

A variation of the technique of establishing one-to-one correspondence is counting the same set in two ways. Since the answer must be the same, we can obtain an identity. Here is an example from number theory. In the chart below, we put a check mark in the box at the intersection of the ith row and the jth column if and only if i is a divisor of j. We now count the check marks in two ways. The check marks on the ith row indicate multiples of i up to 9, and their number is $\lfloor \frac{9}{i} \rfloor$. The check marks on the jth column are the divisors of j, and their number is $\tau(j)$. It follows that

$$\sum_{i=1}^{9} \left\lfloor \frac{9}{i} \right\rfloor = \sum_{j=1}^{9} \tau(j).$$

$j =$	1	2	3	4	5	6	7	8	9
$i = 1$	√	√	√	√	√	√	√	√	√
$i = 2$		√		√		√		√	
$i = 3$			√			√			√
$i = 4$				√				√	
$i = 5$					√				
$i = 6$						√			
$i = 7$							√		
$i = 8$								√	
$i = 9$									√

Examples

Example 0.4.1.

In preparing for a tournament, a chessmaster wants to play at least one game per day, but not more than 132 games over 77 days. Prove that there is a sequence of successive days on which the chessmaster plays a total of exactly 21 games.

Solution:

Let a_i be the number of games played up to the i-th day inclusive. Then $1 \leq a_1 \leq \cdots \leq a_{77} \leq 132$, so that

$$22 \leq a_1 + 21 \leq a_2 + 21 \leq \cdots \leq a_{77} + 21 \leq 153.$$

These 154 numbers a_1, ..., a_{77}, $a_1 + 21$, ..., $a_{77} + 21$ are the pigeons. Each is of value 1 to 153 inclusive, so that the holes are the values 1, 2, ..., 153. Since there are more pigeons than holes, the Pigeonhole Principle guarantees two pigeons in the same hole, that is, two numbers of the same value. Hence there are indices i and j such that $a_j = a_i + 21$. The chessmaster has played exactly 21 games on the days numbered $i + 1$, $i + 2$, ..., j.

Example 0.4.2.

Given 52 integers, prove that two of them can be chosen such that

(a) their difference is divisible by 51;

(b) the difference of their squares is divisible by 100;

(c) their sum or difference is divisible by 100.

Solution:

(a) The pigeons are the 52 numbers. The holes are the 51 possible values of the remainder when a number is divided by 51. Since there are more pigeons than holes, two pigeons will be in the same hole. If two numbers have the same remainder when divided by 51, their difference must be divisible by 51.

(b) The pigeons are the squares of the 52 numbers. Since we are only interested in the remainder when divided by 100, we may assume that each of the 52 given numbers is less than 100. The holes are the possible values of the remainder when the square of a number is divided by 100. There are only 51 possible remainders, since the numbers n^2 and $(100 - n)^2$ give the same remainder. Since there are more pigeons than holes, two pigeons will be in the same hole. If two squares have the same remainder when divided by 100, their difference must be divisible by 100.

(c) The pigeons are the 52 numbers. We construct 51 holes as follows. Hole 0 contains numbers ending in 00, hole 1 contains numbers ending in 01 or 99, hole 2 contains numbers ending in 02 or 98, and so on. Finally, hole 49 contains numbers ending in 49 or 51, and hole 50 contains numbers ending in 50. Since there are more pigeons than holes, two pigeons will be in the same hole. If these two numbers end in the same two digits, their difference will be divisible by 100. If not, their sum will be divisible by 100.

Example 0.4.3.

A partition is said to be *self-conjugate* if its Ferrers graph is symmetric about the line at a 45° angle to the base and passing through the dot at the bottom left corner. In other words, the vertical count and the horizontal count of the dots yield exactly the same expression.

(a) List the self-conjugate partitions of 16.

(b) List the partitions of 16 into distinct odd terms.

(c) Prove that the number of self-conjugate partitions of a positive integer n is equal to the number of partitions of n into distinct odd terms, using the case $n = 16$ for illustration.

Solution:

(a) There are 5 self-conjugate partitions, namely, 8+2+1+1+1+1+1+1, 7+3+2+1+1+1+1, 6+4+2+2+1+1, 5+5+2+2+2 and 4+4+4+4. In (c), the respective Ferrers graphs are shown.

(b) There are 5 partitions into distinct odd parts, namely, 15+1, 13+3, 11+5, 9+7 and 7+5+3+1.

(c) We shall exhibit a one-to-one correspondence between the two types of partitions. For each self-conjugate partition of a positive integer n, separate its Ferrers graph into layers as shown in the diagram below for $n = 16$. This automatically yields a partition into distinct terms. Moreover, the symmetry of the Ferrers graphs implies that the terms are all odd. Thus we have a partition into distinct odd terms.

Conversely, suppose we start with a partition of n into distinct odd terms. Draw an improper Ferrers graphs with a row of dots for each number, starting with the largest on the bottom. Align them so that a vertical line passes through the middle dot in each row. Keeping the right side of the base fixed, bend the left side so that the line is now at a 45° angle to the base. Since the terms in the original partition are distinct, this is a proper Ferrers graph as the columns are in non-ascending order. Moreover, it is still symmetric about the line, so that we have a self-conjugate partition. The case 13+3=7+3+2+1+1+1+1 of $n = 16$ is illustrated in the diagram below.

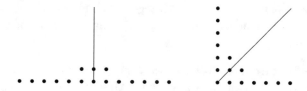

Exercises

1. Several football teams enter a tournament in which each team plays every other team exactly once. Show that at any moment during the tournament there will be two teams which have played, up to that moment, an identical number of games.

2. Eleven students have formed five study groups in a summer camp. Prove that two students can be found, say A and B, such that every study group which includes student A also includes student B.

3. Prove that from ten positive integers, not necessarily distinct, one can always find a nonempty subset such that the sum of its elements is divisible by ten.

4. Prove that from ten distinct two-digit numbers, one can always choose two disjoint nonempty subsets such that their elements have the same sum.

5.   (a) List the partitions of 8 into distinct terms.

     (b) List the partitions of 8 into odd terms.

     (c) Prove that the number of partitions of a positive integer n into distinct terms is equal to the number of partitions of n into odd terms, using the case $n = 8$ for illustration.

6.   (a) List the partitions of 8 into terms which appear at most twice.

     (b) List the partitions of 8 into terms which are not divisible by 3.

     (c) Prove that the number of partitions of a positive integer n into terms which appear at most twice is equal to the number of partitions of n into terms which are not divisible by 3, using the case $n = 8$ for illustration.

Section 0.5. Mathematical Induction

A good problem-solving strategy is to analyze the initial cases and generate some data. This process serves three purposes.

(1) The initial analysis helps in getting a better understanding of the problem.

(2) The initial data may suggest a general approach to the problem.

(3) The initial data may be used to double-check the final result.

This exploration and pattern-recognition process is a form of *inductive* reasoning. It is an art and not a science. As an illustration, suppose that some children have among them 2^n slices of bread. We may select two of the children, with k and ℓ slices respectively, where $k \leq \ell$, and make the second child give k slices to the first. Prove that we can arrange for all 2^n slices to be given to one child eventually.

Let us first examine the case $n = 1$. Here we have 2 slices of bread. They may both belong to one child, in which case there is nothing further to prove. We may also have two children each with one slice, but then we can make one of them give his or her slice to the other. We summarize our finding in the following chart.

In the case $n = 2$, there are 4 slices and five initial distributions.
Distribution 1: One child has all 4 slices.
Distribution 2: One child has 3 slices and another has 1 slice.
Distribution 3: Each of two children has 2 slices.
Distribution 4: One child has 2 slices and each of two other children has 1 slice.
Distribution 5: Each of four children has 1 slice.

Distribution 1 is the desired scenario. Distribution 3 may be converted to Distribution 1, Distribution 2 or 4 into Distribution 3 and Distribution 5 into Distribution 4. Note that Distribution 4 may be converted to itself, but there is no point in doing that. All distributions may eventually be converted to Distribution 1. We summarize our finding in the following chart.

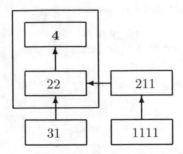

So far, we know that the result is true for $n = 1$ and $n = 2$. However, we cannot expect to solve the problem by continuing to examine all possible initial distributions. Even if we are only interested in a finite value of n, the number of such distributions rises very quickly. However, we still do not have enough data to tell us which way to go.

So we continue with one more case, namely, $n = 3$. This time, there are twenty-two initial distributions, which can all be converted eventually to the desired scenario. We summarize our finding in the following chart.

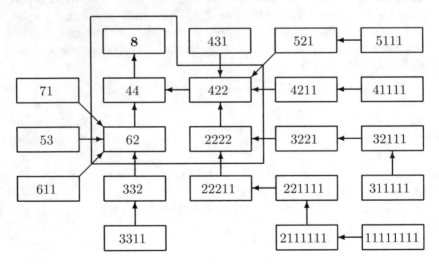

We notice that the desired scenario and the four distributions close to it form a structure just like in the chart for $n = 2$. We highlight this by inserting a box into the chart above. Then we notice that a similar box can be inserted into the chart for $n = 2$.

What is the difference between the distributions inside the boxes and the distributions outside the boxes? It is not hard to see that those inside involve only even numbers, and those outside involve at least one odd number.

This is the key to solving this problem. We are now ready to switch from educated guesses to a logical proof. This process is a science called *deductive* reasoning.

The requirement of our problem may be rephrased as follows. Let $P(n)$ be the statement that we can arrange for all 2^n slices to be given to one child eventually. We wish to prove that this is true for all positive integers n.

A method known as **mathematical induction** is particularly suited to solving this kind of problem, in which we wish to prove that a sequence of statements $P(n)$ is true for *all* positive integers n. Despite its name, mathematical induction is a form of deductive reasoning.

First Principle of Mathematical Induction

Let $P(n)$ be a sequence of statements such that $P(k)$ is true for some positive integer k, and $P(n + 1)$ is true whenever $P(n)$ is true. Then $P(n)$ is true for *all* integers $n \geq k$.

Usually, we have $k = 1$, and wish to conclude that $P(n)$ is true for all positive integers n.

A proof by mathematical induction consists of the following two steps.

(1) **Basis.** Prove that $P(k)$ is true for some positive integer k. Usually, we have $k = 1$.

(2) **Induction.** Assuming that $P(n)$ is true for *some* integer $n \geq k$, prove that $P(n + 1)$ is also true.

Let us carry out the proof by mathematical induction for our problem. For $n = 1$, we have 2 slices. Either some child has both of them already, or two children have one each. In the former case, nothing further is needed. In the latter case, one child can give his or her slice to the other. Suppose the results holds for some $n \geq 1$. Consider the next case where the total number of slices is 2^{n+1}. If each child has an even number of slices, give them some jam and ask them to use all the slices to make sandwiches. Then we have 2^n sandwiches and, by the induction hypothesis, all of them may be given to one child eventually. Suppose some children have odd numbers of slices. Since the total number of slices is even, the number of such children is also even. Thus they can be divided into pairs arbitrarily. Suppose in a pair one child has k slices while the other has ℓ slices, with $k \leq \ell$. After the give-and-take, the former will have $2k$ slices while the latter will have $\ell - k$ slices. Both numbers are even. Thus we can make sure that each child has an even number of slices after one round of give-and-take. We can then bring in the jam to make sandwiches and invoke the induction hypothesis. By mathematical induction, the result holds for all $n \geq 1$.

As another illustration, suppose there are n cars on a circular track. Together they have just enough gas for one car to complete a lap. Let $P(n)$ be the statement that there is a car which can complete a lap by collecting gas from the other cars on its way around. We prove by mathematical induction that $P(n)$ is true for all positive integers n.

The case $n = 1$ is trivial because the only car has enough fuel to go once around. Suppose the result holds for some $n \geq 1$. Consider the next case with $n + 1$ cars. Now one of them must have enough fuel to reach the next car. If this is not the case, then the cars do not have among them enough fuel to go once around the track. So we may assume that car A can reach the next car B. In a parallel universe, we take out car B and transfer all its fuel to car A. Now we have only n cars, and still enough fuel to go once around. By the induction hypothesis, some car can make that trip. Suppose this car is A. In the real world, car A has enough fuel to reach car B, and it will pick up all the fuel from car B. The situation is then the same as in the parallel universe, and car A can complete the trip once around the track. Suppose in the parallel universe, the car that can make the trip is not car A but some car C. At some point in time, car C must be able to reach car A, and this happens in the real world also. Car C will pick up less fuel from car A in the real world than in the parallel universe, but still enough to reach car B when it will then pick up the fuel in car B. At this point, the situation is once again the same in the real world as in the parallel universe. Since car C can complete the trip in the latter, it can also do so in the former. This completes the induction argument.

The traditional application of mathematical induction is to prove algebraic identities. As an illustration, $P(n)$ may be the statement

$$1 + 3 + 5 + \cdots + (2n - 1) = n^2.$$

For $n = 1$, the left side is equal to 1 while the right side is equal to $1^2 = 1$. Assume that the result holds for some $n \geq 1$. Then

$$1 + 3 + 5 + \cdots + (2n - 1) + (2n + 1) = n^2 + 2n + 1 = (n + 1)^2.$$

By mathematical induction, the result holds for all $n \geq 1$. A geometric interpretation is shown below.

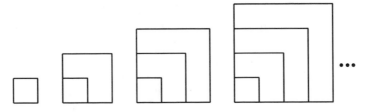

As another illustration, let $P(n)$ be the statement

$$1 - \frac{1}{2} + \frac{1}{3} - \frac{1}{4} + \cdots + \frac{1}{2n-1} - \frac{1}{2n} = \frac{1}{n+1} + \frac{1}{n+2} + \cdots + \frac{1}{2n}.$$

For $n = 1$, the left side is equal to $1 - \frac{1}{2} = \frac{1}{2}$ while the right side is equal to $\frac{1}{1 \times 2} = \frac{1}{2}$. Assume that the result holds for some $n \geq 1$. Then

$$
\begin{aligned}
& 1 - \frac{1}{2} + \frac{1}{3} - \frac{1}{4} + \cdots + \frac{1}{2n+1} - \frac{1}{2n+2} \\
={} & \frac{1}{n+1} + \frac{1}{n+2} + \cdots + \frac{1}{2n} + \frac{1}{2n+1} - \frac{1}{2n+2} \\
={} & \frac{1}{n+2} + \cdots + \frac{1}{2n} + \frac{1}{2n+1} + \frac{2-1}{2n+2} \\
={} & \frac{1}{n+2} + \cdots + \frac{1}{2n} + \frac{1}{2n+1} + \frac{1}{2n+2}.
\end{aligned}
$$

By mathematical induction, the result holds for all $n \geq 1$.

Suppose $P(n)$ is that statement that in every set of n crayons, all are of the same color. Clearly $P(1)$ is true. Suppose $P(n)$ is true for some $n \geq 1$. Consider the next case with $n+1$ crayons. By the induction hypothesis, the first n crayons are of the same color, and the last n crayons are also of the same color. Hence the first crayon and the last crayon must be of the same color as all the other crayons. It follows that $P(n+1)$ is true, so that $P(n)$ is true for all positive integers n.

What is going on? The statement $P(n)$ is blatantly false for $n \geq 2$, and yet we seem to have given a proof that it is true for all positive integers n. The problem is that $P(1)$ does not imply $P(2)$. For $n \geq 3$, $P(n)$ does imply $P(n+1)$, as the first crayon and the last crayon must be of the same color as all the other crayons. However, for $n = 2$, there are no other crayons.

In some problems, the truth of statement $P(n+1)$ may depend on more than just the truth of statement $P(n)$. It may depend on the truth of some or all of $P(1)$, $P(2)$, $\ldots$, $P(n)$. There are other variations too, but this version occurs often enough that we give it a formal statement.

Second Principle of Mathematical Induction.
Let $P(n)$ be a sequence of statements such that $P(k)$ is true for some positive integer k, and $P(n+1)$ is true whenever $P(k)$, $P(k+1)$, $\ldots$, $P(n)$ are true. Then $P(n)$ is true for all integers $n \geq k$.

Usually, when we say we use mathematical induction, we mean the First Principle of Mathematical Induction. We have to say explicitly the Second Principle of Mathematical Induction if we wish to emphasize that this is what we are using.

We give an illustration. Let $P(n)$ be the statement that the positive integer n can be expressed as a product of prime numbers. We wish to prove that $P(n)$ is true for all integers $n > 1$. The case $n = 2$ is trivial since 2 is itself a prime number. Suppose $P(2)$, $P(3)$, $\ldots$, $P(n)$ are true. Consider the positive integer $n + 1$. If it is a prime number, there is nothing further to prove. Otherwise, $n + 1 = ab$ for integers a and b, with $2 \le a \le n$ and $2 \le b \le n$. Then $P(a)$ and $P(b)$ are true, so that each of a and b is a product of prime numbers. Since $n + 1$ is the product of a and b, it is also a product of prime numbers. This completes the induction argument. Actually, 1 may also be considered as an empty product of prime numbers.

Although the Second Principle has a stronger induction hypothesis than the First Principle, they are in fact equivalent to each other. Moreover, they are also equivalent to the Well-Ordering Principle which we considered in Section 0.2. We prove this in three steps.

First, we establish that the First Principle of Mathematical Induction implies the Second Principle of Mathematical Induction. Let $P(n)$ be a sequence of statements such that $P(k)$ is true, and $P(n + 1)$ is true whenever $P(k)$, $P(k + 1)$, $\ldots$, $P(n)$ are true. Let $Q(n)$ be the statement that $P(k)$, $P(k + 1)$, $\ldots$, $P(n)$ are true. Since $P(k)$ is true, $Q(k)$ is true. Suppose $Q(n)$ is true. Then $P(k)$, $P(k+1)$, $\ldots$, $P(n)$ are true, so that $P(n+1)$ is true. It follows that $Q(n+1)$ is also true. By the First Principle of Mathematical Induction, $Q(n)$ is true for all integers $n \ge k$. This implies that $P(n)$ is also true for all integers $n \ge k$.

Next, we establish that the Second Principle of Mathematical Induction implies the Well-Ordering Principle, by showing that a subset S of the positive integers without a smallest element must be empty. Let $P(n)$ denote the statement that n is not an element of S. Now $P(1)$ must be true, as otherwise 1 is an element of S, and will be its smallest element. Suppose $P(1), P(2), \ldots, P(n)$ are all true. Then none of 1, 2, $\ldots$, n belong to S. If $n + 1$ does, it will be the smallest element of S. It follows that $P(n + 1)$ is also true, so that $P(n)$ is true for all positive integers n. This means that S is empty.

Finally, we establish that the Well-Ordering Principle implies the First Principle of Mathematical Induction. Let $P(n)$ be a sequence of statements such that $P(k)$ is true for some positive integer k, and $P(n + 1)$ is true whenever $P(n)$ is true. Let S be the subset of the integers $n \ge k$ such that $P(n)$ is false. Suppose S is non-empty. Then it has a smallest element m. Now $m > k$ since $P(k)$ is true. Hence $m - 1 \ge k$ is not in S, so that $P(m-1)$ is true. However, this implies that $P(m)$ is also true, so that m is not in S. This contradiction shows that S is empty. In other words, $P(n)$ is true for all integers $n \ge k$.

We now give the proof of the Addition Rule deferred from Section 0.1. This rule states that

$$|A_1 \cup A_2 \cup \cdots \cup A_n| = \sum_n |A_i| - \sum_n |A_i \cap A_j| + \cdots + (-1)^{n-1}|A_1 \cap A_2 \cap \cdots \cap A_n|.$$

We use induction on n. For $n = 2$, note that

$$A_1 \cup A_2 = (A_1 - A_1 \cap A_2) \cup (A_1 \cap A_2) \cup (A_2 - A_1 \cap A_2)$$

and the three subsets are pairwise disjoint. Moreover, $A_1 \cap A_2 \subseteq A_1$ and $A_1 \cap A_2 \subseteq A_2$. By the Addition Principle and the Subtraction Rule,

$$
\begin{aligned}
|A_1 \cup A_2| &= |A_1 - A_1 \cap A_2| + |A_1 \cap A_2| + |A_2 - A_1 \cap A_2| \\
&= |A_1| - |A_1 \cap A_2| + |A_1 \cap A_2| + |A_2| - |A_1 \cap A_2| \\
&= |A_1| + |A_2| - |A_1 \cap A_2|.
\end{aligned}
$$

Suppose the result holds for some $n \geq 2$. By the induction hypothesis, we have

$$
\begin{aligned}
&\quad |A_1 \cup A_2 \cup \cdots \cup (A_n \cup A_{n+1})| \\
&= \Sigma_{n-1}|A_i| + |A_n \cup A_{n+1}| \\
&\quad -\Sigma_{n-1}|A_i \cap A_j| - \Sigma_{n-1}|A_i \cap (A_n \cup A_{n+1})| + \cdots \\
&= \Sigma_{n+1}|A_i| - |A_n \cap A_{n+1}| \\
&\quad -\Sigma_{n-1}|A_i \cap A_j| - \Sigma_{n-1}|(A_i \cap A_n) \cup (A_i \cap A_{n+1})| + \cdots \\
&= \Sigma_{n+1}|A_i| - \Sigma_{n+1}|A_i \cap A_j| + \Sigma_{n-1}|A_i \cap A_n \cap A_{n+1}| + \cdots \\
&= \cdots \\
&= \Sigma_{n+1}|A_i| - \Sigma_{n+1}|A_i \cap A_j| + \cdots + (-1)^n|A_i \cap A_2 \cap \cdots \cap A_{n+1}|.
\end{aligned}
$$

Examples

Example 0.5.1.
An arbitrary square of a $2^n \times 2^n$ chessboard is removed. Prove that the remaining part can be covered by copies of the following shape, which when placed on the chessboard, covers exactly three of its squares. The copies may not overlap or stick out beyond the border of the chessboard.

Solution:
The result is trivially true for $n = 0$. Suppose it holds for some $n \geq 0$. Consider a $2^{n+1} \times 2^{n+1}$ chessboard. Divide it into four equal quadrants. Without loss of generality, we may assume that the missing square is in the north-east quadrant. By the induction hypothesis, the rest of this quadrant may be covered by copies of the given shape. We now place a copy of the given shape so that it covers the south-east corner of the north-west quadrant, the north-east corner of the south-west quadrant and the north-west corner of the south-east quadrant. Then each of these quadrant is missing one square. By the induction hypothsis again, each of them may be covered by the copies of the given shape. By mathematical induction, the result holds for all $n \geq 1$. This is due to **Solomon Golomb**.

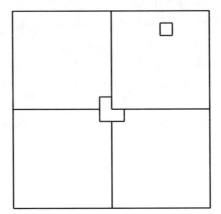

Example 0.5.2.
Use mathematical induction to prove that for all $n \geq 1$,

(a) $\frac{1}{1 \times 2} + \frac{1}{2 \times 3} + \cdots + \frac{1}{n(n+1)} = \frac{n}{n+1}$;

(b) $1^3 + 2^3 + \cdots + n^3 = \frac{n^2(n+1)^2}{4}$.

Solution:

(a) For $n = 1$, the left side is equal to $\frac{1}{1\times2} = \frac{1}{2}$ while the right side is equal to $\frac{1}{1+1} = \frac{1}{2}$. Assume that the result holds for some $n \geq 1$. Then

$$\frac{1}{1 \times 2} + \frac{1}{2 \times 3} + \cdots + \frac{1}{n(n+1)} + \frac{1}{(n+1)(n+2)}$$
$$= \frac{n}{n+1} + \frac{1}{(n+1)(n+2)}$$
$$= \frac{n(n+2)+1}{(n+1)(n+2)}$$
$$= \frac{n+1}{n+2}.$$

By mathematical induction, the result holds for all $n \geq 1$. Compare with Example 0.2.3.

(b) For $n = 1$, the left side is equal to $1^3 = 1$ while the right side is equal to $\frac{1^2 2^2}{4} = 1$. Assume that the result holds for some $n \geq 1$. Then

$$1^3 + 2^3 + \cdots + n^3 + (n+1)^3$$
$$= \frac{n^2(n+1)^2}{4} + (n+1)^3$$
$$= \frac{(n+1)^2}{4}(n^2 + 4n + 4)$$
$$= \frac{(n+1)^2(n+2)^2}{4}.$$

By mathematical induction, the result holds for all $n \geq 1$.

Here is a geometric solution of (b) due to **Solomon Golomb**. Consider the left side as the total volume of cubes of sides 1, 2, ..., n. Cut the $2 \times 2 \times 2$ cube into two $1 \times 2 \times 2$ slabs, cut one of them in halves, and assemble these three pieces along with the $1 \times 1 \times 1$ cube into a $1 \times (1+2) \times (1+2)$ square prism, as shown in the diagram below at the top left corner. Then cut the $3 \times 3 \times 3$ cube into three $1 \times 3 \times 3$ slabs, and use them to expand the existing prism into a $1 \times (1+2+3) \times (1+2+3)$ prism as shown in the diagram below. Continue to cut the remaining cubes into slabs of thickness 1, and for cubes of even side lengths, cut one of the slabs in halves. Eventually, the prism expands into a $1 \times (1+2+\cdots+n) \times (1+2+\cdots+n)$ prism, thus proving that

$$1^3 + 2^3 + \cdots + n^3 = (1 + 2 + \cdots + n)^2 = \frac{n^2(n+1)^2}{4}.$$

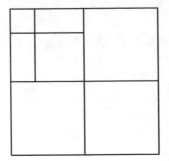

Example 0.5.3.

A merchant wanted to ship 100 valuable coins to a foreign market. A reputable courier charged 500 dollars for shipping each coin. Very Quiet Trucking would only charge 1000 dollars per shipment, regardless of the number of coins in the shipment. However, VQT was unreliable and might simply steal the entire shipment. The salesman from VQT said to the merchant, "You don't have to send your 100 coins with us all at once. You can divide them into various shipments of whatever sizes you choose, and they do not have to be uniform. If we steal a shipment, you can stop using us and go back to the reputable courier to send the remaining coins. You may also decide to do that anyway whenever it is in your interest. If we can make just as much money without stealing, we will not steal from you." Should the merchant use VQT? If so, how should the 100 coins be divided into several shipments, and what would be the net saving for using VQT instead of the reputable courier?

Solution:

More generally, suppose the merchant has n coins. Apply the Second Principle of Mathematical Induction by taking the statement $P(n)$ to mean that the merchant should not use VQT if the number of coins is n. Clearly, $P(1)$ is true as the merchant gets nothing by using VQT with only 1 coin, even if VQT is honest. Suppose $P(1)$, $P(2)$, ..., $P(n)$ are true, that is, the merchant should not use VQT if the number of coins is less than or equal to n. Consider the next case where the number of coins is $n+1$. Using VQT, the merchant must send in the first shipment of k coins for some positive integer $k < n+1$. Then the number of coins left is $(n+1) - k \le n$. VQT knows by the induction hypothesis that the merchant will not use VQT again. Hence VQT will either pocket the 1000 dollars if $k = 1$ or steal the shipment if $k > 1$. Now the merchant must go back to the reputable courier, but would have been better off had this been done in the first place. This completes the induction argument.

Exercises

1. A game is played with a deck of 100 cards numbered from 1 to 100. They are shuffled thoroughly and the top card is turned over. If it is number 1, the game is won. If it is number i where $2 \leq i \leq 100$, then it is inserted into the ith place from the top. Then the new top card is turned over and the same process is applied. Can this game be won eventually, regardless of how the cards are stacked initially?

2. One hundred girls are standing in a row, and a hat is put on each. None of them can see her own hat, but each can see the hats of all the girls with higher numbers than herself. They are told that all are wearing red hats, except for one who has a black hat on. Each is then asked in turn, starting with number 1, whether she can deduce the color of her hat. They answer either "Yes" or "No". Each can hear the answers of all the girls with lower numbers. Prove that exactly one girl will say 'No".

3. Use mathematical induction to prove that for all $n \geq 1$,

   (a) $2 + 2^2 + \cdots + 2^n = 2^{n+1} - 2$;

   (b) $1^2 + 2^2 + \cdots + n^2 = \frac{n(n+1)(2n+1)}{6}$.

4. Use mathematical induction to prove that for all $n \geq 2$,

   (a) $1 \times 2 + 2 \times 3 + \cdots + (n-1)n = \frac{(n-1)n(n+1)}{3}$;

   (b) $\left(1 - \frac{1}{4}\right)\left(1 - \frac{1}{9}\right) \cdots \left(1 - \frac{1}{n^2}\right) = \frac{n+1}{2n}$.

5. Prove that every positive integer can be represented as a sum of several distinct powers of 2.

6. (a) Prove that there exists a positive real number $x \neq 1$ such that $x + \frac{1}{x}$ is an integer.

   (b) Prove that $x^n + \frac{1}{x^n}$ is an integer for any positive integer n, where x is the real number in (a).

Section 0.6 Critical Measures

In applying the Extremal Value Principle, we seek the maximum or minimum value of a certain parameter. In applying the Mean Value Principle, we seek a value of a certain parameter which is not above average, or which is not below average. In applying the Pigeonhole Principle, the parameter is the number of pigeons per hole. We call such a parameter the **critical measure** for the problem.

The concept of critical measures is most suitable for handling problems in which the scenario continues to change. By focusing on a certain parameter, we can avoid being distracted by various aspects of the changing scenario.

As an illustration, consider six evenly spaced trees along a straight road, each with a sparrow perched on it. Whenever a sparrow flies from one tree to another, then at the same time some other sparrow flies from some tree to another the same distance away, but in the opposite direction. Is it possible for all the sparrows to gather on one tree?

Number the trees 1 to 6 from one end to the other. Each sparrow is labeled with the number of the tree on which it is perched. Let S be the sum of the six labels. This is taken to be our critical measure. Initially, we have $S = 1 + 2 + 3 + 4 + 5 + 6 = 21$. When the label of a sparrow increases because of its motion, the label of the other sparrow in motion must decrease by the same amount, so that S is unchanged. If all six sparrows end up on the same tree, S will be a multiple of 6. Since 21 is not a multiple of 6, the situation is impossible.

A critical measure which remains unchanged is called an **invariant**. Here are three more illustrations.

Initially, there are three piles with a, b and c chips, respectively. In one step, you may transfer one chip from any pile with x chips onto any other pile with y chips. You will then be paid $y - x + 1$ coins. Of course, if $y - x + 1 < 0$, it means that you will be fined $x - y - 1$ coins instead. After a while, the three piles have a, b and c chips again. What is the maximum number of coins you may have gained?

Let g be the number of coins you have gained. Initially, $g = 0$. After each move, the value of g changes by $y - x + 1$ from the given condition, whereas $a^2 + b^2 + c^2$ changes by $2(y - x + 1)$ since $(x-1)^2 + (y+1)^2 - x^2 - y^2 = 2(y - x + 1)$. Hence the critical measure $a^2 + b^2 + c^2 - 2g$ is an invariant. If you get back to the original distribution, then g must be zero again.

Note that $(x-1)(y+1) + (y+1)z + z(x-1) - xy - yz - zx = -(y-x+1)$. Hence another invariant is $ab + bc + ca + g$.

There is a heap of 1001 stones on a table. You are allowed to perform the following operation: you choose one of the heaps containing more than 2 stones, throw away 1 stone from that heap and divide the rest of it into two smaller (not necessarily equal) heaps. Is it possible to make all the heaps on the table contain exactly 3 stones?

Let S be the sum of the number of stones and the number of heaps. In each operation, the number of stones decreases by 1 while the number of heaps increases by 1. Hence S is invariant. Initially, $S = 1002$. If each heap contains exactly 3 stones, then S must be a multiple of 4. Since 1002 is not a multiple of 4, the task is impossible.

Each vertex of a regular 12-gon is labeled with a 1 except for one which is labeled with -1. In a move, we may change the signs of the labels of any k adjacent vertices. Is it possible to shift the only -1 to the next vertex in the clockwise direction, if
(a) $k = 3$;
(b) $k = 4$;
(c) $k = 6$?

The answer is negative in all three cases. The proof for all of them follows the same general scheme: we mark some of the vertices such that among any k adjacent vertices, the number of marked vertices is even. In the diagram below on the left, there are two marked vertices among any 3 adjacent vertices. In the diagram below in the center, there are two marked vertices among any 4 adjacent vertices. In the diagram below on the right, there are two marked vertices among any 6 adjacent vertices. Place the -1 on a marked vertex such that the next vertex in the clockwise direction is not marked. For our invariant, we take the product of the numbers on the marked vertices. Initially, it is equal to -1, but if the -1 has been "shifted" to the next vertex in the clockwise direction, it is equal to 1. Hence the task is impossible.

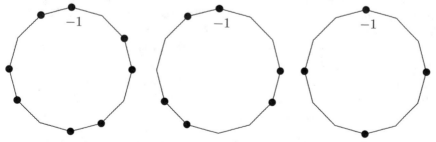

In the majority of cases, the critical measure is not an invariant, but its value is confined to a fixed range, which allows us to draw various conclusions. Perhaps the most common occurence is when the critical measure has a fixed value modulo m, where m is some positive integer. In the special case where $m = 2$, this is called a **parity** argument.

We present two simple illustrations.

The numbers 1, 2, 3, ..., 1989 are written on a blackboard. It is permitted to erase any two of them and replace them with their difference. Can this operation be used to obtain a situation where all the numbers on the blackboard are zeros?

Let S be the sum of all the numbers on the blackboard at the time. Initially, S is odd, since it is the sum of 994 even numbers and 995 odd numbers. Since the difference of two numbers is even if and only if so is their sum, the parity of S is invariant. If all the numbers on the blackboard are zeros, then S will be even. This is impossible.

There are 7 glasses on a table, all upside down. In each move, we may turn over any 4 of them, from upside down to right side up or vice versa. Is it possible to make all the glasses stand right side up?

Let S denote the number of glasses standing right side up. Initially, $S = 0$. Each move changes S by 0, ± 2 or ± 4. Hence the parity of S is invariant, so that $S = 7$ is impossible.

In the next illustration, consider three printing machines. The first accepts a card with any two numbers a and b on it and returns a card with the numbers $a + 1$ and $b + 1$. The second accepts only cards with two even numbers a and b and returns a card with the numbers $\frac{a}{2}$ and $\frac{b}{2}$ on it. The third accepts one card with the numbers a and b, and another card with the numbers b and c, and returns a card with the numbers a and c. All these machines also return the cards they accepted. Is it possible to obtain a card with numbers 1 and 2011, if we originally have only a card with the numbers 5 and 19?

The first machine changes (a, b) to $(a + 1, b + 1)$. The difference between the numbers on the cards is invariant here since $(a + 1) - (b + 1) = a - b$. However, the second machine changes the difference from $a - b$ to $\frac{a}{2} - \frac{b}{2} = \frac{a-b}{2}$. Moreover, the third machine adds the differences on the two cards, yielding $(a - b) + (b - c) = a - c$. Hence the difference between the numbers on the card is not an invariant, but it is clearly the critical measure. Note that $19 - 5 = 14 = 2 \times 7$. The factor 2 may be lost when we invoke the second machine. However, the factor 7 is preserved by all three machines. It follows that the critical measure is always divisible by 7. Since $1 - 2011 = -2010$ is not divisible by 7, the answer is negative.

For another illustration, consider a solitaire game played on an infinite chessboard. Initially, each cell of an $n \times n$ square is occupied by a chip. A move consists of jumping a chip over another chip in a horizontal or vertical direction onto a free cell directly beyond it. The chip jumped over is removed. Find all values of n for which the game can end with one chip left over.

A 1×1 square is already the desired state. It is easy to see how a 2×2 square can be reduced to a single occupied cell. Jump the two chips on the left column over the two chips on the right column. Then one of the surviving chips jumps over the other, and the task is accomplished. We now prove that there is no such reduction of a 3×3 square.

More generally, consider any $n \times n$ square where n is divisible by 3. We color the cells one diagonal at a time, using three colors A, B and C in cyclic order. The diagram below illustrates the case $n = 6$.

C	A	B	C	A	B
B	C	A	B	C	A
A	B	C	A	B	C
C	A	B	C	A	B
B	C	A	B	C	A
A	B	C	A	B	C

Denote the number of occupied cells of colors A, B and C by a, b and c, respectively. Initially, $a = b = c$, so that they have have the same parity. If we make a jump, two of these numbers are decreased by 1, and one is increased by 1. After the jump, all three numbers change parity. In other words, the parities of $a - b$, $b - c$ and $c - a$ are all invariant. However, this condition is violated if only one chip remains.

We now prove that the task is possible for all $n \times n$ squares where n is not divisible by 3. A sequence of jumps, called a T-operation, will be a useful tool. The diagram below shows how we can remove the central column of a T-shaped region with the help of an extra chip on the top bar on one side and an empty cell on the other side. In each step, the chip making the jump is marked black. Note that apart from the removal of the central column, nothing else is changed.

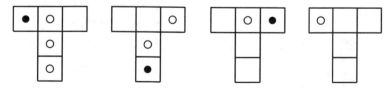

We now apply T-operations to reduce a 4×4 square to an occupied 1×1 square.

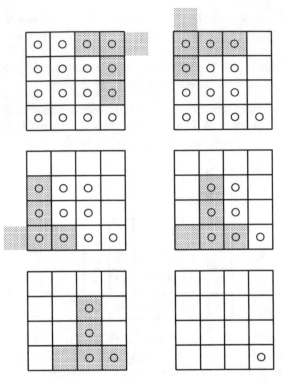

Next, we apply T-operations to reduce a 5×5 square to an occupied 2×2 square.

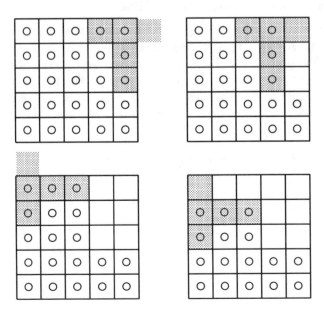

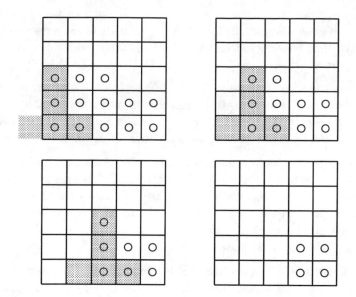

More generally, T-operations may be used to reduce an $n \times n$ square to an $(n - 3) \times (n - 3)$ square in three stages.

1. We keep an occupied 3×3 square at the top left corner and empty the remaining part of the top three rows using a vertical T.

2. We keep an occupied 3×3 square at the bottom left corner and empty the remaining part of the leftmost three columns using a horizontal T.

3. We empty the 3×3 square at the bottom left corner using a vertical T.

Finally, there are problems in which the critical measure is monotonic, and may increase or decrease without bounds. When the critical measure is a decreasing positive integer, the argument is akin to the method of infinite descent.

We start with a simple problem. The numbers 1, 2, 3, ..., 19, 20 are written on a blackboard. In each move, we may erase any two numbers a and b and replace them with $a + b - 1$. Only one number will remain after 19 moves. What is this number?

The critical measure is the sum S of all the numbers on the blackboard. Initially, $S = 210$. In each move, S decreases by 1. Hence after 19 moves, $S = 191$. By then, there is only one number on the blackboard. Hence this number must be 191.

We now conclude with three more illustrations. Each vertex of an n-gon is labeled with a real number. If $(a-d)(b-c) < 0$ for four successive labels a, b, c and d, then we may switch b with c. Prove that no further switching is possible after a while.

We take as the critical measure $S = ab+bc+cd+da$. After the switching operation, we have $S = ac+cb+bd+da$ so that $ab+cd$ is replaced by $ac+bd$. Since $(a-d)(b-c) < 0$, $ab+cd < ac+bd$. Hence S increases. However S can take only finitely many values, and the desired conclusion follows.

Suppose there is a table of integers with 1000 columns. The integers in the first row are chosen arbitrarily. Under each number in the first row, we count the number of its occurrences in the first row and enter it in the corresponding place in the second row. In the same way, we get the third row from the second row, and so on. Prove that we will eventually obtain a row identical to the preceding row.

Consider the j-th column of the table. Suppose the number in the i-th row is k. This means that the number in the $(i-1)$-st row appears k times in that row. Each of these appearances generates a k in the i-th row, so that the number in the $(i+1)$-st row and the j-th column is at least k. It follows that the numbers in the j-th column are in non-descending order. However, none of the numbers can exceed 1000. Hence the numbers are constant beyond a certain point. Since this is true of any column, the desired conclusion follows.

Suppose there is a table of integers with 4 columns. The integers in the first row are chosen arbitrarily, not all four equal. If the numbers in one row are a, b, c and d, then the numbers in the next row will be $a-b$, $b-c$, $c-d$ and $d-a$. Prove that the table eventually contains arbitrarily large integers.

Note that from the second row on, the sum of the four numbers in each row is 0. We take as the critical measure the sum S of the squares of these four numbers. In going from one row to the next, the new value of the critical measure is

$$
\begin{aligned}
S &= (a-b)^2 + (b-c)^2 + (c-d)^2 + (d-a)^2 \\
&= 2(a^2 + b^2 + c^2 + d^2) - 2ab - 2bc - 2cd - 2da.
\end{aligned}
$$

Now $0 = (a+b+c+d)^2 = (a+c)^2 + (b+d)^2 + 2ab + 2ad + 2bc + 2cd$. Adding these two equations, we get

$$
\begin{aligned}
S &= (a-b)^2 + (b-c)^2 + (c-d)^2 + (d-a)^2 \\
&= 2(a^2 + b^2 + c^2 + d^2) + (a+c)^2 + (b+d)^2 \\
&\geq 2(a^2 + b^2 + c^2 + d^2).
\end{aligned}
$$

Hence the value of S increases. Since it is an integer, it increases without bound. It follows that at least one of the labels must become arbitrarily large.

Examples

Example 0.6.1.

The numbers 1, 2, 3, ..., 19, 20 are written on a blackboard. In each move, we may erase any two numbers a and b and replace them with another number. Only one number will remain after 19 moves. What is this number if the replacement number is

(a) $a + b - 1$;

(b) $ab + a + b$?

Solution:

(a) Note that $(a+b-1)-1 = (a-1)+(b-1)$. If we decrease each number on the blackboard by 1, then their sum S is invariant. Initially, we have $S = 190$. Hence the final number on the blackboard must be $S + 1 = 191$.

(b) Note that $ab + a + b + 1 = (a+1)(b+1)$. If we increase each number on the blackboard by 1, then their product P is invariant. Initially, we have $P = 21!$. Hence the final number on the blackboard must be $P - 1 = 21! - 1$.

Example 0.6.2.

There are 20 azure, 21 beige and 22 cyan chameleons.

(a) When two chameleons of different colors meet, they merge into a chameleon of the third color. After several such mergers, only one chameleon remains. What is its color?

(b) When two chameleons of different colors meet, they both change their colors to the third one. Is it possible that after some time all 45 chameleons are the same color?

Solution:

Let a be the difference between the number of beige and cyan chameleons, b the difference between the number of cyan and azure chameleons, and c the difference between the number of azure and beige chameleons.

(a) Initially, $a = -1$, $b = 2$ and $c = -1$. In each move, one of them changes by 0, a second by 2 and the third by -2. Hence their parities are invariant. When the number of chameleons is down to 1, chameleons of the other two colors have vanished, and the difference between their numbers is 0. We can only make $b = 0$, meaning that we have the same number, namely 0. Hence only a beige chameleon remains.

(b) Initially, $a = -1$, $b = -1$ and $c = 2$. In each move, one of them changes by 0, a second by 3 and the third by -3. Hence their values modulo 3 are invariant. When all chameleons are of the same color, chameleons of the other two colors have vanished, and the difference between their numbers is 0. We cannot make any of a, b or c equal to 0, meaning that the desired state cannot be attained.

Example 0.6.3.
Each vertex of a pentagon is labeled with an integer and the sum of the five labels is positive. If x, y and z are the labels on three successive vertices and if $y < 0$, then we replace (x, y, z) by $(x + y, -y, y + z)$. By repeating this step if necessary, can we always obtain five non-negative labels?

Solution:
Let the labels be a, b, c, d and e in cyclic order. We take as the critical measure $(a-c)^2 + (b-d)^2 + (c-e)^2 + (d-a)^2 + (e-b)^2$. Consider a particular replacement. By symmetry, we may assume that we are replacing (a, b, c) by $(a + b, -b, b + c)$. Now the critical measure becomes

$$(a - c)^2 + (-b - d)^2 + (b + c - e)^2 + (d - a - b)^2 + (e + b)^2.$$

The change in the critical measure is given by $2b(a + b + c + d + e) < 0$ since $b < 0$ but $a + b + c + d + e > 0$. However, the critical measure is a positive integer, and cannot decrease forever. However, the replacement will continue as long as at least one label is negative. When the process stops, all five labels are non-negative.

Exercises

1. The world of the amoeba consists of the first quadrant of the plane divided into unit cells. Initially, a solitary amoeba is imprisoned in the south-west left corner cell. The prison consists of the six shaded cells as shown in the diagram below. It is unguarded, and the Great Escape is successful if the entire prison is unoccupied.

   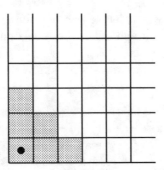

   In each move, an amoeba splits into two, with one going to the cell directly north and one going to the cell directly east. However, the move is not permitted if either of those two cells is already occupied. Can the Great Escape be achieved?

2. The world of the beetles consists of the entire plane divided into unit cells. Initially, all cells south of an inner wall constitute the prison, and each is occupied by a beetle. Freedom lies beyond an outer wall which is four rows north of the inner wall. If any beetle reaches any cell outside the unguarded prison, such as the shaded one in the diagram below, it will trigger the release of all surviving beetles. In that case, the Great Escape is successful.

   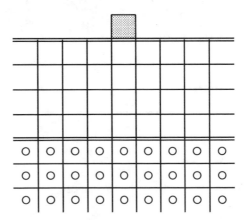

In each move, a beetle can jump over another beetle in an adjacent cell and land on the cell immediately beyond. However, the move is not permitted if that cell is already occupied. The beetle being jumped over is removed, making a sacrifice for the common good. The jump may be northward, eastward or westward. Can the Great Escape be achieved?

3. An Evil Dragon has 100 heads. It can only be killed if all the heads are cut off. The King has a magic sword which can cut off exactly 21 heads of the dragon. Prince Ivan has a magic sword which can cut off exactly 4 heads, but after that the dragon grows 1985 new heads. Prince Igor also has a magic sword. It can cut off exactly 5 heads, but after that the dragon grows 2012 new heads. Either Prince can borrow the King's magic sword. Can the dragon be killed by

   (a) Prince Ivan alone;

   (b) Prince Igor alone;

   (c) both Princes together?

4. A chip moves on an $n \times n$ chessboard so that in one move it can move one cell east, one cell north or one cell south-west. Can it visit all the cells of the board exactly once, and finish its trip on the cell immediately east of its starting cell?

5. Initially, some 1×1 cells of a 10×10 board are infected. In one minute, the cells sharing a common side with at least two infected cells become infected. What is the minimum number of cells initially infected so that the infection can spread to the whole board?

6. A blackboard is initially empty. In each move, one may either add two 1s, or erase two copies of a number n and replace them with $n-1$ and $n+1$. What is the minimum number of moves needed to put 2011 on the blackboard?

Practice Questions

Basic Counting.

1. There are five types of envelopes and four types of stamps in a post office. How many ways are there to buy an envelope or a stamp?

2. A restaurant offers seven main courses, five soups, and two salads. How many ways are there of ordering a complete dinner, consisting of one main course, one soup and one salad?

3. There are four towns A, B, C and D. Six roads go from A to B, and four roads go from B to D. Two roads go from A to C and three roads go from C to D. In how many ways can one go from A to D?

4. There are five different teacups, three saucers, and four teaspoons in a store. How many ways are there to buy two items of different types?

5. How many six-digit numbers have their digits all odd or all even?

6. Each of two novice collectors has 20 stamps and 10 postcards. We call an exchange fair if they exchange a stamp for a stamp or a postcard for a postcard. How many ways are there to carry out one fair exchange between these two collectors?

Inclusion-Exclusion.

7. In the graduating class of 80 students, 55 are trainees with MicroSoft, 35 with Google and 20 with both. How many are not trainees with either company?

8. In a class, 50 students take Swedish, 30 take Hungarian and 20 take both. If 40 students take neither of these two languages, how many students are there in the class?

9. In a cafeteria, 70 ordered soup, 40 ordered salad, 65 ordered sandwiches, 30 ordered both soup and salad, 60 ordered both soup and sandwiches, 20 ordered both salad and sandwiches and 15 ordered all three. How many customers are there in the cafeteria?

10. In a class, every student takes one of physics, chemistry and biology, and 6 take all three. The number of students taking physics is 30, chemistry 20 and biology 25. The number of students taking physics and chemistry is 10. The number of students taking chemistry and biology is at least 8. The number of students taking biology and physics is at most 8. What is the maximum number of students in this class?

11. How many positive integers up to 1000000 are neither squares or cubes of integers?

12. How many positive integers up to 1000000 are divisible by neither 4 nor 5?

Pigeonhole Principle.

13. A library has 70 books in English, 60 books in French, 50 books in German, 40 books in Italian, 30 books in Russian and 20 books in Spanish. How many books must a student take out at random in order to include 6 books in English, 11 books in French, 7 books in German, 4 books in Italian, 20 books in Russian or 8 books in Spanish?

14. A library has 70 books in English, 60 books in French, 50 books in German, 40 books in Italian, 30 books in Russian and 20 books in Spanish. How many books must a student take out at random in order to include 6 books in English, 11 books in French, 7 books in German, 4 books in Italian, 20 books in Russian and 8 books in Spanish?

15. The spider web shown in the diagram below consists of four straight threads each of length 2 meters. They intersect to form a square of side length 1 meter, with 0.5 meter protruding at either end. What is the largest number of spiders which can amicably share this spider web, if a spider will tolerate a neighbour only at a distance of 1.1 meters or more, traveling along the web?

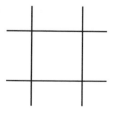

16. What is the largest number of identical kings which can be placed on a chessboard so that no two of them are in adjacent squares in a row, a column or a diagonal?

17. The shape shown in the diagram below is to be placed anywhere on an 8×8 checkerboard, such that it covers exactly three squares. It may be rotated. What is the largest number of the 64 squares which can be colored green so that at least one of the squares covered is not colored green?

18. The shape shown in the diagram above is to be placed anywhere on an 8×8 checkerboard, such that it covers exactly three squares. It may be rotated. What is the smallest number of the 64 squares which must be colored green so that at least one of the squares covered is colored green?

Mathematical Induction.

From small values of n, discover a closed form for each of the following summations and prove the result by mathematical induction:

19. $1 + 3 + \cdots + (2n - 1)$;

20. $1 + 2 + 2^2 + \cdots + 2^n$;

21. $1^3 + 3^3 + \cdots + (2n - 1)^3$;

22. $1 \times 2 \times 3 + 2 \times 3 \times 4 + \cdots + n(n + 1)(n + 2)$;

23. $\frac{1}{2 \times 5} + \frac{1}{5 \times 8} + \cdots + \frac{1}{(3n-1)(3n+2)}$;

24. $1^2 - 2^2 + 3^2 - 4^2 + \cdots + (-1)^{n-1} n^2$.

Answer

(1) 9 (2) 70 (3) 30 (4) 47 (5) 28125 (6) 500 (7) 10 (8) 100 (9) 85 (10) 57
(11) 998910 (12) 600000 (13) 51 (14) 260 (15) 4 (16) 22 (17) 32 (18) 24
(19) n^2 (20) $2^{n+1} - 1$ (21) $n^2(2n^2 - 1)$ (22) $\frac{n(n+1)(n+2)(n+3)}{4}$
(23) $\frac{n}{2(3n+2)}$ (24) $(-1)^{n-1} \frac{n(n+1)}{2}$

Chapter One: Combinations

Section 1.1. Combinations without Repetitions

A **combination** is a selection of elements from a finite set. By the definition of a set, each element appears only once. It is either selected or not, but never more than once. We refer to this as a combination *without repetitions*.

The number of ways of choosing k objects from a set of n objects is denoted by $\binom{n}{k}$, which is verbalized as "n choose k". We begin by making a few simple observations:

- $\binom{n}{k} = 0$ if $n < k$;

- $\binom{n}{0} = 1$;

- $\binom{n}{n} = 1$;

- $\binom{n}{k} = \binom{n}{n-k}$.

In the last item, the number of ways of choosing k objects from a set of n is equal to the number of ways of eliminating $n - k$ objects from the set of n. This is a prototype of a *combinatorial* argument, which we will pursue further. Such an argument is also featured in the proof of the following result.

Pascal's Formula: $\binom{n}{k} = \binom{n-1}{k-1} + \binom{n-1}{k}$.

Proof:
Fix an arbitrary one of the n objects. If we take it, we can choose $k - 1$ more from the remaining $n - 1$. If we leave it, we will be choosing all k from the remaining $n - 1$. The desired result follows from the Addition Principle since we either take it or leave it.

We can compute the value of $\binom{n}{k}$ recursively by building the famous Pascal's Triangle, the first few rows of which are shown below.

$$\binom{0}{0}=1$$

$$\binom{1}{0}=1 \qquad \binom{1}{1}=1$$

$$\binom{2}{0}=1 \qquad \binom{2}{1}=2 \qquad \binom{2}{2}=1$$

$$\binom{3}{0}=1 \qquad \binom{3}{1}=3 \qquad \binom{3}{2}=3 \qquad \binom{3}{3}=1$$

$$\binom{4}{0}=1 \qquad \binom{4}{1}=4 \qquad \binom{4}{2}=6 \qquad \binom{4}{3}=4 \qquad \binom{4}{4}=1$$

It is also possible to express $\binom{n}{k}$ directly in terms of the **factorial** function. For a positive integer n, we define $n! = n(n-1)(n-2)\cdots 3\cdot 2\cdot 1$. We also take $0!=1$.

© The Author(s), under exclusive license to Springer Nature Switzerland AG 2021
S. W. Golomb, A. Liu, *Solomon Golomb's Course on Undergraduate Combinatorics*,
https://doi.org/10.1007/978-3-030-72228-9_2

Factorial Formula.

For integers $n \geq k \geq 0$, $\binom{n}{k} = \frac{n!}{k!(n-k)!}$.

Proof:

We first verify the boundary conditions on Pascal's Triangle. We have $\frac{n!}{0!(n-0)!} = 1 = \binom{n}{0}$ and $\frac{n!}{n!(n-n)!} = 1 = \binom{n}{n}$. Thus the first two rows of Pascal's Triangles conform. We now proceed row by row. In each, the two outside entries have been verified. Each inside entry is the sum of two entries of the preceding row by Pascal's Formula. Indeed,

$$
\binom{n}{k} = \binom{n-1}{k-1} + \binom{n-1}{k}
$$

$$
= \frac{(n-1)!}{(k-1)!(n-k)!} + \frac{(n-1)!}{k!(n-k-1)!}
$$

$$
= \frac{(n-1)!}{k!(n-k)!}(k + (n-k))
$$

$$
= \frac{n!}{k!(n-k)!}.
$$

On the other hand, if we can establish this formula by another means, we can turn the above argument around to obtain a proof of Pascal's Formula.

The diagram below is the map of a small town with nine streets running north to south, and nine avenues running east to west.

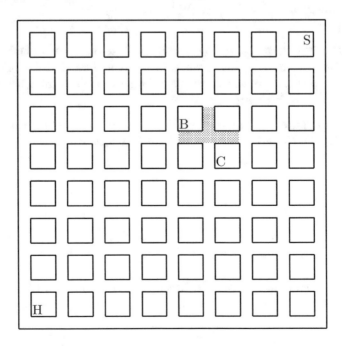

Mary's home H is at the intersection of 1st Street and 1st Avenue, and her school S is at the intersection of 9th Street and 9th Avenue. She goes to school walking 8 blocks east and 8 blocks north. Thus she can follow any of $\binom{16}{8}$ different routes. It is easy to see that the numbers of different routes leading to various points of intersection of the streets and avenues form part of Pascal's Triangle.

One day, 6th Avenue is blocked off between 5th Street and 7th Street, and 6th Street is blocked off between 6th Avenue and 7th Avenue. How many different routes of 16 blocks to school are there on that day?

We may as well assume that 6th Street is also blocked off between 5th Avenue and 6th Avenue, as using that block will take Mary to a dead-end. The total number of routes without obstruction is $\binom{16}{8}$. A blocked route takes Mary to the bank B at the intersection of 5th Street and 6th Avenue, or to the church C at the intersection of 5th Avenue and 6th Street, each in $\binom{9}{4}$ ways. After that, Mary proceeds to the intersection of 6th Avenue and 6th Street. From here, she can go to school in $\binom{6}{3}$ ways. Hence the total number of routes is $\binom{16}{8} - 2\binom{9}{4}\binom{6}{3} = 7830$.

In our next illustration, we turn to a problem in combinatorial number theory. In how many ways can three distinct integers be chosen from 1 to 1000 inclusive if their sum is to be divisible by 7?

Suppose $A = \{1, 8, \ldots, 995\}$, $B = \{2, 9, \ldots, 996\}$, $C = \{3, 10, \ldots, 997\}$, $D = \{4, 11, \ldots, 998\}$, $E = \{5, 12, \ldots, 999\}$, $F = \{6, 13, \ldots, 1000\}$ and take $G = \{7, 14, \ldots, 994\}$. If we choose three integers from the same set, they must all come from G and there are $\binom{142}{3}$ ways. If exactly two of the three integers are from the same set, we may have two from A and one from E, two from B and one from C, two from C and one from A, two from D and one from F, two from E and one from D, or two from F and one from B. The number of ways is $6 \times 143\binom{143}{2}$. If all three integers are from different sets, they may be from (A, B, D), (C, E, F), (A, F, G), (B, E, G) or (C, D, G). The number of ways is $2 \times 143^3 + 3 \times 142 \times 143^2$. The total is $\binom{142}{3} + 858\binom{143}{2} + 2 \times 143^3 + 426 \times 143^2 = 32449416$.

Our final illustration is a problem from combinatorial geometry. In a *convex n-gon*, each diagonal lies entirely inside it. All diagonals are drawn, and no three are concurrent. Many overlapping triangles are formed by the sides, the diagonals and parts of the diagonals. How many such triangles are there?

Let us paint the n vertices of the polygon red, and consider the following cases.

Case 1. Triangles with three red vertices.
Such a triangle corresponds to a set of three red vertices, and conversely. Hence there are $\binom{n}{3}$ of them.

Case 2. Triangle with exactly two red vertices.
Such a triangle corresponds to a set of four red vertices, the other two being at the end of the extensions of two sides past the non-red vertex. On the other hand, each set of four red vertices define four such triangles. The total is $4\binom{n}{4}$.

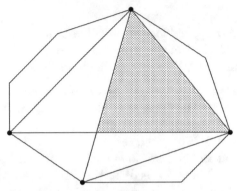

Case 3. Triangles with exactly one red vertex.
Such a triangle corresponds to a set of five red vertices. Two of them are at the end of the extensions of two sides past a non-red vertex, while the other two are at the end of the extensions in both directions of the side opposite to the red vertex. On the other hand, each set of five red vertices define five such triangles. The total is $5\binom{n}{5}$.

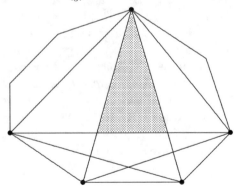

Case 4. Triangle with no red vertices.
Such a triangle corresponds to a set of six red vertices, at the end of the extensions in both directions of all three sides. On the other hand, each set of six red vertices define one such triangle. The total is $\binom{n}{6}$.

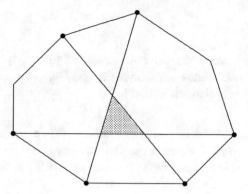

The grand total is therefore $\binom{n}{3} + 4\binom{n}{4} + 5\binom{n}{5} + \binom{n}{6}$.

Examples

Example 1.1.1.
In how many ways can we choose two subsets of $\{1, 2, \ldots, 100\}$ with respective sizes 3 and 5 such that each number in the 3-element set is greater than each element in the 5-element subset?

Solution:
We first choose an arbitrary 8-element subset. Then we partition it into a 3-element subset and a 5-element subset, putting the largest three numbers in the 3-element subset. This yields two subsets with the desired properties. Conversely, given two such subsets, we can merge them into a single 8-element subset. Thus we have a one-to-one correspondence. The number of 8-element subsets is clearly $\binom{100}{8}$.

Example 1.1.2.
In how many ways can three distinct integers be chosen from 1 to 1000 inclusive if their sum is to be divisible by 5?

Solution:
Let $A = \{1, 6, 11, \ldots, 996\}$, $B = \{2, 7, 12, \ldots, 997\}$, $C = \{3, 8, 13, \ldots, 998\}$, $D = \{4, 9, 14, \ldots, 999\}$ and $E = \{5, 10, 15, \ldots, 1000\}$. If we choose three integers from the same set, they must all come from E and there are $\binom{200}{3}$ ways. If exactly two of the three integers are from the same set, we may have two from A and one from C, two from B and one from A, two from C and one from D, or two from D and one from B. The number of ways is $4 \times 200\binom{200}{2}$. If all three integers are from different sets, one of them must come from E, and the other two from A and D or from B and C. The number of ways is 2×200^3. The total number of choices is $\binom{200}{3} + 800\binom{200}{2} + 2 \times 200^3$.

Example 1.1.3.
All diagonals of a convex n-gon are drawn, no three concurrent. Into how many segments do they divide themselves?

Solution:
The number of diagonals is $\binom{n}{2} - n$. The number of points of intersections of the diagonals is $\binom{n}{4}$. Such a point divides each of the two diagonals intersecting there into two. Hence the total number of segments is

$$\binom{n}{2} - n + 2\binom{n}{4}.$$

Exercises

1. Simplify $\binom{n}{2} + \binom{n+1}{2}$.

2. Determine positive integers $n \geq k$ which satisfy $\binom{n}{k+1} = 3\binom{n}{k-1}$ and $\binom{n}{k} = 2\binom{n}{k-1}$.

3. In how many ways can three distinct integers be chosen from 1 to 1000 inclusive if their sum is to be divisible by 3?

4. In how many ways can three distinct integers be chosen from 1 to 1000 inclusive if their sum is to be divisible by 2?

5. On the plane are n lines, no two parallel and no three concurrent. Into how many regions is the plane divided by these n lines?

6. On a circle are n points. All chords determined by these points are drawn, no three concurrent inside the circle. Into how many regions is the interior of the circle divided by these chords?

Section 1.2. Combinations with Repetitions

Suppose we have a multi-set consisting of n distinct elements each with infinite multiplicity. What is the number of ways of choosing a multi-subset of size k? In other words, how can we choose k objects from n if each may be taken any number of times?

For example, let $n = 3$ and $k = 4$. Since the numbers involved are small enough, we can make a list of all 15 multi-subsets as shown below.

1111	1133	2222
1112	1222	2223
1113	1223	2233
1122	1233	2333
1123	1333	3333

When the numbers involved are larger, counting the multi-subsets directly is no easy matter either. We now change each into a binary sequence as follows. Write down a number of 0s equal to the number of 1s in the label. Insert a 1 after this block. Then write down a number of 0s equal to the number of 2s, followed by another 1, and so on. Note that each binary sequence consists of two 1s and four 0s. They are listed below.

000011	001100	100001
000101	010001	100010
000110	010010	100100
001001	010100	101000
001010	011000	110000

It is easy to see that we have a one-to-one correspondence between the multi-subsets and the binary sequences. There are 6 symbols in a row, and we just have to choose four of them to place the 0s. The 1s will take up the remaining places. Thus the desired number is indeed $\binom{6}{4} = 15$.

In general, consider a set with infinite multiplicities for its n elements. The number of multi-subsets of size k is equal to the number of binary sequences with $n-1$ 1s and k 0s. The desired number is $\binom{k+n-1}{n-1} = \binom{k+n-1}{k}$.

Here is an alternative derivation of this formula. Let the elements in the set be 1, 2, ..., n. Let the k elements selected be $a_0 \le a_1 \le \cdots \le a_{k-1}$, Define $b_i = a_i + i$ for $0 \le i \le k - 1$. Then $b_0 < b_1 < \cdots < b_{k-1}$, and the maximum value of b_{k-1} is $n + k - 1$. We have a one-to-one correspondence between the desired multi-subsets on the one hand, and the subsets of the set $\{1, 2, \ldots, n + k - 1\}$ on the other hand. The number of the latter is clearly $\binom{n+k-1}{k}$.

We give yet another derivation of this formula. A boy has a deck of kn cards consisting of k copies of each of $1, 2, \ldots, n$. How many different hands of k cards can be dealt from his deck? This is equivalent to asking for the number of combinations of k objects chosen from among n objects with unlimited repetition.

The boy's fairy godmother also has a deck of cards. It consists of n ordinary cards $1, 2, \ldots, n$. In addition, there are $k-1$ magic cards $1, 2, \ldots, k-1$. The number of different hands of k cards which can be dealt from her deck is clearly $\binom{n+k-1}{k}$. We now establish a one-to-one correspondence between the k-card hands dealt from the two decks.

We first prove that each hand of the fairy godmother may be transformed into a hand of the boy. She arranges the k cards starting with the ordinary cards in increasing order, followed by the magic cards in increasing order. She then waves a wand over them. Each magic card in turn changes into an ordinary card, the magic card numbered j duplicating the j-th card in her hand.

Note that the fairy godmother's hand has at least one ordinary card. The earliest the magic card j can appear in her hand is when all higher numbered magic cards are also in the hand. Even then, this puts the magic card j in the $(j+1)$-st place. By the time it is to be transformed, the j-th card in the hand will be an ordinary card. It follows that each magic card duplicates an ordinary card, which may be an original one or a duplicate itself. We illustrate with the case $n = k = 3$ as shown below. The magic cards are in bold face.

Transformation of her hand		His hand
123	123	123
12**1**	121	112
12**2**	122	122
1**3**1	131	113
1**3**2	133	133
2**3**1	232	223
2**3**2	233	233
1**12**	112	111
2**12**	222	222
3**12**	332	333

We now prove that each hand of the fairy godmother may be obtained by transformation from a hand of the boy. Put the first appearance of each card in the first cycle, the second appearance of any card in the second cycle, and so on. Arrange the cards within each cycle in increasing order. Then put the cycles together starting with the first, followed by the second, and so on.

The transformation this time starts with the last card of the assembled hand. If it is already in the first cycle, there is nothing to do. If not, run through the cards towards the start of the hand until we reach a duplicate of the last card. Denote the place number of this duplicate by j. We then change the last card into the magic card j. We then turn to the second last card, and repeat this process until we reach the first cycle, after which no further transformation takes place. We cannot have duplicate magic cards, as the second appearance should have been replaced by a higher number. We illustrate with an instance of the case $n = k = 9$ as shown below. As before, the magic cards are in bold face.

$$
\begin{array}{ccccccccc}
2 & 2 & 3 & 3 & 3 & 5 & 8 & 8 & 9 \\
(2 & 3 & 5 & 8 & 9) & (2 & 3 & 8) & (\mathbf{3}) \\
2 & 3 & 5 & 8 & 9 & 2 & 3 & 8 & \mathbf{7} \\
2 & 3 & 5 & 8 & 9 & 2 & 3 & \mathbf{4} & \mathbf{7} \\
2 & 3 & 5 & 8 & 9 & 2 & \mathbf{2} & \mathbf{4} & \mathbf{7} \\
2 & 3 & 5 & 8 & 9 & \mathbf{1} & \mathbf{2} & \mathbf{4} & \mathbf{7}
\end{array}
$$

This completes the argument, which is due to **Solomon Golomb**.

We are now in a position to tackle two more cases of the Distribution Problem introduced in Section 0.1. In Case (3), n identical objects are to be distributed into k distinct boxes, with empty boxes allowed. Label each object 1 to k according to which box it is put, and convert each label into a binary sequence as before. Thus the desired answer is $\binom{n+k-1}{k-1}$.

In Case (4) where now empty boxes are not allowed, we first put one object into each box, and then distribute the remaining $n - k$ objects into the k boxes. Thus the desired answer is $\binom{(n-k)+(k-1)}{k-1} = \binom{n-1}{k-1}$.

Here is an alternative argument. Not allowing empty boxes means that no two 0s in the binary sequences may be adjacent. Hence each of the $k - 1$ 1s must be placed in a different one of the $n-1$ spaces between two adjacent 0s. This yields the same answer.

Imagine that you are the owner of a small coffee shop, and you have just imported 6 kilograms of coffee beans in a box. You will use 1 kilogram each day. Each time you open the box, some of the aroma disappears into thin air. Here is a mathematical model which may be used to measure aroma loss. Each kilogram of coffee beans in a box loses 1 aroma point every time the box is opened. Fortunately, you have 1 empty box which helps in reducing future losses. You want to minimize the total number of points lost.

Let the boxes be numbered 1 and 2, with the 6 kilograms of coffee beans in Box 1. At the end of Day 1, we have 5 exposed kilograms, 1 or 2 of which may be shifted to Box 2.

Day	Open Box	Points Lost	Shift to Box 2	Amount in Box 1	Amount in Box 2
1	1	6	1 kg	4 kg	1 kg
2	2	1		4 kg	
3	1	4	1 kg	2 kg	1 kg
4	2	1		2 kg	
5	1	2		1 kg	
6	1	1			
Total =		15			

Day	Open Box	Points Lost	Shift to Box 2	Amount in Box 1	Amount in Box 2
1	1	6	2 kg	3 kg	2 kg
2	2	2		3 kg	1 kg
3	2	1		3 kg	
4	1	3	1 kg	1 kg	1 kg
5	2	1		1 kg	
6	1	1			
Total =		14			

The second strategy is better, and the minimum loss is 14 points. However, this is obviously not the way to tackle the problem if the numbers involved are larger.

Suppose instead we have 66 kilograms of coffee beans and 4 empty boxes. Number the boxes 1, 2, 3, 4 and 5. By symmetry, we can arrange to have no more coffee beans in a box with a higher number than in a box with a lower one. Each day, we always open the non-empty box with the highest number. Thus we never transfer coffee beans from a box with a higher number to a box with a lower one.

Instead of measuring aroma loss day by day, which will depend on what is being done on the day, we measure aroma loss kilogram by kilogram. The number of points lost by each kilogram is equal to the number of times it is exposed. We keep track of this by putting a label on each kilogram. A label is initially empty. Every time the kilogram is exposed while in box i, add an i to the end of its current label.

The label lengthens progressively until the kilogram is used up. Its length at that time is the total number of points lost by this kilogram of coffee bean. Since exactly one kilogram is used each day, no two kilograms can have the same label. What we want is a set of the shortest labels.

As an illustration, let us find the total number of labels of length 7, such that the first term is 1, each subsequent term is 1, 2 or 3, and the terms are in non-descending order. This is equivalent to choosing a multi-subset of size 6 from the multiset $\{1, 2, 3\}$ in which each element has infinite multiplicity. There is a one-to-one correspondence between the labels and the binary sequences of length 8, consisting of 6 0s and 2 1s. The number of ways of putting 2 1s in the 8 places is $\binom{8}{2} = 28$.

Next, we find the total number of such labels of length up to 7. Using the one-to-one correspondence between the labels of length m and the binary sequences of length $m+1$, consisting of $m-1$ 0s and 2 1s, the total we seek is

$$\binom{2}{2} + \binom{3}{2} + \binom{4}{2} + \binom{5}{2} + \binom{6}{2} + \binom{7}{2} + \binom{8}{2}$$

$$= \binom{3}{3} + \left(\binom{4}{3} - \binom{3}{3}\right) + \left(\binom{5}{3} - \binom{4}{3}\right) + \left(\binom{6}{3} - \binom{5}{3}\right)$$

$$+ \left(\binom{7}{3} - \binom{6}{3}\right) + \left(\binom{8}{3} - \binom{7}{3}\right) + \left(\binom{9}{3} - \binom{8}{3}\right)$$

$$= \binom{9}{3}.$$

Here is a quicker approach. For labels shorter than 7, add a number of terms to make the total length 7. Each new term is 4. The corresponding binary sequence will be of length 9, consisting of 6 0s and 3 1s. Hence the total number is $\binom{9}{3}$. In general, the number of labels of length up to m, such that the first term is 1, each subsequent term is 1, 2, $\ldots$, k, and the terms are in non-descending order, is given by $\binom{m+k-1}{k}$.

Returning to our problem, since $\binom{8}{5} = 56 < 66 < 126 = \binom{9}{5}$, we use all 56 labels of length up to 4 and $66 - 56 = 10$ labels of length 5. The total point loss is $1 \times \binom{4}{4} + 2 \times \binom{5}{4} + 3 \times \binom{6}{4} + 4 \times \binom{7}{4} + 5 \times 10 = 246$.

So far, we have considered combination problems with unlimited repetition. Often, we have only limited repetition because the elements in the multi-set in question have finite multiplicities. We illustrate the general method with an example. From the multi-set $\{1,2,2,2,2,3,3,3,3,3,4,4\}$, we wish to choose 7 elements. Let A_i, $1 \le i \le 4$, denote the set of multi-subsets in which the number of copies of i taken exceeds its multiplicity. By the Principle of Inclusion-Exclusion, the number of desired multi-subsets is $|S| - \sum_4 |A_i| + \sum_4 |A_i \cap A_j| - \sum_4 |A_i \cap A_j \cap A_k| + |A_1 \cap A_2 \cap A_3 \cap A_4|$.

We have $|S| = \binom{10}{3}$. In A_1, we count all multi-sets which contain at least two 1s. The other five elements can be chosen arbitrarily without further concern over multiplicities. It follows that $|A_1| = \binom{8}{3}$. Similarly, $|A_2| = \binom{5}{3}$, $|A_3| = \binom{4}{3}$ and $|A_4| = \binom{7}{3}$.

What are we counting in $A_1 \cap A_2$? Here are all multi-subsets which contains at least two 1s and at least five 2s. Hence $|A_1 \cap A_2| = \binom{3}{3}$. The only other non-empty pairwise intersection of the As is $A_1 \cap A_4$ which counts all multi-subsets which contains at least two 1s and at least three 4s. We have $|A_1 \cap A_4| = \binom{5}{3}$. All triple intersections and the quadruple intersection of the As are empty. It follows that the number of desired multi-subsets is $\binom{10}{3} - \binom{8}{3} - \binom{5}{3} - \binom{4}{3} - \binom{7}{3} + \binom{3}{3} + \binom{5}{3} = 26$.

In the next illustration, we turn to a problem in combinatorial number theory. Determine the number of triples of positive integers a, b and c such that $abc = 4000000$ and none of a, b and c is divisible by the fifth power of a prime.

Note that $4000000 = 2^8 5^6$. Let $a = 2^i 5^p$, $b = 2^j 5^q$ and $c = 2^k 5^r$. Then $i + j + k = 8$, $p + q + r = 6$, and none of i, j, k, p, q and r exceeds 4. If there are no restrictions on i, j and k, the number of solutions to $i + j + k = 8$ is $\binom{8+2}{2} = 45$. At most one of i, j and k can exceed 5. Hence number of solutions to be excluded is $3\binom{3+2}{2} = 30$. It follows that there are 15 choices for i, j and k. Similarly, the number of choices for p, q and r is $\binom{6+2}{2} - 3\binom{1+2}{2} = 28 - 9 = 19$. Hence the number of choices for a, b and c is $15 \times 19 = 285$.

Examples

Example 1.2.1.
Consider the multiset $\{a, a, a, a, a, a, a, b, b, b, b, b, b, c, c, c\}$. How many multi-subsets of size 12 does it have?

Solution:
With unlimited repetition of the elements a, b and c, the number of ways is $\binom{12+2}{2} = 91$. Let A be the set of those combinations which include at least 8 copies of a, B be the set of those which include at least 7 copies of b, and C be the set of those which include at least 4 copies of c. Then $|A| = \binom{4+2}{2} = 15$, $|B| = \binom{5+2}{2} = 21$, $|C| = \binom{8+2}{2} = 45$, $|A \cap C| = \binom{0+2}{2} = 1$ and $|B \cap C| = \binom{1+2}{2} = 3$. By the Principle of Inclusion-Exclusion, the number of desired combinations is $91 - 15 - 21 - 45 + 1 + 3 = 14$.

Example 1.2.2.
You have just imported 66 kilograms of coffee beans in a box. You will use 1 kilogram each day. Each kilogram of coffee beans in a box loses 1 aroma point every time the box is opened. You have 3 empty boxes available. What is the minimum number of points lost?

Solution:
Since $\binom{7}{4} = 35 < 66 < 70 = \binom{8}{4}$, we use all 35 labels of length up to 4 and $66 - 35 = 31$ labels of length 5. The total point loss is

$$1 \times \binom{3}{3} + 2 \times \binom{4}{3} + 3 \times \binom{5}{3} + 4 \times \binom{6}{3} + 5 \times 31 = 274.$$

Example 1.2.3.
A company has three shareholders. Determine the number of ways of distributing $2n$ identical shares among them so that any two of them will hold more shares than the third.

Solution:
The total number of ways of distributing the shares is $\binom{2n+3-1}{2n}$. The number of ways in which a particular shareholder has at least n shares is $\binom{n+3-1}{n}$ and the number of ways in which two particular shareholders have at least n shares each is $\binom{0+3-1}{0}$. There are no ways in which all three shareholders have at least n shares each. By the Principle of Inclusion-Exclusion, the number of ways of distributing the shares without any of the three shareholders getting at least n shares is $\binom{2n+3-1}{2n} - 3\binom{n+3-1}{n} + 3\binom{0+3-1}{0} = \binom{n-1}{2}$.

Remark:
The simplicity of the answer suggests that perhaps there is an easier approach. Each shareholder may have at most $n - 1$ shares. Give that many to each at the start. Since the total number of shares is only $2n$, we must take back $n - 3$ of them, and this can be chosen at random. The number of ways of doing so is $\binom{n-3+2}{2} = \binom{n-1}{2}$.

Exercises

1. Determine the number of non-negative integer triples (p, q, r) whose sum is 52,

2. Determine the number of integer triples (p, q, r) whose sum is 25, subject to the constraints $0 \le p \le 5$, $5 \le q \le 25$ and $-5 \le r \le 5$.

3. You have just imported 35 kilograms of coffee beans in a box. You will use 1 kilogram each day. Each kilogram of coffee beans in a box loses 1 aroma point every time the box is opened. You have 2 empty boxes available. What is the minimum number of points lost?

4. You have just imported 53 kilograms of coffee beans in a box. You will use 1 kilogram each day. Each kilogram of coffee beans in a box loses 1 aroma point every time the box is opened. You have 2 empty boxes available. What is the minimum number of points lost?

5. Each of four players has a deck of cards numbered from 1 to 10. Each of them plays one card from his or her deck. In how many ways can it happen that the total of the four numbers is 27?

6. An ordinary six-sided die has the numbers 1, 2, 3, 4, 5 and 6 on its faces. When we roll three such dice which are distinguishable and examine the numbers on their top faces, there are $6 \times 6 \times 6 = 216$ different outcomes. How many different outcomes are there if the three dice are not distinguishable?

Section 1.3. Combinatorial Identities

The numbers $\binom{n}{k}$ satisfy a number of identities apart from Pascal's Formula. Here is a very useful one: $k\binom{n}{k} = n\binom{n-1}{k-1}$. By the Factorial Formula, we have

$$
\begin{aligned}
k\binom{n}{k} &= \frac{k(n!)}{k!(n-k)!} \\
&= \frac{n(n-1)!}{(k-1)!(n-k)!} \\
&= n\binom{n-1}{k-1}.
\end{aligned}
$$

At the end of Section 0.4, we mention a problem-solving technique, that of counting the same set in two different ways. Quite often, the obvious way to count the set is the harder way, so that using the alternative way leads to an easier solution to the problem. This is because counting the same set in two different ways should yield the same answer. When the equality of the two answers is not immediately apparent, we have discovered and established an identity.

We now give a combinatorial proof of $k\binom{n}{k} = n\binom{n-1}{k-1}$. Suppose we have a pool of n players from whom we would choose a team of size k. Then we choose a captain among the team players. Since the team can be chosen in $\binom{n}{k}$ ways and the captain can be chosen in k ways, the total number of ways is $k\binom{n}{k}$. However, we could have chosen the captain first in n ways, and the remaining team players in $\binom{n-1}{k-1}$ ways. The desired identity now follows.

Here is a more complicated identity, $\binom{\binom{n}{2}}{2} = 3\binom{n+1}{4}$. This can be proved as before by the Factorial Formula, but a combinatorial proof is more interesting.

Suppose we have n tennis players from which we wish to choose a pair to play in a tournament on Saturday and a different pair to play in a tournament on Sunday. Now there are $\binom{n}{2}$ possible pairs, and the total number of choosing two different pairs is $\binom{\binom{n}{2}}{2}$.

We now count the choices in a different way. Suppose the two pairs involve four different players. They can be chosen in $\binom{n}{4}$ ways, and partitioned into two pairs in 3 ways. Suppose the two pairs have one player in common, so that only three different players are involved. They can be chosen in $\binom{n}{3}$ ways. The player on both pairs can be chosen in 3 ways. Hence the overall total is $3\binom{n}{4} + 3\binom{n}{3} = 3\binom{n+1}{4}$ by Pascal's Formula. It follows that

$$
\binom{\binom{n}{2}}{2} = 3\binom{n+1}{4}.
$$

The simplicity of the term $3\binom{n+1}{4}$ suggests that there should be a simpler argument, and indeed there is. Let the coach be available for selection. We can now choose 4 players in $\binom{n+1}{4}$ ways, and partition them into two pairs in 3 ways. If the coach happens to be chosen, let the other three players be a, b and c. Then the three combinations will be (a,b) with (a,c), (b,c) with (b,a), and (c,a) with (c,b). This neat idea is due to **Solomon Golomb**.

We now turn our attention to identities involving summations. In our first illustration, we wish to prove that

$$\binom{n}{k} - \binom{n}{k-1} + \binom{n}{k-2} - \cdots + (-1)^k\binom{n}{0} = \binom{n-1}{k}.$$

The desired result follows from Pascal's Formula as the left side may be rewritten as

$$\left(\binom{n-1}{k} + \binom{n-1}{k-1}\right) - \left(\binom{n-1}{k-1} + \binom{n-1}{k-2}\right) - \cdots + (-1)^k\binom{n-1}{0}.$$

Here is an alternative approach, also using Pascal's Formula. We have

$$\binom{n-1}{k} = \binom{n}{k} - \binom{n-1}{k-1}$$

$$= \binom{n}{k} - \left(\binom{n}{k-1} - \binom{n-1}{k-2}\right)$$

$$= \binom{n}{k} - \binom{n}{k-1} + \left(\binom{n}{k-2} - \binom{n-1}{k-3}\right)$$

$$= \cdots$$

$$= \binom{n}{k} - \binom{n}{k-1} + \binom{n}{k-2} - \cdots + (-1)^k\binom{n}{0}.$$

This motivates the following combinatorial argument. Suppose we wish to choose a team of size k from a pool of n players, but one of the players is injured and cannot be chosen. Now $\binom{n}{k}$ counts the number of all possible teams, including those in which the injured player is chosen. From these, we exclude the $\binom{n}{k-1}$ where $k-1$ players are chosen from the whole pool, and the injured player is then added. However, the injured player may appear twice on the team. So we must add back the teams in which $k-2$ players are chosen, and the injured player added twice. Continuing to include and exclude alternately, the number of desirable teams is

$$\binom{n}{k} - \binom{n}{k-1} + \binom{n}{k-2} - \cdots + (-1)^k\binom{n}{0}.$$

However, the number of such selections is clearly $\binom{n-1}{k}$.

In our next illustration, we wish to prove that

$$\binom{n}{0}\binom{m}{k} - \binom{n}{1}\binom{m-1}{k} + \cdots + (-1)^n \binom{n}{n}\binom{m-n}{k} = \binom{m-n}{k-n}$$

for $m \geq k \geq n$. We use induction on n. For $n = 0$, both sides are equal to $\binom{m}{k}$. Suppose the identity holds for some $n \geq 0$. Consider the next case. We have

$$\binom{n+1}{0}\binom{m}{k} - \binom{n+1}{1}\binom{m-1}{k} + \cdots$$

$$+ (-1)^{n+1}\binom{n+1}{n+1}\binom{m-n-1}{k}$$

$$= \binom{n}{0}\binom{m}{k} - \left(\binom{n}{0} + \binom{n}{1}\right)\binom{m-1}{k} + \cdots$$

$$+ (-1)^n \left(\binom{n}{n-1} + \binom{n}{n}\right)\binom{m-n}{k} + (-1)^{n+1}\binom{n}{n}\binom{m-n-1}{k}$$

$$= \binom{n}{0}\binom{m}{k} - \binom{n}{1}\binom{m-1}{k} + \cdots + (-1)^n\binom{n}{n}\binom{m-n}{k}$$

$$- \binom{n}{0}\binom{m-1}{k} + \binom{n}{1}\binom{m-2}{k} - \cdots + (-1)^{n+1}\binom{n}{n}\binom{m-n-1}{k}$$

$$= \binom{m-n}{k-n} - \binom{m-n-1}{k-n}$$

$$= \binom{m-n-1}{k-n-1}.$$

Here is a combinatorial proof. Suppose we have a pool of m players with n captains. We choose a team of size k and all captains must be included. There are $\binom{m-n}{k-n}$ ways of choosing the rest of the team. On the other hand, the team can be chosen without regard to the captains in $\binom{n}{0}\binom{m}{k}$ ways. The number of ways of excluding one captain is $\binom{n}{1}\binom{m-1}{k}$. The number of ways of excluding two captains is $\binom{n}{2}\binom{m-2}{k}$, and so on. The result follows from the Principle of Inclusion-Exclusion.

In our final illustration, we wish to prove that

$$\binom{n}{0}\binom{n^2}{n} - \binom{n}{1}\binom{n^2-n}{n} + \cdots + (-1)^{n-1}\binom{n}{n-1}\binom{n}{n} = n^n.$$

Suppose each of n provinces send n players to the national camp. A national team of size n is to be chosen, and each province must be represented. Since there are n possible choices of players from each of the n provinces, the total number of permissible teams is n^n.

On the other hand, let S denote the set of all possible teams of size n, and for $1 \leq k \leq n$, let A_k denote the sets of teams of size n in which the kth province is not represented. By the Principle of Inclusion-Exclusion, the number of permissible teams is given by

$$
\begin{aligned}
n^n &= |\overline{A_1} \cap \overline{A_2} \cap \cdots \overline{A_n}| \\
&= |S| - \sum_n |A_i| + \sum_n |A_i \cap A_j| - \cdots + (-1)^n |A_1 \cap A_2 \cap \cdots \cap A_n| \\
&= \binom{n}{0}\binom{n^2}{n} - \binom{n}{1}\binom{n^2-n}{n} + \cdots + (-1)^{n-1}\binom{n}{n-1}\binom{n}{n}.
\end{aligned}
$$

Examples

Example 1.3.1.
Prove that $\binom{n+1}{2} + 2\binom{n+1}{3} + \cdots + n\binom{n+1}{n+1} = (n-1)2^n + 1$.

Solution:
We have $n + 1$ players wearing jerseys numbered consecutively from 1 to $n + 1$. We choose a team of arbitrary size, and a captain from among the team members, with only the stipulation that the captain must not have the lowest jersey number on the team. We count the number of ways of choosing team and captain by two methods. Let the team size be $k + 1$. We must have $1 \le k \le n$, and such a team can be chosen in $\binom{n+1}{k+1}$ ways. Moreover, the captain may be chosen in k ways since anyone but the player with the lowest jersey number on the team is in the running. Hence the total count is $\binom{n+1}{2} + 2\binom{n+1}{3} + \cdots + n\binom{n+1}{n+1}$. Without the stipulation, we can choose the captain in $n + 1$ ways and the rest of the team in 2^n ways. From $(n + 1)2^n$, we subtract the $2^{n+1} - 1$ non-empty teams in which the player with the lowest jersey number is the captain, yielding $(n-1)2^n + 1$. Equating the two answers yields the desired result.

Example 1.3.2.
Prove that $\binom{n}{0} + \binom{n+1}{1} + \binom{n+2}{2} + \cdots + \binom{n+m}{m} = \binom{n+m+1}{m}$.

Solution:
We have

$$
\begin{aligned}
& \binom{n}{0} + \binom{n+1}{1} + \binom{n+2}{2} + \cdots + \binom{n+m}{m} \\
= {} & \binom{n+1}{0} + \left(\binom{n+2}{1} - \binom{n+1}{0}\right) + \left(\binom{n+3}{2} - \binom{n+2}{1}\right) + \cdots \\
& + \left(\binom{n+m+1}{m} - \binom{n+m}{m-1}\right) \\
= {} & \binom{n+m+1}{m}.
\end{aligned}
$$

Here is a combinatorial proof. Let there be n boys and $m + 1$ girls. We wish to choose a team of size $n + 1$, and the number of ways of doing so is clearly $\binom{n+m+1}{n+1} = \binom{n+m+1}{m}$. However, we establish the following procedure. Let the girls be numbered from 1 to $m+1$. Since there is at least one girl on the team, let the highest number among the girls on the team be $k+1$. Then $0 \le k \le m$. For each value of k, the remaining n players may be chosen from the boys as well as the girls numbered from 1 to k, and the number of choices is $\binom{n+k}{n}$. Summing from $k = 0$ to $k = m$ yields the desired result.

Example 1.3.3.

Prove that $\binom{n}{0}\binom{m}{k} + \binom{n}{1}\binom{m}{k-1} + \cdots + \binom{n}{k}\binom{m}{0} = \binom{m+n}{k}$.

Solution:

We first give a combinatorial proof. Suppose we wish to choose a team of k players from among m girls and n boys. The total number of ways, without regard to gender, is $\binom{m+n}{k}$. If we break down according to gender, then there are $\binom{n}{0}\binom{m}{k}$ teams with 0 boys, $\binom{n}{1}\binom{m}{k-1}$ teams with 1 boy, and so on. Now we use induction on m. For $m = 0$, we have $k = 0$ as well, and the result is trivial. Suppose the identity holds for some $m \geq 0$. Consider the next case. We have

$$\binom{n}{0}\binom{m+1}{k} + \binom{n}{1}\binom{m+1}{k-1} + \cdots + \binom{n}{k}\binom{m+1}{0}$$

$$= \binom{n}{0}\left(\binom{m}{k-1} + \binom{m}{k}\right) + \binom{n}{1}\left(\binom{m}{k-2} + \binom{m}{k-1}\right)$$

$$\quad + \cdots + \binom{n}{k}\binom{m}{0}$$

$$= \binom{n}{0}\binom{m}{k-1} + \binom{n}{1}\binom{m}{k-2} + \cdots + \binom{n}{k-1}\binom{m}{0}$$

$$\quad + \binom{n}{0}\binom{m}{k} + \binom{n}{1}\binom{m}{k-1} + \cdots + \binom{n}{k}\binom{m}{0}$$

$$= \binom{m+n}{k-1} + \binom{m+n}{k}$$

$$= \binom{m+n+1}{k}.$$

Exercises

1. (a) Prove that $\binom{n}{1} + 2\binom{n}{2} + \cdots + n\binom{n}{n} = n2^{n-1}$.

    (b) Prove that $\binom{n}{1} - 2\binom{n}{2} + \cdots + (-1)^{n-1}n\binom{n}{n} = 0$ for $n \geq 2$.

2. (a) Prove that $\binom{n}{1} + 2^2\binom{n}{2} + \cdots + n^2\binom{n}{n} = n(n+1)2^{n-2}$.

    (b) Prove that $\binom{n}{1} - 2^2\binom{n}{2} + \cdots + (-1)^{n-1}n^2\binom{n}{n} = 0$ for $n \geq 3$.

3. Prove that $\binom{n+1}{1} + 2\binom{n+2}{2} + \cdots + m\binom{n+m}{m} = \frac{m(n+1)}{n+2}\binom{n+m+1}{m}$

    (a) using Pascal's Formula;

    (b) using a combinatorial argument.

4. Prove that $\binom{n+1}{1} + 2^2\binom{n+2}{2} + \cdots + m^2\binom{n+m}{m} = \frac{m(n+1)(nm+2m+1)}{(n+2)(n+3)}\binom{n+m+1}{n+1}$

    (a) using Pascal's Formula;

    (b) using a combinatorial argument.

5. Prove that $\binom{n}{0}\binom{n}{k} + \binom{n}{1}\binom{n-1}{k-1} + \cdots + \binom{n}{k}\binom{n-k}{0} = 2^k\binom{n}{k}$.

6. Prove that $\binom{n}{1}^2 + 2\binom{n}{2}^2 + \cdots + n\binom{n}{n}^2 = n\binom{2n-1}{n-1}$.

Section 1.4. Binomial Theorem

The number $\binom{n}{k}$ is called a **binomial coefficient** because it is the coefficient of the x^k-th term in the expansion of the n-th power of the binomial expression $1+x$. For instance, $(1+x)^0 = 1 = \binom{0}{0}$, $(1+x)^1 = 1+x = \binom{1}{0}+\binom{1}{1}x$ and $(1+x)^2 = 1+2x+x^2 = \binom{2}{0}+\binom{2}{1}x+\binom{2}{2}x^2$. The following is the general result.

Binomial Theorem. For any non-negative integer n and any complex number x,

$$(1+x)^n = \sum_{k=0}^{n} \binom{n}{k} x^k.$$

Proof:

We use mathematical induction on n. The basis has been established. Suppose the result holds for some $n \geq 0$. For the next case, we have

$$
\begin{aligned}
(1+x)^{n+1} &= (1+x)(1+x)^n \\
&= (1+x)\sum_{k=0}^{n}\binom{n}{k}x^k \\
&= \sum_{k=0}^{n}\binom{n}{k}x^k + \sum_{k=0}^{n}\binom{n}{k}x^{k+1} \\
&= \sum_{k=0}^{n}\binom{n}{k}x^k + \sum_{k=1}^{n+1}\binom{n}{k-1}x^k \\
&= \binom{n}{0} + \sum_{k=1}^{n}\binom{n}{k}x^k + \sum_{k=1}^{n}\binom{n}{k-1}x^k + \binom{n}{n}x^{n+1} \\
&= \binom{n+1}{0} + \sum_{k=1}^{n}\left(\binom{n}{k}+\binom{n}{k-1}\right)x^k + \binom{n+1}{n+1}x^{n+1} \\
&= \sum_{k=0}^{n+1}\binom{n+1}{k}x^k.
\end{aligned}
$$

This completes the induction argument.

Although x is a formal variable, we may assign it numerical values which yield various identities. For instance, if we put $x = 1$, we have

$$\sum_{k=0}^{n}\binom{n}{k} = 2^n.$$

If we put $x = -1$, we have, for any $n > 0$,

$$\sum_{k=0}^{n}(-1)^k\binom{n}{k} = 0.$$

This last identity is equivalent to the Principle of Inclusion-Exclusion. Assuming first the Principle of Inclusion-Exclusion, we take $S = A_i = \{x\}$ for $1 \leq i \leq n$. Then the intersection of any of the As is also $\{x\}$. It follows that

$$
\begin{aligned}
0 &= |\overline{A_1} \cap \overline{A_2} \cap \cdots \cap \overline{A_n}| \\
&= |S| - \Sigma_n |A_i| + \Sigma_n |A_i \cap A_j| - \cdots \\
&= \sum_{k=0}^{n} (-1)^k \binom{n}{k}.
\end{aligned}
$$

Assuming now the above identity, consider the sum

$$
|S| - \Sigma_n |A_i| + \Sigma_n |A_i \cap A_j| - \cdots.
$$

An element belonging to 0 of the As will be counted exactly once in $|S|$ and nowhere else. An object belonging to exactly k of the As will be counted once in $|S|$, $\binom{k}{1}$ times in $\Sigma_n |A_i|$, $\binom{k}{2}$ times in $\Sigma_n |A_i \cap A_j|$, and so on. Since $k > 0$, we have $\sum_{i=0}^{k} (-1)^i \binom{k}{i} = 0$. It follows that we are simply counting elements in $\overline{A_1} \cap \overline{A_2} \cap \cdots \cap \overline{A_n}$.

We now use the Binomial Theorem to solve some problems involving summation identities. In our first illustration, we wish to find a closed form for $\sum_{i=0}^{n} \frac{k^i}{i+1} \binom{n}{i}$. We have $\int_0^k (1+x)^n dx = \left[\frac{(1+x)^{n+1}}{n+1} \right]_0^k = \frac{(1+k)^{n+1} - 1}{n+1}$. Integrating term by term in the expansion of the binomial, we have

$$
\left[\sum_{i=0}^{n} \frac{x^{i+1}}{i+1} \binom{n}{i} \right]_0^k = k \left(\sum_{i=0}^{n} \frac{k^i}{i+1} \binom{n}{i} \right).
$$

Hence the desired closed form is $\frac{(1+k)^{n+1}-1}{k(n+1)}$.

Here is an alternative argument which also makes use of the Binomial Theorem. It follows from $(i+1)\binom{n+1}{i+1} = (n+1)\binom{n}{i}$ that the given expression is equal to

$$
\sum_{i=0}^{n} \frac{k^i}{n+1} \binom{n+1}{i+1} = \frac{1}{k(n+1)} \sum_{i=1}^{n+1} k^i \binom{n+1}{i} = \frac{(1+k)^{n+1} - 1}{k(n+1)}.
$$

In our next illustration, we wish to prove that for $m \geq k \geq n$,

$$
\sum_{i=0}^{n} (-1)^i \binom{n}{i} \binom{m-i}{k} = \binom{m-n}{k-n}.
$$

We have $x^n = ((1+x) - 1)^n = \sum_{i=0}^{n} (-1)^i \binom{n}{i} (1+x)^{n-i}$. Multiply both sides

by $(1+x)^{m-n}$, $x^n (1+x)^{m-n} = \sum_{i=0}^{n} (-1)^i \binom{n}{i} (1+x)^{m-i}$. Consider now the

coefficient of the term x^k. On the left side, it is the same as the coefficient
of x^{k-n} in $(1+x)^{m-n}$, namely, $\binom{m-n}{k-n}$. On the right side, it is

$$\sum_{i=0}^{n} (-1)^i \binom{n}{i} \binom{m-i}{k}.$$

This yields the desired result.

In our final illustration, we prove that $\sum_{i=0}^{n-1} (-1)^i \binom{n}{i} \binom{n^2 - in}{n} = n^n$.
Define $f(x) = (1 - (1+x)^n)^n$ We have

$$
\begin{aligned}
f(x) &= \sum_{i=0}^{n} (-1)^i \binom{n}{i} (1+x)^{in} \\
&= \sum_{i=0}^{n} (-1)^i \binom{n}{i} \sum_{j=0}^{i} n \binom{in}{j} x^j \\
&= \sum_{j=0}^{n^2} x^j \sum_{i=m}^{n} (-1)^i \binom{n}{i} \binom{in}{j},
\end{aligned}
$$

where m is the greatest integer not exceeding $\frac{j}{n}$. Thus the coefficient of x^n in
$f(x)$ is $\sum_{j=1}^{n} (-1)^j \binom{n}{j} \binom{jn}{n}$. On the other hand, $f(x) = \left(1 - \sum_{i=0}^{n} \binom{n}{i} x^i\right)^n$.
Thus the coefficient of x^n in $f(x)$ is $(-1)^n n^n$. Equating the two results, we
have

$$\sum_{j=1}^{n} (-1)^j \binom{n}{j} \binom{kn}{n} = (-1)^n n^n.$$

Canceling $(-1)^n$, we have

$$\sum_{j=1}^{n} (-1)^{j-n} \binom{n}{n-j} \binom{jn}{n} = n^n.$$

Setting $i = n - j$, we have

$$\sum_{i=0}^{n-1} (-1)^i \binom{n}{i} \binom{n^2 - in}{n} = n^n.$$

Examples

Example 1.4.1.

Prove that $\displaystyle\sum_{i=1}^{n} i\binom{n+1}{i+1} = (n-1)2^n + 1$ by the Binomial Theorem.

Solution:

Note that $x(x+1)^{n-1} = ((x+1)-1)(x+1)^{n-1} = (x+1)^n - (x+1)^{n-1}$.

It follows that $(x+1)^n - (x+1)^{n-1} = \displaystyle\sum_{i=0}^{n-1}\binom{n-1}{i}x^{i+1}$. Integration yields

$$\frac{(x+1)^{n+1}}{n+1} - \frac{(x+1)^n}{n} + C = \sum_{i=0}^{n-1}\binom{n-1}{i}\frac{x^{i+2}}{i+2}.$$ Setting $x = 0$, the integra-

tion constant C assumes the value $\frac{1}{n(n+1)}$. Setting $x = 1$, we have

$$\sum_{i=0}^{n-1}\frac{1}{i+2}\binom{n-1}{i} = \frac{(n-1)2^n + 1}{n(n+1)}.$$

Since $\frac{n(n+1)}{i+2}\binom{n-1}{i} = (i+1)\binom{n+1}{i+2}$, $\displaystyle\sum_{i=0}^{n-1}(i+1)\binom{n+1}{i+2} = (n-1)2^n + 1$, which

is equivalent to $\displaystyle\sum_{i=1}^{n} i\binom{n+1}{i+1} = (n-1)2^n + 1$.

Here is an alternative argument which also uses the Binomial Theorem. Writing i as $(i+1) - 1$ and using $(i+1)\binom{n+1}{i+1} = (n+1)\binom{n}{i}$, we have

$$\begin{aligned}
\sum_{i=1}^{n} i\binom{n+1}{i+1} &= \sum_{i=1}^{n}(i+1)\binom{n+1}{i+1} - \sum_{i=1}^{n}\binom{n+1}{i+1}\\
&= (n+1)\sum_{i=1}^{n}\binom{n}{i} - \sum_{i=2}^{n+1}\binom{n+1}{i}\\
&= (n+1)(2^n - 1) - (2^{n+1} - 1 - (n+1))\\
&= (n-1)2^n + 1.
\end{aligned}$$

Compare with Example 1.4.1.

Example 1.4.2.

Prove that $\displaystyle\sum_{i=0}^{m}\binom{n+i}{i} = \binom{n+m+1}{m}$ by the Binomial Theorem.

Solution:

The given sum is the coefficient of x^n in

$$\begin{aligned}
S &= \sum_{i=0}^{m}(1+x)^{n+i}\\
&= (1+x)^n\frac{(1+x)^{m+1} - 1}{x}
\end{aligned}$$

$$= \frac{1}{x}(1+x)^{n+m+1} - \frac{1}{x}(1+x)^n.$$

The coefficient of x^n in the first term is the same as the coefficient of x^{n+1} in $(1+x)^{n+m+1}$ or $\binom{n+m+1}{m}$. The coefficient of x^n in the second term is 0. Hence the coefficient of x^n in S is $\binom{n+m+1}{m}$. Compare with Example 1.3.2.

Example 1.4.3.

Prove that $\displaystyle\sum_{i=0}^{k} \binom{n}{i}\binom{m}{k-i} = \binom{m+n}{k}$ by the Binomial Theorem.

Solution:

We have $(1+x)^{m+n} = (1+x)^n(1+x)^m$. The coefficient of the term x^k on the left side is $\binom{m+n}{k}$ while that on the right side is $\displaystyle\sum_{i=0}^{k} \binom{n}{i}\binom{m}{k-i}$. Compare with Example 1.3.3.

Exercises

1. Prove by the Binomial Theorem that

   (a) $\displaystyle\sum_{i=1}^{n} i\binom{n}{i} = n2^{n-1}$;

   (b) $\displaystyle\sum_{i=1}^{n} (-1)^i i\binom{n}{i} = 0$ for $n \geq 2$.

2. Prove by the Binomial Theorem that

   (a) $\displaystyle\sum_{i=1}^{n} i^2\binom{n}{i} = n(n+1)2^{n-2}$;

   (b) $\displaystyle\sum_{i=1}^{n} (-1)^i i^2\binom{n}{i} = 0$ for $n \geq 3$.

3. Use the Binomial Theorem to prove that

$$\sum_{i=1}^{m} i\binom{n+i}{i} = \frac{m(n+1)}{n+2}\binom{n+m+1}{m}.$$

4. Use the Binomial Theorem to prove that

$$\sum_{i=1}^{m} i^2\binom{n+1}{i} = \frac{m(n+1)(nm+2m+1)}{(n+2)(n+3)}\binom{n+m+1}{n+1}.$$

5. Prove that $\displaystyle\sum_{i=0}^{k} \binom{n}{i}\binom{n}{k-i} = 2^k\binom{n}{k}$ by the Binomial Theorem.

6. Prove that $\displaystyle\sum_{i=1}^{n} i\binom{n}{i}^2 = n\binom{2n-1}{n-1}$ by the Binomial Theorem.

Section 1.5. Multinomial Theorem

The Binomial Theorem in Section 1.4 is a special case of a result known as the Multinomial Theorem. To provide motivation for this generalization, we reexamine the related concepts in a symmetric form.

Instead of regarding $\binom{n}{k}$ as the number of selections of k objects from n objects, we may regard it as the number of distributions of n objects into two boxes, with k objects in the first box. It goes without saying that there will be $n - k$ objects in the second box. Thus a more symmetric notation is $\binom{n}{k,n-k}$. This is still known as the binomial coefficient. However, if we distribute the n objects into three boxes, with t_1 in the first box and t_2 in the second, then the number of distributions is $\binom{n}{t_1,t_2,t_3}$ where $t_1 + t_2 + t_3 = n$. This is an example of a **trinomial** coefficient.

One of the basic properties of the binomial coefficients may be restated in the symmetric form as $\binom{n}{0,n} = \binom{n}{n,0} = 1$. Pascal's Formula, restated in the symmetric form

$$\binom{n}{k,\,n-k} = \binom{n-1}{k-1,\,n-k} + \binom{n-1}{k,\,n-k-1},$$

is particularly illuminating. We are just reducing in turns each of k and $n - k$ by 1. The corresponding properties of the trinomial coefficients are $\binom{n}{0,0,n} = \binom{n}{0,n,0} = \binom{n}{n,0,0} = 1$ and

$$\binom{n}{t_1,t_2,t_3} = \binom{n-1}{t_1-1,t_2,t_3} + \binom{n-1}{t_1,t_2-1,t_3} + \binom{n-1}{t_1,t_2,t_3-1},$$

provided that $t_1 - 1$, $t_2 - 1$ and $t_3 - 1$ are non-negative. Using these results, we can construct the first five layers of Pascal's Tetrahedron.

$$\binom{4}{0,4,0}$$

$$\binom{4}{1,3,0} \qquad \binom{4}{0,3,1}$$

$$\binom{4}{2,2,0} \qquad \binom{4}{1,2,1} \qquad \binom{4}{0,2,2}$$

$$\binom{4}{3,1,0} \qquad \binom{4}{2,1,1} \qquad \binom{4}{1,1,2} \qquad \binom{4}{0,1,3}$$

$$\binom{4}{4,0,0} \qquad \binom{4}{3,0,1} \qquad \binom{4}{2,0,2} \qquad \binom{4}{1,0,3} \qquad \binom{4}{0,0,4}$$

The numerical values are

$$
\begin{array}{ccccccccc}
 & & & & 1 & & & & \\
 & & & & 1 & & & & \\
 & & & 4 & & 4 & & & \\
 & & & & 1 & & & & \\
 & & & 3 & & 3 & & & \\
 & & 6 & & 12 & & 6 & & \\
 & & & & 1 & & & & \\
 & & & 2 & & 2 & & & \\
 & & 3 & & 6 & & 3 & & \\
 & 4 & & 12 & & 12 & & 4 & \\
 & & & & 1 & & & & \\
 & & & 1 & & 1 & & & \\
 & & 1 & & 2 & & 1 & & \\
 & 1 & & 3 & & 3 & & 1 & \\
1 & & 4 & & 6 & & 4 & & 1 \\
\end{array}
$$

In its symmetric form, the Binomial Theorem expands $(x_1 + x_2)^n$ into

$$\binom{n}{n,0}x_1^n + \binom{n}{n-1,1}x_1^{n-1}x_2 + \binom{n}{n-2,2}x_1^{n-2}x_2^2 + \cdots + \binom{n}{0,n}x_2^n.$$

The usual form is obtained by setting $x_1 = 1$ and $x_2 = x$. We now state the general result.

Multinomial Theorem.
We have

$$(x_1 + x_2 + \cdots + x_k)^n = \sum \binom{n}{t_1, t_2, \ldots, t_k} x_1^{t_1} x_2^{t_2} \cdots x_k^{t_k},$$

where the summation ranges over all non-negative integers $t_1, t_2, \ldots, t_k$ whose sum is n.

This may be proved by induction, and we omit the details. The numbers $\binom{n}{t_1, t_2, \ldots, t_k}$ are called **multinomial coefficients**.

As an illustration, we have

$$(x_1 + x_2 + x_3)^3 = \binom{3}{3,0,0}x_1^3 + \binom{3}{0,3,0}x_2^3 + \binom{3}{0,0,3}x_3^3$$
$$+ \binom{3}{2,1,0}x_1^2x_2 + \binom{3}{0,2,1}x_2^2x_3 + \binom{3}{1,0,2}x_3^2x_1$$
$$+ \binom{3}{2,0,1}x_1^2x_3 + \binom{3}{1,2,0}x_2^2x_1 + \binom{3}{0,1,2}x_3^2x_2$$
$$+ \binom{3}{1,1,1}x_1x_2x_3$$
$$= x_1^3 + x_2^3 + x_3^3 + 3x_1^2x_2 + 3x_2^2x_3 + 3x_3^2x_1$$
$$+ 3x_1^2x_3 + 3x_2^2x_1 + 3x_3^2x_2 + 6x_1x_2x_3.$$

The number of distinct terms in the above expansion is 10. This can be determined in advance as follows. We change each term into a binary sequence as follows. Write down a number of 0s equal to the number of x_1s in the term. Insert a 1 after this block. Then write down a number of 0s equal to the number of x_2s, followed by another 1. Finally, write down a number of 0s equal to the number of x_3s. Note that each binary sequence consists of 2 1s and 3 0s, and there are $\binom{2+3}{2} = 10$ such binary sequences.

The Multinomial Theorem also provides an alternative solution to Case (1) of the Distribution Problem in Section 0.1. Consider $\sum \binom{n}{t_1, t_2, \ldots, t_k}$ where the summation ranges over all non-negative integers t_1, t_2, ..., t_k whose sum is n. This expression represents all possible ways of distributing n distinct objects into k distinct boxes. Setting $x_1 = x_2 = \cdots = x_k = 1$ in the Multinomial Theorem, it reduces to k^n.

The answer to Case (2) of the Distribution Problem, where empty boxes are not allowed, is given by $\sum \binom{n}{t_1, t_2, \ldots, t_k}$. Here the summation ranges over all *positive* integers t_1, t_2, ..., t_k whose sum is n. For instance, the number of distributions of 4 distinct objects into 2 distinct boxes, with empty boxes not allowed, is given by $\binom{4}{3,1} + \binom{4}{2,2} + \binom{4}{1,3} = 4 + 6 + 4 = 14$.

Examples

Example 1.5.1.
Determine the coefficient of the term $x_2 x_3 x_4^2$ in the expression

$$(x_1 + 2x_2 + 3x_3 + 4x_4)^4.$$

Solution:
The desired term is $\binom{4}{0,1,1,2}(2x_2)(3x_3)(4x_4)^2$ and the desired coefficient is $12 \times 96 = 1152$.

Example 1.5.2.
Determine the number of distinct terms in the expression

$$(x_1 + x_2 + x_3 + x_4)^4.$$

Solution:
We change each term into a binary sequence as follows. Write down a number of 0s equal to the number of x_1s in the term. Insert a 1 after this block. Then write down a number of 0s equal to the number of x_2s, followed by another 1. Next, write down a number of 0s equal to the number of x_3s, followed by yet another 1. Finally, write down a number of 0s equal to the number of x_4s. Note that each binary sequence consists of 3 1s and 4 0s, and there are $\binom{3+4}{3} = 35$ such binary sequences.

Example 1.5.3.
Determine the number of distributions of 5 distinct objects into 3 distinct boxes, with empty boxes not allowed.

Solution:
The number of desired distributions is given by

$$\binom{5}{3,1,1} + \binom{5}{1,3,1} + \binom{5}{1,1,3} + \binom{5}{2,2,1} + \binom{5}{2,1,2} + \binom{5}{1,2,2}$$
$$= 3\binom{5}{3,1,1} + 3\binom{5}{2,2,1}$$
$$= 3 \times 20 + 3 \times 30$$
$$= 150.$$

Exercises

1. Determine the coefficient of the term $x_2 x_3 x_4$ in the expression

$$(x_1 + 2x_2 + 3x_3 + 4x_4)^3.$$

2. Determine the coefficient of the term $x_1 x_2 x_3^2$ in the expression

$$(x_1 + 2x_2 + 3x_3)^4.$$

3. Determine the number of distinct terms in the expression

$$(x_1 + x_2 + x_3 + x_4)^3.$$

4. Determine the number of distinct terms in the expression

$$(x_1 + x_2 + x_3)^4.$$

5. Determine the number of distributions of 6 distinct objects into 3 distinct boxes, with empty boxes not allowed.

6. Determine the number of distributions of 6 distinct objects into 4 distinct boxes, with empty boxes not allowed.

Section 1.6. Probability

A **probability** problem consists of a pair of enumeration problems. First, we enumerate the **sample space**, that is, we count the total number of possible outcomes. Second, we enumerate the **specific event**, that is, we count the number of outcomes in a specific subset. The probability of this event occurring is given by the size of the latter divided by the size of the former.

For instance, we may be rolling an ordinary six-sided die, and seek the probability that the number on the top face is even. The sample space is of size six, namely, 1, 2, 3, 4, 5 and 6. The even event is of size three, namely, 2, 4 and 6. Hence the probability is $\frac{3}{6} = \frac{1}{2}$.

We should emphasize that what we are dealing with is often called **a priori** probability. In our example, we presuppose that the six outcomes appear with equal probability, namely, $\frac{1}{6}$. In practice, if we roll a die 100 times and the number 4 shows up 72 times, it would be unwise to continue to assume that we have a fair die. We must continuously revise our assumption according to available data.

This is often called **empirical** probability. A typical example is the batting average of a professional baseball player. In our discussion, we will focus on a priori probability, which is often used as a mathematical model for studying empirical probability. In the absence of concrete data, we have to make a priori assumptions which facilitate the computation of probabilities.

We begin with a simple problem. Robin and Kelly were siblings. Robin said, "If you pick one of my siblings at random, the probability that you have picked a boy is $\frac{1}{3}$." Kelly said, "If you pick one of my siblings at random, the probability that you have picked a girl is $\frac{1}{2}$." How many siblings were there altogether?

First note that Robin is a boy and Kelly is a girl. Let there be m boys and n girls. Then Robin's statement translates into $\frac{m-1}{m+n-1} = \frac{1}{3}$ and Kelly's statement translates into $\frac{n-1}{m+n-1} = \frac{1}{2}$. These simplify into $2m - n = 2$ and $n - m = 1$. Addition yields $m = 3$, so that $n = 4$ and there are 7 siblings altogether.

Five of the numbers 0, 1, 2, 3, 4, 5, 6, 7, 8 and 9 are drawn. What is the probability that the sum of any three of the numbers drawn is greater than the sum of the other two? There are $\binom{10}{5} = 252$ possible outcomes. The only way the sum of the smallest three numbers can exceed the sum of the largest two numbers is for the five numbers to be 5, 6, 7, 8 and 9. Hence the desired probability is $\frac{1}{252}$.

Three distinct integers are chosen at random from 1 to 100 inclusive. What is the probability that their sum is divisible by 2? The total number of choices is $\binom{100}{3} = 161700$. Let $A = \{1, 3, 5, \ldots, 99\}$ and $B = \{2, 4, 6, \ldots, 100\}$. In order for the sum of three numbers to be divisible by 2, we can either choose all three from B, or two from A and one from B. The total number of such choices is $\binom{50}{3} + 50\binom{50}{2} = 80850$. Hence the desired probability is $\frac{80850}{161700} = \frac{1}{2}$.

The above result may also be arrived at by an argument using a one-to-one correspondence. Such may not be the case if we want the sum of three numbers to be divisible by 3 instead of 2. Let $A = \{1, 4, 7, \ldots, 100\}$, $B = \{2, 5, 8, \ldots, 98\}$ and $C = \{3, 6, 9, \ldots, 99\}$. We can either choose all three from the same set or one from each set. The total number of choices is $\binom{34}{3} + 2\binom{33}{3} + 34 \times 33^2 = 20262$. Hence the desired probability is $\frac{20262}{161700} = \frac{921}{7350}$.

Probability is full of surprises. Here is a simple illustration. There is a marble inside an urn. It is either red or green. A green marble is added to the urn, and a marble is then drawn at random. If the marble drawn is green, what is the probability that the marble still inside is also green?

Initially, the marble inside the urn is green with probability $\frac{1}{2}$. Since a green marble is added and a green marble is withdrawn, it would appear that the marble still inside should be green with the same probability of $\frac{1}{2}$. However, the marble drawn is not necessarily the one added. We may have drawn the added green ball and left behind the original red ball, we may have drawn the added green ball and left behind the original green ball, or we may have drawn the original green ball and left behind the added green ball. Hence the desired probability is $\frac{2}{3}$.

Each of two families has two children. In the first family, one of the children is a girl. In the second family, the older child is a girl. It is assumed that a boy is equally likely to be born as a girl. For each family, what is the probability that both children are girls?

Casual reflection may suggest that the answer should be the same, but it is not the case. With two children, there are four equally likely outcomes: boy-boy, boy-girl, girl-boy and girl-girl. For each family, we seek the probability of the fourth outcome. In the first family, the given condition eliminates the first outcome. Hence the desired probability is $\frac{1}{3}$. However, in the second family, the given condition eliminates the first and the second outcomes. Hence the desired probability rises to $\frac{1}{2}$.

Suppose we are on the fifth floor of a building with six floors and a single elevator. It is assumed that the elevator may be on any floor at any particular time. Hence when it comes to the fifth floor, it is going down $\frac{1}{6}$ of the time.

Suppose a second elevator is installed, and the same assumption applies to the new elevator. Consider the first elevator to arrive at the fifth floor. Is it still going down with probabilities $\frac{1}{6}$ as before? It would appear that this should be the case, but once again, we are misled by intuition.

Let us analyse the situation carefully. There are four scenarios. In the first, both elevators are below the fifth floor, so that the first elevator to arrive at the fifth floor is going up. In the second, both elevators are above the fifth floor, so that the first elevator to arrive at the fifth floor is going down. This occurs with probability $(\frac{1}{6})^2 = \frac{1}{36}$. In the third scenario, one elevator is above the fifth floor and the other is below the fourth floor, so that the first elevator to arrive at the fifth floor is going down. This occurs with probability $\frac{1}{6} \times \frac{4}{6} + \frac{4}{6} \times \frac{1}{6} = \frac{8}{36}$. In the fourth scenario, one elevator is above the fifth floor and the other is between the fourth and the fifth floor. Now either elevator may arrive first at the fifth floor, with equal probability. This occurs with probability $2(\frac{1}{6})^2 = \frac{2}{36}$, so that the first elevator to arrive at the fifth floor is going down $\frac{1}{36}$ of the time. Taking all four scenarios into consideration, the first elevator to arrive at the fifth floor is going down $\frac{1}{36} + \frac{8}{36} + \frac{1}{36} = \frac{5}{18}$ of the time.

In fact, if the number of elevators continue to increase, the first elevator to arrive at the fifth floor will be going down about half of the time. More precisely, suppose we are on the $(n-1)$st floor of a building with n floors and k elevators. Then for any n and any positive real number ϵ, there exists a k such that the first elevator to arrive at the $(n-1)$st floor is going down with a probability exceeding $\frac{1}{2} - \epsilon$.

Historically, the interest in probability arose from its applications to gambling, and this has not abated with time. Lotteries, casinos and horse racing come to mind.

We discuss here the probabilities associated with the game Poker. It is played with a standard deck of 52 cards, consisting of the Ace, 2, 3, 4, 5, 6, 7, 8, 9, 10, Jack, Queen and King of each of the four suits Spades (♠), Hearts (♡), Diamonds (♢) and Clubs (♣). The Ace also ranks above the King.

A poker hand consists of 5 of the 52 cards, and is classified into nine categories. A flush consists of five cards of the same suit. A straight consists of five consecutive cards, from A, 2, 3, 4 and 5 to 10, J, Q, K and A. A straight flush is both a straight and a flush. A full house consists of three of a kind and a pair, which means two of a kind. Hands which occur in an earlier category are excluded from later categories. The total number of poker hands is $\binom{52}{5} = 2598960$. We now calculate the probability that the hand is in a particular category.

- **Straight Flush**. Example: ♠8, ♠9, ♠10, ♠J, ♠Q.
  The suit may be chosen in 4 ways and the lowest card in 10 ways, for a total of 40 hands. The probability is $\frac{40}{2598960} = \frac{1}{64974}$.

- **Four of a Kind**. Example: ♠4, ♡4, ◇4, ♣4, ♡7.
  The set of four may be chosen in 13 ways and the fifth card in 12×4 ways. The probability is $\frac{624}{2598960} = \frac{1}{4165}$.

- **Full House**. Example: ♠6, ◇6, ♣6, ♡2, ♣2.
  The set of three may be chosen in 13×4 ways and the pair in $12 \times \binom{4}{2}$ ways. The probability is $\frac{3744}{2598960} \approx \frac{1}{694}$.

- **Flush**. Example: ◇3, ◇7, ◇Q, ◇K, ◇A.
  The suit may be chosen in 4 ways and the cards in $\binom{13}{5}$ ways. We then exclude the 40 straight flushes. The probability is $\frac{5108}{2598960} \approx \frac{1}{509}$.

- **Straight**. Example: ♠5, ♡6, ♠7, ♡8, ♣9.
  The lowest card may be chosen in 10×4 ways and the other four cards in 4^4 ways. We then exclude the 40 straight flushes. The probability is $\frac{10200}{2598960} \approx \frac{1}{255}$.

- **Three of a Kind**. Example: ♠A, ♡A, ♣A, ♡5, ♡9.
  The set of three may be chosen in 13×4 ways, and the other two cards in $\binom{12}{2} \times 4^2$ ways. The probability is $\frac{54912}{2598960} \approx \frac{1}{47}$.

- **Two Pairs**. Example: ♡10, ◇10, ♠K, ♣K, ♣8.
  The two pair may be chosen in $\binom{13}{2} \times \binom{4}{2}^2$ ways and the fifth card in 44 ways. The probability is $\frac{123552}{2598960} \approx \frac{1}{21}$.

- **One Pair**. Example: ♡J, ◇J, ♠2, ♡3, ♣7.
  The pair can be chosen in $13 \times \binom{4}{2}$ ways and the other three cards in $\binom{12}{3} \times 4^3$ ways. The probability is $\frac{1098240}{2598960} \approx \frac{2}{5}$.

- **Bust**. Example: ♣3, ♣5, ◇8, ♣10, ♡Q.
  Five cards of different kinds may be chosen in $\binom{13}{5} \times 4^5$ ways. We then exclude the 40 straight flushes, the 5108 flushes and the 10200 straights. The probability is $\frac{1302540}{2598960} \approx \frac{1}{2}$.

To conclude this section, we analyse a gambling game which is also played with a standard deck of 52 cards. You pay 4 dollars, choose a positive integer n and draw n cards from the deck. You receive n dollars if no Aces appear, but nothing if at least one Ace appears. What is the best choice for n, and on the average, do you win or lose money?

If you choose $n \le 4$, we will get back at most what you pay, and this is no certainty. Hence you must choose $n > 4$. Suppose you choose $n = 5$. There are $\binom{52}{5}$ hands of which $\binom{48}{5}$ are winning hands. Your expected return is

$$5 \times \frac{\binom{48}{5}}{\binom{52}{5}} = 5 \times \frac{48!47!5!}{52!43!5!} = 5 \times \frac{48 \times 47 \times 46 \times 45 \times 44}{52 \times 51 \times 50 \times 49 \times 48} = \frac{38916}{10829} \approx 3.59.$$

Thus if you choose $n = 5$, you will lose on the average 41 cents each time you play.

If you choose $n = 6$, you multiply the above result by $\frac{6}{5} \times \frac{43}{47} = \frac{258}{235} > 1$, which means that $n = 6$ is a better choice than $n = 5$. If you increase n to 10, the last multiplier is $\frac{10}{9} \times \frac{39}{43} = \frac{130}{129} > 1$, but if you increase n to 11, the last multiplier is $\frac{11}{10} \times \frac{38}{42} = \frac{209}{210} < 1$. It follows that your best choice is $n = 10$, and your expected return is

$$10 \times \frac{\binom{48}{10}}{\binom{52}{10}} = 5 \times \frac{48!42!5!}{52!38!5!} = 10 \times \frac{48 \times 47 \times \cdots \times 39}{52 \times 51 \times \cdots \times 43} = \frac{492}{119} \approx 4.13.$$

You will win on the average 13 cents each time you play.

It turns out that you still win if you choose $n = 8$, 9, 11 or 12, but your expected return will be less. For instance, if you choose $n = 11$, you will win on the average only 11 cents each time you play.

Examples

Example 1.6.1.

Tom has a fair cubical die with the numbers 1, 1, 5, 5, 9 and 9 on its faces. Dick has a fair cubical die with the numbers 2, 2, 6, 6, 7 and 7 on its faces. Harry has a fair cubical die with the numbers 3, 3, 4, 4, 8 and 8 on its faces. All three roll their dice, and the one with the highest number on the top face wins. Find the winning probability for each of Tom, Dick and Harry.

Solution:

Tom can win in two ways. First, he rolls a 9, which occurs with probability $\frac{1}{3}$. Alternatively, he rolls a 5 while Dick rolls a 2 and Harry rolls a 3 or a 4. This occurs with probability $\frac{1}{3} \times \frac{1}{3} \times \frac{2}{3} = \frac{2}{27}$. Hence Tom's winning probability is $\frac{1}{3} + \frac{2}{27} = \frac{11}{27}$. Harry can win in two ways. First, he rolls an 8 while Tom rolls a 1 or a 5. This occurs with probability $\frac{1}{3} \times \frac{2}{3} = \frac{2}{9}$. Alternatively, he rolls a 3 or a 4 while Tom rolls a 1 and Dick rolls a 2. This occurs with probability $\frac{2}{3} \times \frac{1}{3} \times \frac{1}{3} = \frac{2}{27}$. Hence Harry's winning probability is $\frac{2}{9} + \frac{2}{27} = \frac{8}{27}$. Then Dick's winning probability is $1 - \frac{11}{27} - \frac{8}{27} = \frac{8}{27}$.

Example 1.6.2.

In a game show, the climax is when the contest winner may get an additional bonus of a car. The host shows three doors on stage, and states that there is exactly one car behind one of them. The contest winner chooses one of the doors. Then the host opens one of the remaining doors, and there is no car behind it. The contest winner is then given the chance to switch to the other door instead. Should the contest winner make the switch?

Solution:

If the show host never offers the contest winner the option of switching unless the car is behind the chosen door, clearly the contest winner should not switch. Hence we assume that the offer is always made. Suppose the show host opens one of the two doors which are not chosen at random. Then it may happen that the car is revealed, which leads to an anti-climax and switching becomes a moot point. Hence we assume that the show host chooses a door with no car behind it, and this is always possible since there is only one car, and two doors are left. Now that the ground rules are set, we claim that the contest winner should always switch. Before the show host opens a door, the probability that the car is behind the chosen door is clearly $\frac{1}{3}$. Hence the probability that it is behind either of the other two doors is $\frac{2}{3}$. After the show host opens a door with no car behind it, the probability that the car is behind the chosen door is still $\frac{1}{3}$ since the opening of the door with no car behind is an action that can always be carried out. On the other hand, since there is only one other door left, the probability that the car is behind it is now $\frac{2}{3}$. Thus the contest winner should switch.

Example 1.6.3.

The game Craps is played with two fair cubical dice with the numbers 1, 2, 3, 4, 5 and 6 on the faces of each of them. A player rolls the dice and determines the sum of the numbers on the top faces of them. Let this sum be s on the first roll. If s is 7 or 11, the player wins immediately. If s is 2, 3 or 12, the player loses immediately. Otherwise, the player continues to roll the two dice until the sum is either s again, or 7. In the former case, the player wins. In the latter case, the player loses. Determine the probability that the player wins the game.

Solution:

When these two dice are rolled, the sum of the numbers on the top faces of them may be any of 2, 3, 4, 5, 6, 7, 8, 9, 10, 11 and 12. The number of times these sums occur are 1, 2, 3, 4, 5, 6, 5, 4, 3, 2 and 1 respectively, and their total is 36. The player may win in any of the following ways.

- $s = 7$. This occurs with a probability of $\frac{6}{36} = \frac{1}{6}$.

- $s = 11$. This occurs with a probability of $\frac{2}{36} = \frac{1}{18}$.

- $s = 4$, and reappears before 7. This occurs with a probability of $\frac{3}{36} \times \frac{3}{3+6} = \frac{1}{36}$.

- $s = 10$, and reappears before 7. This occurs with a probability of $\frac{3}{36} \times \frac{3}{3+6} = \frac{1}{36}$.

- $s = 5$, and reappears before 7. This occurs with a probability of $\frac{4}{36} \times \frac{4}{4+6} = \frac{2}{45}$.

- $s = 9$, and reappears before 7. This occurs with a probability of $\frac{4}{36} \times \frac{4}{4+6} = \frac{2}{45}$.

- $s = 6$, and reappears before 7. This occurs with a probability of $\frac{5}{36} \times \frac{5}{5+6} = \frac{25}{396}$.

- $s = 8$, and reappears before 7. This occurs with a probability of $\frac{5}{36} \times \frac{5}{5+6} = \frac{25}{396}$.

Thus the winning probability is $\frac{1}{6} + \frac{1}{18} + 2(\frac{1}{36} + \frac{2}{45} + \frac{25}{396}) = \frac{244}{495}$.

Exercises

1. You have to play a three-game match against a master and a novice. The rule is that to win the match, you must win two games in a row. Your probability of winning a game is $\frac{1}{3}$ against the master and $\frac{2}{3}$ against the novice. You play them in alternate games. You may choose to start the match against the master or against the novice. What should your choice be in order to maximize your probability of winning the match?

2. You are given ten red marbles, ten green marbles and two identical urns. You may distribute the twenty marbles in any way between the two urns. The next day, the two urns with their contents intact are presented to you, and you choose one of them at random. Then you draw a marble at random from the urn. How should you distribute the marbles in order to maximize your chances of drawing a green marble?

3. Each of two apartment buildings has two floors. On each floor, some residents keep pets. The probability that a randomly chosen pet is a cat is higher on the lower floor of the first apartment building than on the lower floor of the second apartment building. The probability that a randomly chosen pet is a cat is also higher on the upper floor of the first apartment building than on the upper floor of the second apartment building. Does it follow that the probability that a randomly chosen pet is a cat is higher in the first apartment building than in the second apartment building?

4. One fair cubical die has the numbers 1, 1, 5, 5, 9 and 9 on its faces. A second one has 2, 2, 6, 6, 7 and 7, and a third one has 3, 3, 4, 4, 8 and 8. Peter chooses one of the dice, and then Mary chooses either of the remaining dice. They then roll their dice, and whoever has the higher number on the top face of the die wins. Who has the advantage in this game, Peter or Mary?

5. The game Chuck-a-luck is played with three fair cubical dice with the numbers 1, 2, 3, 4, 5 and 6 on the faces of each of them. A player bets $1 on one of the six numbers and then rolls the dice. If the number does not appear on any of the top faces, the player loses his bet. If the number does appear, the player gets back the $1 and receives an additional $1 for each appearance of that number. On the average, does the player win or lose money?

6. The game Roulette is played with a wheel divided into 38 equal sectors. The wheel is spun, and when it comes to rest, a fixed pointer will point at one of the sectors. The sectors are numbered from 1 to 36, as well as 0 and 00. These two special squares are green. Half of the other 36 sectors are red and the other half black. A player may bet $1 on either a number from 1 to 36 inclusive, or the color red or black. If the chosen number or color matches that of the pointed sector, the player gets back the $1 and receives an additional $1 if he is betting on red or black, and $35 if he is betting on a number. Otherwise, he loses his bet. Which is the better bet for the player, and on the average, does he win or lose money?

Practice Questions

Combinations without Repetitions

1. One student has 6 math books, and another has 8 books. How many ways are there to exchange 3 books belonging to the first student with 3 books belonging to the second?

2. There are 2 girls and 7 boys in a chess club. A team of four persons must be chosen for a tournament, and there must be at least 1 girl on the team. In how many ways can this be done?

3. Ten points are marked on a straight line, and 11 points are marked on another line, parallel to the first one. How many triangles are there with vertices at these points?

4. A set of 15 different words is given. In how many ways is it possible to choose a subset of no more than 5 words?

5. Pete and John are the stars on a basketball team with 8 other players. How many ways are there to choose 5 starters so that Pete and John are not both starters?

6. How many ways are there to divide 10 players into two basketball teams of 5 players each?

Combinations with Repetitions

7. There 8 red balls, 7 yellow balls and 6 blue balls. How many ways are there to choose 6 of these balls?

8. There 5 red balls, 4 yellow balls and 3 blue balls. How many ways are there to choose 6 of these balls?

9. How many solutions (p, q, r, s) in non-negative integers are there to the equation $p + q + r + s = 10$?

10. How many solutions (p, q, r, s, t) in non-negative integers are there to the equation $p + q + r + s + t = 10$?

11. How many solutions (x, y, z) in integers are there to the equation $x + y + z = 15$ with $0 \leq x \leq 10, 1 \leq y \leq 4$ and $3 \leq z \leq 7$?

12. How many solutions (x, y, z) in integers are there to the equation $x + y + z = 25$ with $3 \leq x \leq 13, 0 \leq y \leq 14$ and $2 \leq z \leq 15$?

Combinatorial Summations

Evaluate the following sums without computing the value of any individual term:

13. $\binom{7}{2} + 2\binom{7}{3} + 3\binom{7}{4} + 4\binom{7}{5} + 5\binom{7}{6} + 6\binom{7}{7}$;

14. $\binom{7}{0} + 3\binom{7}{1} + 5\binom{7}{2} + 7\binom{7}{3} + 5\binom{9}{4} + 11\binom{7}{5} + 13\binom{7}{6} + 15\binom{7}{7}$;

15. $\binom{7}{1} + 2^2\binom{7}{2} + 3^2\binom{7}{3} + 4^2\binom{7}{4} + 5^2\binom{7}{5} + 6^2\binom{7}{6} + 7^2\binom{7}{7}$;

16. $\binom{7}{1}^2 + 2\binom{7}{2}^2 + 3\binom{7}{3}^2 + 4\binom{7}{4}^2 + 5\binom{7}{52} + 6\binom{7}{6}^2 + 7\binom{7}{7}^2$;

17. $\binom{7}{0}\binom{7}{4} + \binom{7}{1}\binom{7}{3} + \binom{7}{2}\binom{7}{2} + \binom{7}{3}\binom{7}{1} + \binom{7}{4}\binom{7}{0}$;

18. $\binom{7}{0}\binom{7}{4} + \binom{7}{1}\binom{6}{3} + \binom{7}{2}\binom{5}{2} + \binom{7}{3}\binom{4}{1} + \binom{7}{4}\binom{3}{0}$.

Probability

19. Joe flips ten fair coins and Moe flips eleven fair coins. What is the probability that Moe gets more heads than Joe?

20. Each of Joe and Moe rolls an ordinary cubical die. What is the probability that Joe rolls a higher number than Moe?

21. Two ordinary cubical dice are rolled. What is the probability that the total of the two numbers rolled is 6 or less?

22. Three cards are drawn at random from a shuffled deck consisting of the four Kings and the four Queens. If the Queen of Spades is drawn, what is the probability that at least one of the other two cards is another Queen?

23. Three cards are drawn at random from a shuffled deck consisting of the four Kings and the four Queens. If at least one Queen is drawn, what is the probability that at least one of the other two cards is another Queen?

24. In a bag are three coins each with two heads, four coins each with a head and a tail, and five coins each with two tails. One of them is drawn at random and tossed. It is equally likely to land on either side. What is the probability that it lands heads?

Answers

(1)1120 (2) 91 (3) 1045 (4) 4944 (5) 196 (6) 126 (7) 28 (8) 18
(9) 286 (10) 1001 (11) 19 (12) 154 (13) 321 (14) 1024 (15) 1792 (16) 12012
(17) 1001 (18) 560 (19) $\frac{1}{2}$ (20) $\frac{5}{12}$ (21) $\frac{5}{12}$ (22) $\frac{5}{7}$ (23) $\frac{7}{13}$ (24) $\frac{5}{12}$

Chapter Two: Generating Functions and Recurrence Relations

Section 2.1. Generating Functions of Sequences

The **generating function** of a sequence $\{a_n\}$ is defined to be

$$G(x) = a_0 + a_1 x + a_2 x^2 + \cdots.$$

We usually seek a closed form of $G(x)$. We do not worry about the convergence of the power series since we are primarily interested in x as a formal variable. If we want to substitute a numerical value for x, then of course we will have to check that it falls within the interval of convergence of $G(x)$.

Given a function $G(x)$, we can write down its Maclaurin's series

$$G(x) = \sum_{n=0}^{\infty} \frac{G^{(n)}(0)}{n!} x^n,$$

where $G^{(n)}(x)$ is the n-th derivative of $G(x)$. Hence $G(x)$ is the generating function of the sequence $\{\frac{G^{(n)}(0)}{n!}\}$.

As an example, let $G(x) = (1+x)^m$, where m is not necessarily a nonnegative integer. Then $G^{(n)}(0) = m(m-1)\cdots(m-n+1)$. We define

$$\binom{m}{n} = \frac{m(m-1)\cdots(m-n+1)}{n!}$$

with $\binom{m}{0} = 1$. These are called **Newton's binomial coefficients**. Then $(1+x)^m = \sum_{n=0}^{\infty} \binom{m}{n} x^n$. This is known as **Newton's Binomial Theorem**.

Note that if m is a nonnegative integer, we have the usual binomial coefficients and the usual Binomial Theorem. In particular, $(1+x)^m$ is the generating function of the sequence $\{\binom{m}{n}\}$. Since $(1+x)^m$ is a polynomial, the sequence generated must be finite. This is indeed the case as all terms beyond $n = m$ are equal to 0.

Suppose $m = -k$ where k is a positive integer. Then

$$\binom{-k}{n} = \frac{(-k)(-k-1)\cdots(-k-n+1)}{n!} = (-1)^n \binom{k+n-1}{n}.$$

It follows that

$$\frac{1}{(1-x)^k} = (1+(-x))^{-k} = \sum_{n=0}^{\infty} \binom{-k}{n} (-x)^n = \sum_{n=0}^{\infty} \binom{k+n-1}{n} x^n,$$

so that it generates the sequence $\{\binom{k+n-1}{n}\}$.

© The Author(s), under exclusive license to Springer Nature Switzerland AG 2021
S. W. Golomb, A. Liu, *Solomon Golomb's Course on Undergraduate Combinatorics*,
https://doi.org/10.1007/978-3-030-72228-9_3

In particular, if $m = -1$, we obtain $\frac{1}{1-x} = 1+x+x^2+\cdots$, which generates the constant sequence of 1s. Replacing x with kx in $\frac{1}{1-x} = 1+x+x^2+\cdots$, we have $\frac{1}{1-kx} = 1 + kx + k^2x^2 + \cdots$, which generates the sequence of the powers of k. In the particular case where $k = -1$, $\frac{1}{1+x} = 1 - x + x^2 - \cdots$ generates the alternating sequence of 1s and -1s.

What is the sequence generated by the function $\frac{1}{1-x^2}$? If we replace x by x^2 in $\frac{1}{1-x} = 1 + x + x^2 + \cdots$, we have $\frac{1}{1-x^2} = 1 + x^2 + x^4 + \cdots$, so that it generates the sequence of alternating 1s and 0s.

We use an alternative method which features the important tool of partial fractions. We write

$$\frac{1}{1 - x^2} = \frac{A}{1 - x} + \frac{B}{1 + x},$$

where A and B are constants to be determined.

Clearing fractions, we have $1 = A(1 + x) + B(1 - x)$. Setting $x = 1$, we have $1 = 2A$ so that $A = \frac{1}{2}$. Setting $x = -1$, we have $1 = 2B$ so that $B = \frac{1}{2}$ also. Now

$$\begin{aligned}
\frac{1}{1 - x^2} &= \frac{\frac{1}{2}}{1 - x} + \frac{\frac{1}{2}}{1 + x} \\
&= \frac{1}{2}(1 + x + x^2 + \cdots) + \frac{1}{2}(1 - x + x^2 - \cdots) \\
&= 1 + x^2 + x^4 + \cdots.
\end{aligned}$$

Note also that multiplying $\frac{1}{1-x^2}$ by x, we shift all the terms forward one place, so that $\frac{x}{1-x^2}$ generates the alternating sequence of 0s and 1s.

The function $\frac{1}{1-x}$ is known as a **summation operator** in that if $G(x)$ generates the sequence $\{a_n\}$, then $\frac{G(x)}{1-x}$ generates its sequence of partial sums $\{a_0, a_0 + a_1, a_0 + a_1 + a_2, \ldots\}$. This follows from the following formal multiplication:

$$\begin{aligned}
\frac{G(x)}{1 - x} &= G(x) \cdot \frac{1}{1 - x} \\
&= (a_0 + a_1 x + a_2 x^2 + \cdots)(1 + x + x^2 + \cdots) \\
&= a_0 + (a_0 + a_1)x + (a_0 + a_1 + a_2)x^2 + \cdots.
\end{aligned}$$

For example, $G(x) = \frac{1}{1+x}$ generates the sequence $1, -1, 1, -1, \ldots$. Hence $\frac{G(x)}{1-x}$ generates $1, 1-1=0, 1-1+1=1, 1-1+1-1=0, \ldots$. We have yet another proof that $\frac{1}{1-x^2} = \frac{G(x)}{1-x} = 1 + x^2 + x^4 + \cdots$.

We now tackle the converse process of finding the generating function of a given sequence. As an illustration, which function generates the sequence $\{\sin \frac{n\pi}{2}\}$? The sequence starts with $\{0, 1, 0, -1\}$ and repeats itself thereafter. It is a periodic sequence with period 4. The generating function is clearly $G(x) = x - x^3 + x^5 - x^7 + \cdots$. Hence $x^4 G(x) = x^5 - x^7 + x^9 - x^{11} + \cdots$ so that $(1 - x^4)G(x) = x - x^3$. It follows that $G(x) = \frac{x-x^3}{1-x^4} = \frac{x(1-x^2)}{(1+x^2)(1-x^2)} = \frac{x}{1+x^2}$. Similarly, the generating function of $\{\cos \frac{n\pi}{2}\}$ is $\frac{1}{1+x^2}$.

Examples

Example 2.1.1.
Express $\binom{\frac{1}{2}}{n}$ in terms of an ordinary binomial coefficient.

Solution:

We have
$$
\binom{\frac{1}{2}}{n} = \frac{(\frac{1}{2})(\frac{1}{2}-1)\cdots(\frac{1}{2}-n+1)}{n!}
$$
$$
= \frac{1}{2}\left(-\frac{1}{2}\right)^{n-1}\frac{1\cdot 3\cdots(2n-3)}{n!}
$$
$$
= \frac{1}{2}\left(-\frac{1}{2}\right)^{n-1}\frac{(2n-2)!}{n!(n-1)!2^{n-1}}
$$
$$
= \frac{1}{2n}\left(-\frac{1}{4}\right)^{n-1}\binom{2n-2}{n-1}.
$$

Example 2.1.2.
What sequence is generated by the function $\frac{1}{(1-x)(1-x^2)(1-x^3)}$?

Solution:

Let
$$
\frac{1}{(1-x)(1-x^2)(1-x^3)}
$$
$$
= \frac{A}{1-x} + \frac{B}{(1-x)^2} + \frac{C}{(1-x)^3} + \frac{D}{1+x} + \frac{E+Fx}{1+x+x^2}.
$$
Clearing fractions, we have
$$
\begin{aligned}
1 ={}& A(1-x)^2(1+x)(1+x+x^2) + B(1-x)(1+x)(1+x+x^2)\\
&+C(1+x)(1+x+x^2) + D(1-x)^3(1+x+x^2)\\
&+(E+Fx)(1-x)^3(1+x)\\
={}& A(1-x^2-x^3+x^5) + B(1+x-x^3-x^4)\\
&+C(1+2x+2x^2+x^3) + D(1-2x+x^2-x^3+2x^4-x^5)\\
&+E(1-2x+2x^3-x^4) + F(x-2x^2+2x^4-x^5).
\end{aligned}
$$

Setting $x = 1$, we have $1 = 6C$ so that $C = \frac{1}{6}$. Setting $x = -1$, we have $1 = 8D$ so that $D = \frac{1}{8}$. Comparing the coefficients of the x^5 and x^2 terms, we have $A - F = \frac{1}{8}$ and $A + 2F = \frac{11}{24}$. Hence $A = \frac{17}{72}$ and $F = \frac{1}{9}$. Comparing the coefficients of the x^0 and x^1 terms, we have $B + E = \frac{17}{36}$ and $B - 2E = -\frac{7}{36}$. Hence $B = \frac{1}{4}$ and $E = \frac{2}{9}$. Note that $\frac{E+Fx}{1+x+x^2} = \frac{2-x-x^2}{9(1-x^3)}$ generates the cycle $\frac{2}{9}, -\frac{1}{9}, -\frac{1}{9}, \frac{2}{9}, -\frac{1}{9}, -\frac{1}{9}, \ldots$. Hence the desired sequence is

$$
\frac{17}{72} + \frac{1}{4}\binom{n+1}{1} + \frac{1}{6}\binom{n+2}{2} + \frac{1}{8}(-1)^n + \begin{cases} \frac{2}{9} & \text{if } 3\mid n;\\ -\frac{1}{9} & \text{if } 3\nmid n. \end{cases}
$$

Note that $\frac{1}{4}\binom{n+1}{1} + \frac{1}{6}\binom{n+2}{2} = \frac{(n+1)(n+5)}{12}$, and the sum of the remaining terms is non-negative and strictly less than 1. Hence the final answer may be put in the more compact form $\left\lceil \frac{(n+1)(n+5)}{12} \right\rceil$.

Example 2.1.3.

(a) Determine the generating function of the sequence $\{\sin \frac{n\pi}{3}\}$.

(b) Determine the generating function of the sequence $\{\cos \frac{n\pi}{3}\}$.

Solution:

(a) The sequence is $\{0, \frac{\sqrt{3}}{2}, \frac{\sqrt{3}}{2}, 0, -\frac{\sqrt{3}}{2}, -\frac{\sqrt{3}}{2}\}$, and repeats itself thereafter. Hence the numerator of its generating function is

$$\frac{\sqrt{3}}{2}(x + x^2 - x^4 - x^5) = \frac{\sqrt{3}}{2}x(1+x)(1-x^3)$$

while its denominator is

$$1 - x^6 = (1+x^3)(1-x^3) = (1+x)(1-x+x^2)(1-x^3).$$

It follows that the generating function is $\frac{\frac{\sqrt{3}}{2}x}{1-x+x^2}$.

(b) The sequence is $\{1, \frac{1}{2}, -\frac{1}{2}, -1, -\frac{1}{2}, \frac{1}{2}\}$, and repeats itself thereafter. Hence the numerator of its generating function is

$$\frac{1}{2}(2 + x - x^2 - 2x^3 - x^4 + x^5) = \frac{1}{2}(2-x)(1+x)(1-x^3)$$

while its denominator is

$$1 - x^6 = (1+x^3)(1-x^3) = (1+x)(1-x+x^2)(1-x^3).$$

It follows that the generating function is $\frac{1-\frac{1}{2}x}{1-x+x^2}$.

Exercises

1. Express $\binom{-\frac{1}{2}}{n}$ in terms of an ordinary binomial coefficient.

2. Express $\binom{\frac{3}{2}}{n}$ in terms of an ordinary binomial coefficient.

3. Which sequence is generated by the function

   (a) $\frac{3x}{(1-x)(1+2x)}$;

   (b) $\frac{-8x^2}{(1-x)(1+x)(1-3x)}$;

   (c) $\frac{2-3x+3x^2}{(1+x)(1-x)^2}$?

4. Which sequence is generated by the function

   (a) $\frac{3}{(1+x)(1-2x)}$;

   (b) $\frac{1-4x}{(1-x)(1-2x)(1+2x)}$;

   (c) $\frac{2-x+x^2}{(1-x)(1+x^2)}$?

5. (a) Determine the generating function of the sequence $\{\sin \frac{n\pi}{4}\}$.

   (b) Determine the generating function of the sequence $\{\cos \frac{n\pi}{4}\}$.

6. (a) Determine the generating function of the sequence $\{\sin \frac{n\pi}{6}\}$.

   (b) Determine the generating function of the sequence $\{\cos \frac{n\pi}{6}\}$.

Section 2.2. Direct Counting with Generating Functions

Generating functions are very useful in solving a large class of combination problems. We begin with a simple illustration. We wish to know how many subsets there are of a three-element set, and how many subsets there are of each size.

Let the three elements be a, b and c. The approach may be divided into three stages. First, we construct a generating function for each individual element, using the Addition Principle. For the element a, we either leave it or take it. In other words, we may take 0 copies or 1 copy. We denote taking 0 copies of a by $a^0 = 1$ and taking 1 copy of a by $a^1 = a$. The "or" scenario yields the generating function $1 + a$. Similarly, those for the elements b and c are $1 + b$ and $1 + c$ respectively.

In the second stage, we use the Multiplication Principle to obtain the overall generating function. Since the subsets involve a, b and c, the "and" scenario yields the overall generating function $(1 + a)(1 + b)(1 + c)$.

In the third stage, we extract the answers from the overall generating function. Expansion yields $1 + a + b + c + ab + bc + ca + abc$, meaning that there are eight subsets, namely, $\emptyset$, $\{a\}$, $\{b\}$, $\{c\}$, $\{a, b\}$, $\{b, c\}$, $\{c, a\}$ and $\{a, b, c\}$.

We can also count the number of subsets of any specific sizes, but there is a simpler approach. The current formulation provides more information than is needed. If we replace each of a, b and c by x, we have $(1 + x)^3 = \binom{3}{0} + \binom{3}{1}x + \binom{3}{2}x^2 + \binom{3}{3}x^3$. Thus we have recreated a special case of the Binomial Theorem.

We do not in general intend to assign any numerical value to x. It merely serves as a counter. In the example above, the number of subsets of size 0 is given by the constant or x^0 term, the number of subsets of size 1 is given by the coefficient of the linear or x^1 term, the number of subsets of size 2 is given by the coefficient of the quadratic or x^2 term, and the number of subsets of size 3 is given by the coefficient of the cubic or x^3 term. We sometimes refer to x as a **formal variable**.

Suppose we wish to count the number of ways of choosing n elements from $\{a, b, c\}$, with unlimited repetition. The generating function for each of a, b and c is $1 + x + x^2 + x^3 + \cdots = \frac{1}{1-x}$, and the overall generating function is $\left(\frac{1}{1-x}\right)^3$. Appealing to Newton's Binomial Theorem, we have

$$(1 - x)^{-3} = \sum_{n=0}^{\infty} \binom{n+2}{2} x^n.$$

Hence the number of ways is $\binom{n+2}{2}$, a result we have obtained in the last chapter.

Consider the multiset $\{a, a, a, a, a, a, a, b, b, b, b, b, b, c, c, c\}$. Suppose we wish to count the number of its subsets of size 12. The generating function for a, b and c are $1+x+x^2+x^3+x^4+x^5+x^6+x^7$, $1+x+x^2+x^3+x^4+x^5+x^6$ and $1+x+x^2+x^3$, respectively. We want the coefficient of the x^{12} term in their product.

The generating functions being polynomials, we can carry out the multiplication and obtain

$$1 + 3x + 6x^2 + 10x^3 + 14x^4 + 18x^5 + 22x^6 + 25x^7 + 26x^8$$

$$+25x^9 + 22x^{10} + 18x^{11} + 14x^{12} + 10x^{13} + 6x^{14} + 3x^{15} + x^{16}.$$

It follows that there are 14 subsets of size 12 of the multiset.

Alternatively, we can compute

$$\frac{1-x^8}{1-x} \cdot \frac{1-x^7}{1-x} \cdot \frac{1-x^4}{1-x}$$
$$= (1 - x^4 - x^7 - x^8 + x^{11} + x^{12} + x^{15} - x^{19})(1-x)^{-3}$$
$$= (1 - x^4 - x^7 - x^8 + x^{11} + x^{12} + x^{15} - x^{19}) \sum_{n=0}^{\infty} \binom{n+2}{2} x^n.$$

To extract the coefficient of x^{12}, note that only the first six terms in the first factor are relevant. They are complemented by the terms x^{12}, x^8, x^5, x^4, x and 1 in the second factor. Thus the coefficient of x^{12} is given by $\binom{14}{2} - \binom{10}{2} - \binom{7}{2} - \binom{6}{2} + \binom{3}{2} + \binom{2}{2} = 14$. This is the exact expression obtained when we solve this problem by the Principle of Inclusion-Exclusion in Example 1.2.1.

Our next illustration is a problem in combinatorial number theory. Determine the number of triples of positive integers a, b and c such that $abc = 4000000$ and none of a, b and c is divisible by the fifth power of a prime.

Note that $4000000 = 2^8 5^6$. Let $a = 2^i 5^p$, $b = 2^j 5^q$ and $c = 2^k 5^r$. Then $i + j + k = 8$, $p + q + r = 6$, and none of i, j, k, p, q and r exceeds 4. The generating function for each of i, j and k is $1 + x + x^2 + x^3 + x^4 = \frac{1-x^5}{1-x}$. Hence the composite generating function is

$$\left(\frac{1-x^5}{1-x}\right)^3 = (1-x^5)^3 \sum_{n=0}^{\infty} \binom{-3}{n} (-x)^n$$
$$= (1 - 3x^5 + 3x^{10} - x^{15}) \sum_{n=0}^{\infty} \binom{n+2}{2} x^n.$$

The number of choices for i, j and k is the coefficient of the x^8 term, namely, $\binom{8+2}{2} - 3\binom{3+2}{2} = 15$. Similarly, the number of choices for p, q and r is $\binom{6+2}{2} - 3\binom{1+2}{2} = 19$. Hence the number of choices for a, b and c is $15 \times 19 = 285$.

We now consider an example involving rolling dice. An ordinary six-sided die has the numbers 1, 2, 3, 4, 5 and 6 on its faces. When we roll a pair of such dice and add the two numbers which show up on top, the possible outcomes are 2, 3, 4, 5, 6, 7, 8, 9, 10, 11 and 12. We wish to determine their respective frequencies of occurrence among the $6 \times 6 = 36$ possible outcomes.

When we roll a single ordinary die, the outcome is 1, 2, 3, 4, 5 or 6. This suggests representing the die by $x + x^2 + x^3 + x^4 + x^5 + x^6$. When we roll a pair of them, the generating function is

$$
\begin{aligned}
&(x + x^2 + x^3 + x^4 + x^5 + x^6)^2 \\
= \; & x^2 + 2x^3 + 3x^4 + 4x^5 + 5x^6 + 6x^7 + 5x^8 + 4x^9 + 3x^{10} + 2x^{11} + x^{12}.
\end{aligned}
$$

Thus the answers are 1, 2, 3, 4, 5, 6, 5, 4, 3, 2 and 1 respectively. These frequencies become probability when divided by their sum, namely 36.

A wheel is divided into 4 equal sectors, each labeled with a non-negative integer. The labels need not be distinct. The wheel is spun, and when it comes to a stop, the label of one of the sectors will show through a display window. Does there exist a pair of wheels such that the sum of the displayed numbers is between 0 and 15 inclusive, with each appearing exactly once?

The generating function $1 + x + x^2 + \cdots + x^{15}$ represents the desired outcome. It may be factored as $(1+x)(1+x^2)(1+x^4)(1+x^8)$. When we put $x = 1$, each factor is equal to 2. Hence the four factors may be partitioned into two products, representing the two wheels each with 4 sectors.

The products $(1+x)(1+x^2)$ and $(1+x^4)(1+x^8)$ yield the wheels with labels 0, 1, 2 and 3 on one and 0, 4, 8 and 12 on the other. The products $(1+x)(1+x^4)$ and $(1+x^2)(1+x^8)$ yield the wheels with labels 0, 1, 4 and 5 on one and 0, 2, 8 and 10 on the other. The products $(1+x)(1+x^8)$ and $(1+x^2)(1+x^4)$ yield the wheels with labels 0, 1, 8 and 9 on one and 0, 2, 4 and 6 on the other.

We are now in a position to tackle two more cases of the Distribution Problem introduced in Section 0.1. In Case (7), n identical objects are to be distributed into k identical boxes, with empty boxes allowed. This is equivalent to partitioning n into at most k parts. We have seen in Chapter 0 that this is equal to the number of partitions of n into 1s, 2s, ..., ks.

Consider the case $k = 3$. Since we can use any number of 1s and obtain any positive integer, the generating function for 1 is $\frac{1}{1-x}$. Likewise, we can use any number of 2s, but only even numbers may be obtained. Hence the generating function for 2 is $\frac{1}{1-x^2}$. Similarly, the generating function for 3 is $\frac{1}{1-x^3}$. The overall generating function is $\frac{1}{(1-x)(1-x^2)(1-x^3)}$. By Example 2.1.2, the sequence generated is $\lceil \frac{(n+1)(n+5)}{12} \rceil$.

Case (8) of the Distribution Problem does not allow empty boxes. Since the objects are identical, we may put one in each of the k identical boxes and then distribute the remaining $n - k$ objects without further restriction.

Examples

Example 2.2.1.

A company has three shareholders. Determine the number of ways of distributing $2n$ identical shares among them so that any two of them will hold more shares than the third.

Solution:

The maximum number of shares held by one shareholder is $n-1$. Hence the minimum number is 2. It follows that each shareholder may be represented by the generating function $x^2 + x^3 + \cdots + x^{n-1}$. The overall generating function is

$$\left(\frac{x^2 - x^n}{1 - x}\right)^3 = (x^6 - 3x^{n+4} + 3x^{2n+2} - x^{3n}) \sum_{n=0}^{\infty} \binom{n+2}{2} x^n.$$

To extract the coefficient of x^{2n}, note that only the first two terms in the first factor are relevant. They are complemented respectively by the terms x^{2n-6} and x^{n-4} in the second factor. Thus the coefficient of x^{2n} is given by $\binom{2n-4}{2} - 3\binom{n-2}{2} = \binom{n-1}{2}$. Alternatively, we may represent each shareholder by $1 + x + \cdots + x^{n-1}$. This of course changes the overall generating function, but will not affect the value of the coefficient of the x^{2n} term. This is because if no shareholder has more than $n - 1$ shares, then it is impossible for any of them to have less than 2. Now the overall generating function is

$$\left(\frac{1 - x^n}{1 - x}\right)^3 = (1 - 3x^n + 3x^{2n} - x^{3n}) \sum_{n=0}^{\infty} \binom{n+2}{2} x^n.$$

To extract the coefficient of x^{2n}, note that only the first three terms in the first factor are relevant. They are complemented by the terms x^{2n}, x^n and 1 in the second factor. It follows that the coefficient of x^{2n} is given by $\binom{2n+2}{2} - 3\binom{n+2}{2} + 3\binom{2}{2} = \binom{n-1}{2}$. This is the exact expression obtained in Example 1.2.3.

Example 2.2.2.

Design a pair of six-sided dice, different from a pair of ordinary ones, such that when these dice are rolled, the outcomes and frequencies are exactly the same as those for a pair of ordinary dice. Only positive integers may appear. The two dice do not have to be identical. The same positive integer may appear more than once on the same die.

Solution:

The desired frequencies are obtained from the generating function

$$(x + x^2 + x^3 + x^4 + x^5 + x^6)^2,$$

which may be factored as

$$x^2(1 + x)^2(1 + x + x^2)^2(1 - x + x^2)^2.$$

To design a new pair of dice means a partition of these eight factors into two products, each representing one of the dice. The condition that only positive integers may be used means that each of the two copies of x must belong to a different die. Otherwise, one product will have a constant term, which means putting the number 0 on a face of that die. The condition that the dice are six-sided means that each of the two copies $1+x$ and each of the two copies of $1 + x + x^2$ must belong to a different die. This is because the number of faces of a die is equal to the value of its expression when $x = 1$. For instance, $x + x^2 + x^3 + x^4 + x^5 + x^6 = 6$ when $x = 1$. Now $1 + x = 2$ and $1 + x + x^2 = 3$ when $x = 1$, so that the dice have $2 \times 3 = 6$ faces. If we split the two copies of $1 - x + x^2$ as well, we will just get a pair of ordinary dice. It follows that if there is indeed an alternative, this pair of dice must be represented by

$$x(1 + x)(1 + x + x^2) = x + 2x^2 + 2x^3 + x^4$$

and

$$x(1 + x)(1 + x + x^2)(1 - x + x^2)^2 = x + x^3 + x^4 + x^5 + x^6 + x^8.$$

Thus one die has the numbers 1, 2, 2, 3, 3 and 4 on its faces, while the other has 1, 3, 4, 5, 6 and 8 on its faces. The discovery of this pair of dice is credited to the American mathematician **George Sicherman**, and these dice are known as the Sicherman dice. The chart below shows the outcomes for this pair of dice, and it is easy to verify that the frequencies are indeed as desired.

+	1	3	4	5	6	8
1	2	4	5	6	7	9
2	3	5	6	7	8	10
2	3	5	6	7	8	10
3	4	6	7	8	9	11
3	4	6	7	8	9	11
4	5	7	8	9	10	12

Example 2.2.3.
Use generating functions to prove that there is a unique partition of any positive integer n into a sum of distinct powers of 2.

Solution:
The generating function for the partition into distinct powers of 2 is

$$G(x) = (1 + x)(1 + x^2)(1 + x^4)(1 + x^8) \cdots.$$

Now

$$
\begin{aligned}
(1-x)G(x) &= (1-x)(1+x)(1+x^2)(1+x^4)(1+x^8)\cdots \\
&= (1-x^2)(1+x^2)(1+x^4)(1+x^8)\cdots \\
&= (1-x^4)(1+x^4)(1+x^8)\cdots \\
&= (1-x^8)(1+x^8)\cdots \\
&= \cdots \\
&= 1.
\end{aligned}
$$

Hence $G(x) = \frac{1}{1-x} = 1 + x + x^2 + x^3 + \cdots$. The desired conclusion follows immediately.

Exercises

1. Determine the number of integer triples (p, q, r) whose sum is 25, subject to the constraints $0 \le p \le 5$, $5 \le q \le 25$ and $-5 \le r \le 5$. Use the method of generating functions.

2. Each of four players has a deck of cards numbered from 1 to 10. Each of them plays one card from his or her deck. In how many ways can it happen that the total of the four numbers is 27? Use the method of generating functions.

3. A wheel is divided into n equal sectors, each labeled with a positive integer. The labels need not be distinct. The wheel is spun, and when it comes to a stop, the label of one of the sectors will show through a display window. Design a pair of wheels, one with $n = 3$ and one with $n = 12$, such that the sum of the displayed numbers appears with exactly the same frequencies as with two standard cubical dice. Find all solutions.

4. A wheel is divided into n equal sectors, each labeled with a positive integer. The labels need not be distinct. The wheel is spun, and when it comes to a stop, the label of one of the sectors will show through a display window. Design a pair of wheels, one with $n = 4$ and one with $n = 9$, such that the sum of the displayed numbers appears with exactly the same frequencies as with two standard cubical dice. Find all solutions.

5. Use generating functions to prove that the number of partitions of a positive integer n into distinct terms is equal to the number of partitions of n into odd terms.

6. Use generating functions to prove that the number of partitions of a positive integer n into terms which appear at most twice is equal to the number of partitions of n into terms which are not divisible by 3.

Section 2.3. Recurrence Relations and Iterations

There are at least two mathematical ways of describing a given sequence. For instance, consider the sequence $\{a_n\}$ where a_n is the n-th positive odd integer. We can give a **general formula** for the sequence. In this case, we have $a_n = 2n - 1$. Alternatively, we can give a **recurrence relation** for the sequence.

Since the terms of the sequence are the positive odd integers, each is 2 greater than the preceding one. Hence we have $a_n = a_{n-1} + 2$. Instead of giving the value explicitly, the recurrence relation computes each term based on the values of the earlier terms. This way, we can compute the values of all subsequent terms. A recurrence relation is analogous to the inductive step in the method of mathematical induction.

Note that, by itself, a recurrence relation does not define a unique sequence. For instance, the recurrence relation $a_n = a_{n-1} + 2$ is also satisfied by the sequence of positive even numbers. To get $a_n = a_{n-1} + 2$ going, we need to know the value of a_1, which is 1. We call $a_1 = 1$ the **initial condition** for this recurrence relation. We can also define $a_0 = -1$ artificially to make it fit the recurrence relation.

While a recurrence relation together with appropriate initial conditions defines a sequence uniquely, it is not the most desirable way of computing the values of its terms. The process of obtaining the general formula from the recurrence relation is called *solving the recurrence relation*. There is the companion process of *setting up the recurrence relation* when we solve problems using this approach.

The recurrence relation $a_n = a_{n-1} + 2$ is called a **one-step** recurrence relation because only the value of the preceding term is used. If we have a **multi-step** recurrence relation, we will need several initial conditions, usually the values of the first few terms. In general, a recurrence relation is a function which defines a_n in terms of $a_1, a_2, \ldots, a_{n-1}$.

A one-step recurrence relation can be solved by the **method of iteration**. In the above example, we have

$$
\begin{aligned}
a_n &= a_{n-1} + 2 \\
&= (a_{n-2} + 2) + 2 \\
&= \cdots \\
&= (a_1 + 2) + 2(n - 2) \\
&= 2n - 1.
\end{aligned}
$$

We may also present the computation in a tabular form as follows.

$$
\begin{array}{rcccc}
a_n & = & a_{n-1} & + & 2, \\
a_{n-1} & = & a_{n-2} & + & 2, \\
\cdots & = & \cdots & + & \cdots, \\
a_2 & = & a_1 & + & 2; \\
\hline
a_n & = & a_1 & + & 2(n-1).
\end{array}
$$

Hence $a_n = 2n - 1$. We sometimes describe the massive cancellation of the terms $a_2, a_3, \ldots, a_{n-1}$ as a **telescoping sum**.

The remainder of this section is devoted to the process of setting up recurrence relations. Unlike the companion solving process which is a science, this is an art which can only be learnt by experience.

In our first example, a bug begins at vertex B of a regular hexagon $ABCDEF$. In each minute, it moves from its current position to an adjacent vertex. Once it reaches vertex F, it ends there. What is the number of different paths the bug can take in order to finish at F after exactly $2n$ minutes?

Let b_n denote the number of different paths the bug can take in order to finish at F after exactly $2n$ minutes. Then $b_0 = 0$ and $b_1 = 1$, the unique path being $B - A - F$. Suppose $n \geq 2$. Consider the moves the bug makes in the first two minutes. There are three parallel cases: $B - A - B$, $B - C - B$ and $B - C - D$. If the bug ends up back at B, there are b_{n-1} different ways to complete its path. If it ends up at D, the number of different ways to complete its path is still b_{n-1} by symmetry. It follows that $b_n = 3b_{n-1}$ for $n \geq 2$. Iteration yields

$$
b_n = 3b_{n-1} = 3^2 b_{n-2} = 3^3 b_{n-3} = \cdots = 3^{n-2} b_2 = 3^{n-1} b_1 = 3^{n-1}
$$

for $n \geq 1$, along with $b_0 = 0$.

Our next example is a famous one known as the Problem of the Tower of Hanoi. There are n disks of different sizes. Each has a hole in the middle so that the disk can be put around a peg. There are three numbered pegs, and all the disks are on top one another around peg number 1, forming a tower with the disks in ascending order of size from top to bottom. In each move, one may move the top disk around one peg and put it on top of the tower, possibly empty, around another peg, provided the moving disk is smaller than the top disk around the destination peg. What is the minimum number of moves required to move the tower to peg number 3?

Let the minimum number of moves be a_n. We have $a_0 = 0$ and $a_1 = 1$. To set up a recurrence relation for $\{a_n\}$, we analyse the sequential process of the problem.

The bottom disk should move just once. When this happens, all the other $n-1$ disks must form a tower around peg number 2. This takes a_{n-1} moves. After the move of the bottom disk, we have to move the tower of $n-1$ disks on top on it, and this takes another a_{n-1} moves. Hence $a_n = 2a_{n-1} + 1$.

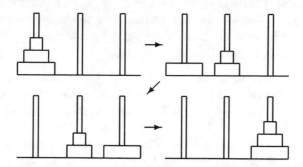

We now solve this recurrence relation as follows.

$$
\begin{aligned}
a_n &= 2a_{n-1} &+& 1, \\
2a_{n-1} &= 2^2 a_{n-2} &+& 2, \\
2^2 a_{n-2} &= 2^3 a_{n-3} &+& 2^2 \\
\cdots &= \cdots &+& \cdots, \\
2^{n-1} a_1 &= 2^n a_0 &+& 2^{n-1}; \\
\hline
a_n &= 2^n a_0 &+& 2^n - 1.
\end{aligned}
$$

Since $a_0 = 0$, $a_n = 2^n - 1$. We can make the iteration easier by defining $b_n = a_n + 1$. Then we have $b_n - 1 = 2(b_{n-1} - 1) + 1$ so that $b_n = 2b_{n-1}$. Since $a_1 = 1$, $b_1 = 2$ so that $b_n = 2^n$ and $a_n = 2^n - 1$.

In the next example, with the patronage of King Arthur, Merlin concocted n bottles of Elixir of equal capacity. He shared them with King Arthur, using a cup with capacity half that of a bottle. Each day, Merlin would open one bottle and pour some Elixir into this cup. King Arthur would choose whether to drink from the bottle or from the cup, and Merlin would drink from the vessel not chosen by King Arthur. During the course of these n days, Merlin was allowed to choose instead of King Arthur, but just once. However, King Arthur would decide when Merlin would exercise this option, after Merlin had done the pouring. Naturally, both King Arthur and Merlin would like to have as much Elixir as possible. What was the maximum amount out of the n bottles that Merlin could get?

Let the capacity of each bottle be 1. Let a_n be the total amount Merlin gets. For $n = 1$, since Merlin is allowed to choose once, and this is the only chance, he will get to do it. Thus he will pour nothing into the cup, and drinks the whole bottle. Thus $a_1 = 1$.

For $n = 2$, let Merlin pour an amount c into the cup, leaving an amount b in the first bottle, where $b + c = 1$ and $b \geq c$. If King Arthur chooses the bottle, Merlin will get the entire second bottle. If King Arthur lets Merlin choose, Merlin will divide the second bottle evenly and get half of that. Hence the possible values of a_2 are $c + 1$ or $b + \frac{1}{2}$. King Arthur decides which value Merlin gets, but Merlin has control over the values c and b. Since $c + 1 + b + \frac{1}{2} = \frac{5}{2}$ is constant, Merlin can get $\frac{5}{4}$ by setting $c + 1 = b + \frac{1}{2}$, with $c = \frac{1}{4}$ and $b = \frac{3}{4}$. It follows that $a_2 = \frac{5}{4}$.

Let us try to express a_n in terms of a_{n-1}. Let Merlin pour an amount c from the first bottle into the cup, leaving an amount b behind. If King Arthur lets Merlin choose, Merlin will divide each of the remaining n bottles evenly. His total take would be $b + \frac{n-1}{2}$. If King Arthur chooses instead, Merlin gets c from the first bottle. Since he has not chosen yet, we have now the same problem but with only $n - 1$ bottles. Hence Merlin's total take would be $c + a_{n-1}$. As before, he wants these two values to be equal, so that $a_n = \frac{1}{2}(b + \frac{n-1}{2} + c + a_{n-1}) = \frac{1}{2}a_{n-1} + (n+1)(\frac{1}{2})^2$. We have

$$
\begin{aligned}
a_n &= \tfrac{1}{2}a_{n-1} &+& (n+1)\left(\tfrac{1}{2}\right)^2, \\
\tfrac{1}{2}a_{n-1} &= \left(\tfrac{1}{2}\right)^2 a_{n-2} &+& n\left(\tfrac{1}{2}\right)^3, \\
\left(\tfrac{1}{2}\right)^2 a_{n-2} &= \left(\tfrac{1}{2}\right)^3 a_{n-3} &+& (n-1)\left(\tfrac{1}{2}\right)^4, \\
\cdots &= \cdots &+& \cdots, \\
\left(\tfrac{1}{2}\right)^{n-2} a_2 &= \left(\tfrac{1}{2}\right)^{n-1} a_1 &+& 3\left(\tfrac{1}{2}\right)^n; \\
\hline
a_n &= \left(\tfrac{1}{2}\right)^{n-1} a_1 &+& S,
\end{aligned}
$$

where $S = (n+1)\left(\frac{1}{2}\right)^2 + n\left(\frac{1}{2}\right)^3 + (n-1)\left(\frac{1}{2}\right)^4 + \cdots + 3\left(\frac{1}{2}\right)^n$. Doubling this expression, we have $2S = \frac{n+1}{2} + n\left(\frac{1}{2}\right)^2 + (n-1)\left(\frac{1}{2}\right)^3 + \cdots + 3\left(\frac{1}{2}\right)^{n-1}$. Subtraction yields

$$
\begin{aligned}
S &= \frac{n+1}{2} - \left(\frac{1}{2}\right)^2 - \left(\frac{1}{2}\right)^3 - \cdots - \left(\frac{1}{2}\right)^{n-1} - 3\left(\frac{1}{2}\right)^n \\
&= \frac{n}{2} + 1 - \left(1 - \frac{1}{2^{n-1}}\right) - \frac{3}{2^n} \\
&= \frac{n}{2} - \frac{1}{2^n}.
\end{aligned}
$$

Since $a_1 = 1$, $a_n = \frac{1}{2^{n-1}} + \frac{n}{2} - \frac{1}{2^n} = \frac{n}{2} + \frac{1}{2^n}$.

Let us see if we could have made the iteration simpler, by substituting $b_n = a_n - \frac{1}{2^n}$. Then we have $b_n + \frac{1}{2^n} = \frac{1}{2}(b_{n-1} + \frac{1}{2^{n-1}}) + \frac{n+1}{4}$. Simplification yields $b_n = \frac{1}{2}b_{n-1} + \frac{n+1}{4}$, which has exactly the same form as before, just with a different initial value. This is no help at all.

On the other hand, we could try $b_n = a_n - \frac{n}{2}$. Then

$$b_n + \frac{n}{2} = \frac{1}{2}\left(b_{n-1} + \frac{n-1}{2}\right) + \frac{n+1}{4}.$$

Simplification leads to $b_n = \frac{1}{2}b_{n-1}$ this time, which yields $b_n = \frac{1}{2^n}$ readily by iteration. Hence $a_n = b_n + \frac{n}{2} = \frac{n}{2} + \frac{1}{2^n}$ as before.

We turn now to a problem in combinatorial geometry. All diagonals of a convex n-gon are drawn, no three concurrent. Many overlapping triangles are formed by the sides, the diagonals and parts of the diagonals. How many such triangles are there?

Let the n-gon be $A_1 A_2 \ldots A_n$ and let the desired number be a_n. Then the interior of $A_1 A_2 \ldots A_{n-1}$ is divided into a_{n-1} regions. To obtain a_n from a_{n-1}, we first add the triangle $A_1 A_{n-1} A_n$. For $2 \le k \le n-2$, the diagonal $A_k A_n$ carves out a number of new regions equal to 1 plus the number of points of intersection of $A_k A_n$ with the diagonals of $A_1 A_2 \ldots A_{n-1}$. Since $k-1$ vertices are on one side of this diagonal and $n-k-1$ are on the other side,

$$
\begin{aligned}
a_n ={}& a_{n-1} + 1 + \sum_{k=2}^{n-2}\left((k-1)(n-k-1)+1\right) \\
={}& a_{n-1} + 1 + \sum_{k=2}^{n-2}\left((n-1)(k-1) - k^2 + k + 1\right) \\
={}& a_{n-1} + 1 + \frac{(n-3)(n-2)(n-1)}{2} - \left(\frac{(n-2)(n-1)(2n-3)}{6} - 1\right) \\
& + \left(\frac{(n-2)(n-1)}{2} - 1\right) + (n-3) \\
={}& a_{n-1} + \binom{n-2}{1} + \binom{n-1}{3}.
\end{aligned}
$$

Iteration yields

$$
\begin{array}{rcccc}
a_n &=& a_{n-1} &+& \binom{n-2}{1} &+& \binom{n-1}{3}; \\
a_{n-1} &=& a_{n-2} &+& \binom{n-3}{1} &+& \binom{n-2}{3}; \\
a_{n-2} &=& a_{n-3} &+& \binom{n-4}{1} &+& \binom{n-3}{3}; \\
\cdots &=& \cdots &+& \cdots &+& \cdots, \\
a_4 &=& a_3 &+& \binom{2}{1} &+& \binom{3}{3}; \\
\hline
a_n &=& a_3 &+& \left(\binom{n-1}{2} - 1\right) &+& \binom{n}{4}.
\end{array}
$$

Since $a_3 = 1$, we have $a_n = \binom{n-1}{2} + \binom{n}{4} = \frac{(n-1)(n-2)(n^2-3n+12)}{24}$ for $n \ge 3$.

We conclude with a problem involving a system of simultaneous recurrence relations.

(a) Determine the number of quaternary sequences of length n with an even number of 0s and an even number of 1s.

(b) Determine the number of quaternary sequences of length n with an odd number of 0s and an odd number of 1s.

(c) Determine the number of quaternary sequences of length n with an even number of 0s and an odd number of 1s.

(d) Determine the number of quaternary sequences of length n with an odd number of 0s and an even number of 1s.

We solve all four parts together. Let a_n be the number of quaternary sequences of length n with an even number of 0s and an even number of 1s. Then $a_0 = 1$. Let b_n be the number of quaternary sequences of length n with an odd number of 0s and an odd number of 1s. Then $b_0 = 0$. Let c_n be the number of quaternary sequences of length n with an even number of 0s and an odd number of 1s. Then $c_0 = 0$. Let d_n be the number of quaternary sequences of length n with an odd number of 0s and an even number of 1s. Then $d_0 = 0$. We have

$$
\begin{align}
a_n &= 2a_{n-1} + c_{n-1} + d_{n-1}; \tag{1}\\
b_n &= 2b_{n-1} + c_{n-1} + d_{n-1}; \tag{2}\\
c_n &= 2c_{n-1} + a_{n-1} + b_{n-1}; \tag{3}\\
d_n &= 2d_{n-1} + a_{n-1} + b_{n-1}; \tag{4}\\
4^{n-1} &= a_{n-1} + b_{n-1} + c_{n-1} + d_{n-1}. \tag{5}
\end{align}
$$

Subtracting (5) from (1), we have $a_n - 4^{n-1} = a_{n-1} - b_{n-1}$, so that we have $b_{n-1} = 4^{n-1} + a_{n-1} - a_n$. Subtracting (5) from (2), $b_n - 4^{n-1} = b_{n-1} - a_{n-1}$. Substituting the preceding result into here, we have

$$4^n + a_n - a_{n+1} - 4^{n-1} = 4^{n-1} + a_{n-1} - a_n - a_{n-1},$$

which simplifies to $a_{n+1} = 2a_n + 2^{2n-1}$. To make this recurrence work all the way, we have to redefine $a_0 = \frac{3}{4}$. Hence

$$
\begin{align}
a_n &= 2a_{n-1} &+& \; 2^{2n-3}; \\
2a_{n-1} &= 2^2 a_{n-2} &+& \; 2^{2n-4}; \\
2^2 a_{n-2} &= 2^3 a_{n-3} &+& \; 2^{2n-5}; \\
\cdots &= \cdots &+& \; \cdots; \\
2^{n-1} a_1 &= 2^n a_0 &+& \; 2^{n-2}.
\end{align}
$$

Addition yields $a_n = 3 \cdot 2^{n-2} + 2^{n-2}(2^n - 1) = 4^{n-1} + 2^{n-1}$. Hence we have $a_{n+1} = 4^n + 2^n$. It follows that $b_n = 4^n + a_n - a_{n+1} = 4^{n-1} - 2^{n-1}$, so that $a_n + b_n = 2 \cdot 4^{n-1}$. Hence $c_n + d_n = 2 \cdot 4^{n-1}$ also. By (3) and (4), we have $c_n = d_n$, Hence each is 4^{n-1}.

Examples

Example 2.3.1.

Solve the recurrence relation $a_n - a_{n-1} = 2^n + 2n$ with the initial condition $a_0 = 3$.

Solution:

We have

$$
\begin{array}{rcrcr}
a_n & - & a_{n-1} & = & 2^n & + & 2n, \\
a_{n-1} & - & a_{n-2} & = & 2^{n-1} & + & 2(n-1), \\
a_{n-2} & - & a_{n-3} & = & 2^{n-2} & + & 2(n-2), \\
\cdots & - & \cdots & = & \cdots & + & \cdots, \\
a_1 & - & a_0 & = & 2 & + & 2, \\
\hline
a_n & - & 3 & = & S_1 & + & 2S_2,
\end{array}
$$

where

$$S_1 = 2^n + 2^{n-1} + 2^{n-2} + \cdots + 2 = 2^{n+1} - 2$$

and

$$S_2 = n + (n-1) + (n-2) + \cdots + 1 = \frac{n(n+1)}{2}.$$

Hence $a_n = S_1 + 2S_2 + 3 = 2^{n+1} + n^2 + n + 1$.

Example 2.3.2.

An Italian pizza is to be divided into pieces by straight cuts perpendicular to its base. What is the maximum number of pieces we can obtain with n cuts?

Solution:

We first consider a very simple problem. A French bread is to be divided into pieces by straight cuts perpendicular to its length. Each cut divides an existing piece into two. As long as no two cuts are in the same position, we must have $n+1$ pieces after n cuts. Now let the maximum number of pieces we can obtain from an Italian pizza with n cuts be denoted by a_n. We have $a_0 = 1$. In order to maximize the number of pieces, every two cuts should intersect at a distinct point. After $n - 1$ such cuts, we have a_{n-1} pieces. The n-th cut will intersect the other cuts at $n - 1$ distinct points. By the French Bread Problem, it is divided into n segments, each of which divides an existing piece into two. Hence $a_n = a_{n-1} + n$. Iterating this recurrence, we have $a_n = a_0 + (1 + 2 + \cdots + n) = \frac{n^2+n+2}{2}$.

Example 2.3.3.

Determine the number of ternary sequences of length n

(a) with an even number of 0s;

(b) with an odd number of 0s.

Solution:

We solve both parts of the problem together. Let the number of sequences with an even number of 0s be a_n and the number of those with an odd number of 0s be b_n. Then $a_0 = 1$ and $b_0 = 0$. We can obtain a sequence counted in a_n by adding a 1 or a 2 to any sequence counted in a_{n-1}, as well as adding a 0 to any sequence counted in b_{n-1}. Hence $a_n = 2a_{n-1} + b_{n-1}$. Clearly, $a_n + b_n = 3^n$. It follows that $a_n = 2a_{n-1} + b_{n-1} = a_{n-1} + 3^{n-1}$. Iteration yields

$$
\begin{aligned}
a_n &= a_{n-1} + 3^{n-1} \\
&= a_{n-2} + 3^{n-2} + 3^{n-1} \\
&= \cdots \\
&= a_0 + 1 + 3 + \cdots + 3^{n-1} \\
&= \frac{1}{2}(3^n + 1).
\end{aligned}
$$

Similarly, we have $b_n = \frac{1}{2}(3^n - 1)$.

Exercises

1. Solve the recurrence relation $a_n - 2a_{n-1} = n2^{n+1}$ with the initial condition $a_0 = 1$. Use the method of iterations.

2. Solve the recurrence relation $a_n - a_{n-1} = 2\cos\frac{n\pi}{3}$ with the initial condition $a_0 = 1$. Use the method of iterations.

3. A Belgian cheese is to be divided into pieces by straight cuts, all passing through the center of the block. What is the maximum number of pieces we can obtain with n cuts?

4. A Dutch cheese is to be divided into pieces by straight cuts. What is the maximum number of pieces we can obtain with n cuts?

5. Determine the number of quaternary sequences of length n with an even number of 0s and at least one 3.

6. Determine the number of quaternary sequences of length n with an even number of 0s and at most one 3.

Section 2.4. The Method of Characteristic Equations

A **homogeneous linear recurrence relation** for a sequence $\{a_n\}$ is an expression of the form $c_0 a_n + c_1 a_{n-1} + \cdots + c_k a_{n-k} = 0$. We take the coefficients c_i to be constants. For instance, $a_n - 2a_{n-1} - 3a_{n-2} = 0$ is a homogeneous linear recurrence relation with constant coefficients. To go with $c_0 a_n + c_1 a_{n-1} + \cdots + c_k a_{n-k} = 0$, we must have k **initial values** $a_0, a_1, \ldots, a_{k-1}$. We assume that they are not all zero, so that the sequence $\{a_n\}$ is not the sequence in which every term is zero.

To solve such a recurrence relation, we turn to a process with which we are familiar, namely, that of solving a polynomial equation. Suppose for now that we have $a_n = x^n$ for some $x \neq 0$. Then the recurrence relation may be rewritten as $c_0 x^n + c_1 x^{n-1} + \cdots + c_k x^{n-k} = 0$. Dividing throughout by x^{n-k}, we have $c_0 x^k + c_1 x^{k-1} + \cdots + c_k = 0$. This is called the **characteristic equation** of the recurrence relation, and the roots of this equation are called the **characteristic roots**.

The characteristic equation of $a_n - 2a_{n-1} - 3a_{n-2} = 0$ is $x^2 - 2x - 3 = 0$. This may be factored into $(x+1)(x-3) = 0$. Hence the characteristic roots are $x = -1$ and $x = 3$.

For any characteristic root r, $a_n = r^n$ is indeed a solution of the recurrence relation. Thus in our example $a_n - 2a_{n-1} - 3a_{n-2} = 0$, we have $(-1)^n - 2(-1)^{n-1} - 3(-1)^{n-2} = (-1)^n(1+2-3) = 0$ so that $a_n = (-1)^n$ is a solution. We also have $3^n - 2 \times 3^{n-1} - 3 \times 3^{n-2} = 3^{n-1}(3-2-1) = 0$ so that $a_n = 3^n$ is another solution.

Suppose the characteristic roots r_1, r_2, $\ldots$, r_k are all distinct, as in our example. Consideration in linear algebra shows that r_1^n, r_2^n, $\ldots$, r_k^n are k linearly independent solutions of the recurrence relation. It follows that the general solution is given by $a_n = C_1 r_1^n + C_2 r_2^n + \cdots + C_k r_k^n$ for some constants C_1, C_2, $\ldots$, C_k which can be determined from the initial conditions. In our example, $a_n = C_1(-1)^n + C_2 3^n$. Suppose the initial values are $a_0 = 1$ and $a_1 = 2$. For $n = 0$, $1 = a_0 = C_1 + C_2$. For $n = 1$, $2 = a_1 = -C_1 + 3C_2$. Addition yields $3 = 4C_2$. It follows that $C_2 = \frac{3}{4}$, $C_1 = \frac{1}{4}$ and $a_n = \frac{1}{4}(3^{n+1} + (-1)^n)$.

Suppose our recurrence relation is $a_n - 2a_{n-1} + a_{n-2} = 0$ instead. The characteristic equation is $x^2 - 2x + 1 = 0$ and the characteristic roots are $x = 1$ and $x = 1$, which are not distinct. Clearly, $a_n = C_1 + C_2$ is not the general solution since it can be collapsed into $a_n = C$ where $C = C_1 + C_2$. We need something else.

For the general recurrence relation $c_0 a_n + c_1 a_{n-1} + \cdots + c_k a_{n-k} = 0$, suppose all characteristic roots are equal to r. We claim that $a_n = n^m r^n$ is a solution of the recurrence relation for $0 \leq m \leq k-1$. For $a_n - 2a_{n-1} + a_{n-2} = 0$, let $a_n = n$. Then $n - 2(n-1) + (n-2) = 0$.

To prove the general result, define $F(x) = c_0 x^k + c_1 x^{k-1} + \cdots + c_k$,

$$F_0(x) = x^{n-k} F(x) = c_0 x^n + c_1 x^{n-1} + \cdots + c_k x^{n-k}$$

and for $m \geq 1$, $F_m(x) = x F'_{m-1}(x)$, where differentiation is taken with respect to x. For instance,

$$
\begin{aligned}
F'_0(x) &= c_0 n x^{n-1} + c_1(n-1)x^{n-2} + \cdots + c_k(n-k)x^{n-k-1}; \\
F_1(x) &= c_0 n x^n + c_1(n-1)x^{n-1} + \cdots + c_k(n-k)x^{n-k}; \\
F'_1(x) &= c_0 n^2 x^{n-1} + c_1(n-1)^2 x^{n-2} + \cdots + c_k(n-k)^2 x^{n-k-1}; \\
F_2(x) &= c_0 n^2 x^n + c_1(n-1)^2 x^{n-1} + \cdots + c_k(n-k)^2 x^{n-k};
\end{aligned}
$$

and so on.

In general, $F_m(x) = c_0 n^m x^n + c_1(n-1)^m x^{n-1} + \cdots + c_k(n-k)^m x^{n-k}$. We use induction on m. For $m = 0$, $F_0(x) = c_0 x^n + c_1 x^{n-1} + \cdots + c_k x^{n-k}$. Suppose the result holds for some $m \geq 0$. Then we have

$$F_{m+1}(x) = c_0 n^{m+1} x^n + c_1(n-1)^{m+1} x^{n-1} + \cdots + c_k(n-k)^{m+1} x^{n-k}$$

by straight-forward differentiation. This completes the inductive argument.

We claim that $F_m(x)$ has r as a root of multiplicity $k - m$. Again we use induction on m. For $m = 0$, r is indeed a root of $F_0(x)$ of multiplicity $k - 0$. Suppose the result holds for some $m \geq 0$, so that $F_m(x) = (x - r)^{k-m} G(x)$ for some polynomial $G(x)$ with $G(r) \neq 0$. Then we have

$$F_{m+1}(x) = x(x-r)^{k-m} G'(x) + x(k-m)(x-r)^{k-(m+1)} G(x).$$

Both terms are divisible by $(x - r)^{k-(m+1)}$. The first term is divisible by $(x - r)^{k-m}$ but the second term is not. Hence r is a root of $F_{m+1}(x)$ of multiplicity $k - (m + 1)$. This completes the inductive argument.

It follows that $F_m(r) = 0$ for $0 \leq m \leq k - 1$, so that $a_n = n^m r^n$ is a solution to the recurrence relation for $0 \leq m \leq k - 1$. As before, consideration in linear algebra shows that these solutions are linearly independent. It follows that the general solution of the recurrence relation is $a_n = C_1 r^n + C_2 n r^n + \cdots + C_k n^{k-1} r^n$ for some constants $C_1, C_2, \ldots, C_k$ which can be determined from the initial conditions.

Returning to our example $a_n - 2a_{n-1} + a_{n-2} = 0$, still with the initial conditions $a_0 = 1$ and $a_1 = 2$, we have $a_n = C_1 + C_2 n$. For $n = 0$, $1 = a_0 = C_1$. For $n = 1$, $2 = a_1 = C_1 + C_2$. It follows that $C_1 = 1$ and $C_2 = 1$, so that $a_n = 1 + n$.

It may happen that the characteristic roots are not all distinct and not all identical. We illustrate with an example.

Suppose $a_n - 3a_{n-1} + 4a_{n-3} = 0$, with $a_0 = 0$, $a_1 = -1$ and $a_2 = 5$. The characteristic equation is $x^3 - 3x^2 + 4 = (x+1)(x-2)^2 = 0$ and the characteristic roots are -1, 2 and 2. Hencethe general solution is $a_n = C_1(-1)^n + C_2 2^n + C_3 n 2^n$. The unique solution of

$$
\begin{aligned}
C_1 + C_2 &= 0, \\
-C_1 + 2C_2 + 2C_3 &= -1, \\
C_1 + 4C_2 + 8C_3 &= 5
\end{aligned}
$$

is $(C_1, C_2, C_3) = (1, -1, 1)$. Hence $a_n = (-1)^n - 2^n + n 2^n$.

Another complication which may arise is when the characteristic roots are not real, such as in the recurrence relation $a_n - 2a_{n-1} + 2a_{n-2} = 0$, with initial conditions $a_0 = a_1 = 2$. The characteristic equation is $x^2 - 2x + 2 = 0$ and the characteristic roots are $1 + i$ and $1 - i$. Hence the general solution is $a_n = C_1(1+i)^n + C_2(1-i)^n$. For $n = 0$, we have $2 = C_1 + C_2$. For $n = 1$, we have $2 = C_1(1+i) + C_2(1-i) = (C_1 + C_2) + i(C_1 - C_2)$. Substituting the first equation into the second, we have $0 = i(C_1 - C_2)$ so that $C_1 = C_2 = 1$. It follows that $a_n = (1+i)^n + (1-i)^n$.

For every n, a_n is clearly a real number, in fact an integer. We can interpret the general solution above as follows. Note that $(1 \pm i)^2 = \pm 2i$ and $(1 \pm i)^4 = -4$. It follows that

$$
a_n = \begin{cases}
(-4)^k(1+1) & = 2(-4)^k & \text{for } n = 4k; \\
(-4)^k(1+1) & = 2(-4)^k & \text{for } n = 4k+1; \\
(-4)^k(2i - 2i) & = 0 & \text{for } n = 4k+2; \\
(-4)^k(-2 + 2i - 2 - 2i) & = (-4)^{k+1} & \text{for } n = 4k+3.
\end{cases}
$$

The first few terms are 2, 2, 0, -4, -8, -8, 0, 16, 32, 32, 0 and -64.

It is possible to eliminate references to complex numbers by making use of their polar forms and a result known as **De Moivre's Formula**.

A complex number $x + yi$, where x and y are real numbers, may be represented in the coordinate plane by the point (x, y). The length of the line segment joining (x, y) to the origin $(0,0)$ is $\sqrt{x^2 + y^2}$ by Pythagoras' Theorem. Denote this by r, and denote by θ the angle made by this line segment with the positive x-axis, so that $\tan \theta = \frac{y}{x}$. Then (r, θ) are the polar coordinates of the point (x, y), where $x = r \cos \theta$ and $y = r \sin \theta$.

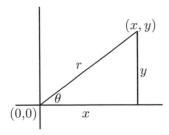

De Moivre's Formula states that for any angle θ and any positive integer n,

$$(\cos\theta \pm i\sin\theta)^n = \cos n\theta \pm i\sin n\theta.$$

We prove this by induction on n. The base $n = 1$ is an identity. Suppose the result holds for some $n \geq 1$. In the next case, we have

$$
\begin{aligned}
(\cos\theta \pm i\sin\theta)^{n+1} &= (\cos\theta \pm i\sin\theta)(\cos\theta \pm i\sin\theta)^n \\
&= (\cos\theta \pm i\sin\theta)(\cos n\theta \pm i\sin n\theta) \\
&= \cos\theta\cos n\theta - \sin\theta\sin n\theta \\
&\quad \pm i(\cos\theta\sin n\theta + \sin\theta\cos n\theta) \\
&= \cos(\theta + n\theta) \pm i\sin(\theta + n\theta) \\
&= \cos(n+1)\theta \pm i\sin(n+1)\theta.
\end{aligned}
$$

Continuing with our example, the characteristic roots $1 \pm i$ have the polar form $(\sqrt{2}, \pm\frac{\pi}{4})$. It follows that

$$
\begin{aligned}
a_n &= K_1 2^{\frac{n}{2}}\left(\cos\frac{\pi}{4} + i\sin\frac{\pi}{4}\right)^n + K_2 2^{\frac{n}{2}}\left(\cos\frac{\pi}{4} - i\sin\frac{\pi}{4}\right)^n \\
&= K_1 2^{\frac{n}{2}}\left(\cos\frac{n\pi}{4} + i\sin\frac{n\pi}{4}\right) + K_2 2^{\frac{n}{2}}\left(\cos\frac{n\pi}{4} - i\sin\frac{n\pi}{4}\right) \\
&= C_1 2^{\frac{n}{2}}\cos\frac{n\pi}{4} + C_2 2^{\frac{n}{2}}\sin\frac{n\pi}{4}.
\end{aligned}
$$

Note that $C_1 = K_1 + K_2$ and $C_2 = i(K_1 - K_2)$. We have $2 = a_0 = C_1$ and $2 = a_1 = C_1\sqrt{2}(\frac{1}{\sqrt{2}}) + C_2\sqrt{2}(\frac{1}{\sqrt{2}})$. Substituting the first equation into the second, we have $C_2 = 0$. It follows that

$$
a_n = 2^{\frac{n+2}{2}}\cos\frac{n\pi}{4} =
\begin{cases}
2^{2k+1}(-1)^k & \text{for } n = 4k; \\
2^{2k+1}(-1)^k & \text{for } n = 4k + 1; \\
0 & \text{for } n = 4k + 2; \\
2^{2k+2}(-1)^{k+1} & \text{for } n = 4k + 3.
\end{cases}
$$

The first few terms are 2, 2, 0, −4, −8, −8, 0, 16, 32, 32, 0 and −64.

We consider another recurrence relation $a_n + a_{n-1} + a_{n-2} = 0$ with initial conditions $a_0 = a_1 = 1$. The characteristic equation is $x^2 + x + 1 = 0$ and the characteristic roots are $\omega = \frac{-1+\sqrt{3}i}{2}$ and $\omega^2 = \frac{-1-\sqrt{3}i}{2}$. Hence the general solution is $a_n = C_1\omega^n + C_2\omega^{2n}$. For $n = 0$, we have $1 = C_1 + C_2$. For $n = 1$, we have $1 = C_1\omega + C_2\omega^2$. Multiplying the first equation by ω^2 and subtracting from this the second equation, we have $\omega^2 - 1 = C_1(\omega^2 - \omega)$. Canceling $\omega - 1$ from both sides, we have $-\omega^2 = \omega + 1 = C_1\omega$ so that $C_1 = -\omega$. Multiplying the first equation by ω and subtracting from this the second equation, we have $\omega - 1 = C_2(\omega - \omega^2)$. Canceling $\omega - 1$ from both sides, we have $\omega^3 = 1 = C_2(-\omega)$ so that $C_2 = -\omega^2$. It follows that $a_n = -\omega^{n+1} - \omega^{2(n+1)}$.

For $n = 3k$, $a_n = -\omega - \omega^2 = 1$. For $n = 3k + 1$, $a_n = -\omega^2 - \omega = 1$. Finally, for $n = 3k + 2$, $a_n = -1 - 1 = -2$. Hence the sequence is periodic with a period of length 3 consisting of 1, 1 and -2. This is because ω and ω^2 lie on the unit circle in the complex plane. In the earlier example, $1 + i$ and $1 - i$ lie on a larger circle. Hence the absolute values of the terms increase without bound.

In polar coordinates, we have $\omega = -\frac{1}{2} + \frac{\sqrt{3}}{2}i = \cos\frac{2\pi}{3} + i\sin\frac{2\pi}{3}$ while $\omega^2 = -\frac{1}{2} - \frac{\sqrt{3}}{2}i = \cos\frac{2\pi}{3} - i\sin\frac{2\pi}{3}$. It follows that

$$
\begin{aligned}
a_n &= K_1\left(\cos\frac{2\pi}{3} + i\sin\frac{2\pi}{3}\right)^n + K_2\left(\cos\frac{2\pi}{3} - i\sin\frac{2\pi}{3}\right)^n \\
&= K_1\left(\cos\frac{2n\pi}{3} + i\sin\frac{2n\pi}{3}\right) + K_2\left(\cos\frac{2n\pi}{3} - i\sin\frac{2n\pi}{3}\right) \\
&= C_1\cos\frac{2n\pi}{3} + C_2\sin\frac{2n\pi}{3}.
\end{aligned}
$$

Note that $C_1 = K_1 + K_2$ and $C_2 = i(K_1 - K_2)$. For $n = 0$, we have $1 = C_1$. For $n = 1$, we have $1 = C_1(-\frac{1}{2}) + C_2\frac{\sqrt{3}}{2}$. Substituting the first equation into the second, we have $C_2 = \sqrt{3}$. It follows that $a_n = \cos\frac{2n\pi}{3} + \sqrt{3}\sin\frac{2n\pi}{3}$.

For $n = 3k$, $a_n = 1 + 0 = 1$. For $n = 3k + 1$, $a_n = -\frac{1}{2} + \frac{3}{2} = 1$. Finally, for $n = 3k + 2$, $a_n = -\frac{1}{2} - \frac{3}{2} = -2$. Hence the sequence is periodic with a period of length 3 consisting of 1, 1 and -2. Note that if θ cannot be expressed as a simple multiple of π, then the polar form may not be easy to use.

Consider now the recurrence relation $a_n - 2a_{n-1} = 1$ with the initial condition $a_0 = 1$. Our convention has been to put all terms involving the sequence $\{a_n\}$ on the left side, and when the term on the right side is 0, we have called it homogeneous. Thus this example is a non-homogeneous linear recurrence relation with constant coefficients.

In the method of characteristic equations, we put $a_n = x^n$. In a homogeneous recurrence relation, we can cancel factors of x so that the resulting characteristic equation is of a fixed degree. In the present instance, $x^n - 2x^{n-1} = 1$ does not allow for any cancellation, and we have a characteristic equation of degree n, where n is unspecified. If we wish to use the same method as before, we must remove the non-homogeneous term 1.

Subtracting $a_{n-1} - 2a_{n-2} = 1$ from $a_n - 2a_{n-1} = 1$, we have

$$a_n - 3a_{n-1} + 2a_{n-2} = 0.$$

We have accomplished homogenization, at the expense of turning a one-step recurrence relation into a two-step recurrence relation. Now the characteristic equation is $x^2 - 3x + 2 = 0$ and the characteristic roots are 1 and 2. Hence the general solution is $a_n = C_1 + C_2 2^n$.

We are given only one initial value $a_0 = 1$, but we can iterate the recurrence relation to get $a_1 = 2a_0 + 1 = 3$. Now we have $1 = a_0 = C_1 + C_2$ and $3 = a_1 = C_1 + 2C_2$. Subtracting the first equation from the second, we have $C_2 = 2$. Substituting this back into either equation, we have $C_1 = -1$. Hence the specific solution is $a_n = 2^{n+1} - 1$.

In general, a non-homogeneous linear recurrence relation with constant coefficients is one of the form

$$c_0 a_n + c_1 a_{n-1} + \cdots + c_k a_{n-k} = b_n, \tag{1}$$

where b_n is not identically zero. The method of characteristic roots will work if the sequence $\{b_n\}$ also satisfies a homogeneous linear recurrence relation with constant coefficients

$$d_0 b_n + d_1 b_{n-1} + \cdots + d_h b_{n-h} = 0. \tag{2}$$

Combining (1) and (2), we obtain a homogeneous recurrence relation satisfied by the sequence $\{a_n\}$, namely,

$$\begin{aligned}
0 = \; & d_0(c_0 a_n + c_1 a_{n-1} + \cdots + c_k a_{n-k}) \\
& + d_1(c_0 a_{n-1} + c_1 a_{n-2} + \cdots + c_k a_{n-k-1}) \\
& + \cdots \\
& + d_h(c_0 a_{n-h} + c_1 a_{n-h-1} + \cdots + c_k a_{n-h-k}).
\end{aligned} \tag{3}$$

The characteristic equation of (3) is

$$(c_0 x^k + c_1 x^{k-1} + \cdots + c_k)(d_0 x^h + d_1 x^{h-1} + \cdots + d_h) = 0,$$

and the set of characteristic roots is the union of that of (2) and

$$c_0 a_n + c_1 a_{n-1} + \cdots + c_k a_{n-k} = 0. \tag{4}$$

The last equation is called the **associated** homogeneous recurrence relation of (1).

The general solution of (3) has $k + h$ terms. The $k + h$ constants can be found by first generating h additional initial conditions by iterating (1). However, we can make things simpler by setting aside the k terms which constitute the general solution of (4), with their constants to be determined later from the given initial conditions. The remaining h terms constitute a particular solution of (1). The constants can be determined by substituting this particular solution into (1).

We illustrate our approach by giving another solution to the example above. The associated homogeneous recurrence relation is $a_n - 2a_{n-1} = 0$, with a characteristic root 2. Now $b_n = 1$ is associated with the characteristic root 1. Hence a particular solution has the form $a_n = A$, where A is an undetermined coefficient.

Substituting this into the original recurrence relation, we have $A - 2A = 1$ so that $A = -1$. The general solution is therefore $a_n = C2^n - 1$. For $n = 0$, $1 = a_0 = C - 1$. Hence $C = 2$ and $a_n = 2^{n+1} - 1$.

The advantage of this approach is in the numerical work. Instead of solving a system of $k + h$ linear equations, it is much better to divide and conquer, namely, solving a system of k linear equations and a system of h linear equations separately. The hard part in this approach is picking out the form of the particular solution.

Let us try just finding a particular solution to $a_n - 2a_{n-1} = n$. Let $a_n = An$. Then we have $n = An - 2(A(n - 1)) = -An + 2A$. Equating the coefficients of the linear term yields $1 = -A$ or $A = -1$, but equating the constant terms yields $0 = 2A$ or $A = 0$. What is going on? The non-homogeneous term n is a linear polynomial whose constant term happens to be 0. The particular solution must be a general linear polynomial, that is, $a_n = An + B$. Now $n = An + B - 2(A(n - 1) + B) = -An + (2A - B)$. Hence $A = -1$ and $B = -2$, so that $a_n = -n - 2$.

To solve the non-homogeneous recurrence relation, $a_n - 2a_{n-1} = 2^n$, let $a_n = A2^n$. Then we have $2^n = A2^n - 2(A2^{n-1}) = 0$. What is going on? That we obtain 0 indicates that $a_n = A2^n$ is a solution to the associated homogeneous recurrence relation. The non-homogeneous term must be treated as a second copy of the same. Hence we must have $a_n = An2^n$. Now $2^n = An2^n - 2(A(n - 1)2^{n-1}) = A2^n$. Hence $A = 1$, so that $a_n = n2^n$.

The initial difficulty in each of these two examples comes from improper handling of repeated characteristic roots, all on the same side or some on each side.

Examples

Example 2.4.1.

Solve the recurrence relation $a_n - 5a_{n-1} + 8a_{n-2} - 8a_{n-3} + 7a_{n-4} - 3a_{n-5} = 0$ with the initial conditions $a_0 = 3$, $a_1 = 4$, $a_2 = 9$, $a_3 = 28$ and $a_4 = 83$.

Solutions:

The characteristic equation is $x^5 - 5x^4 + 8x^3 - 8x^2 + 7x - 3 = 0$, which may be factored as $(x^2 + 1)(x - 1)^2(x - 3) = 0$. Hence the general solution is $a_n = C_1 \sin \frac{n\pi}{2} + C_2 \cos \frac{n\pi}{2} + C_3 + C_4 n + C_5 3^n$. From the initial conditions, we have

$$
\begin{array}{rcrcrcrcl}
        & C_2 & + & C_3 &   &   & + & C_5   & = & 3 \\
C_1   &   & + & C_3 & + & C_4 & + & 3C_5 & = & 4 \\
        & -C_2 & + & C_3 & + & 2C_4 & + & 9C_5 & = & 9 \\
-C_1   &   & + & C_3 & + & 3C_4 & + & 27C_5 & = & 28 \\
        & C_2 & + & C_3 & + & 4C_4 & + & 81C_5 & = & 83
\end{array}
$$

Eliminating C_1 and C_2 simultaneously, we have $2C_3 + 2C_4 + 10C_5 = 12$, $2C_3 + 4C_4 + 30C_5 = 32$ and $2C_3 + 6C_4 + 90C_5 = 92$. Eliminating C_3, we have $2C_4 + 20C_5 = 20$ and $2C_4 + 60C_5 = 60$. Eliminating C_4, we have $C_5 = 1$. Back substituting yields $C_4 = 0$, $C_3 = 1$, $C_2 = 1$ and $C_1 = 0$. Hence we have $a_n = \cos \frac{n\pi}{2} + 1 + 3^n$.

Example 2.4.2.

Solve the recurrence relation $a_n - a_{n-1} = 2^n + 2n$ with the initial condition $a_0 = 3$.

Solution:

We seek a particular solution of the form $a_n = A2^n + Bn^2 + Dn$. Then

$$
\begin{aligned}
2^n + 2n &= a_n - a_{n-1} \\
&= A2^n + Bn^2 + Dn - A2^{n-1} - B(n^2 - 2n + 1) - D(n-1) \\
&= A2^{n-1} + 2Bn + (D - B).
\end{aligned}
$$

It follows that $A = 2$, $B = 1$ and $D = 1$. The general solution is

$$
a_n = C + 2^{n+1} + n^2 + n.
$$

From $a_0 = 3$, we have $C = 1$ so that $a_n = 2^{n+1} + n^2 + n + 1$. Compare with Example 2.3.1.

Example 2.4.3.

Determine the number of ternary sequences of length n with no two 1s adjacent and no two 2s adjacent. Set up a recurrence relation and solve it by the method of characteristic equations.

Solution:
Let a_n denote the number of such ternary sequences, b_n denote those ending in 1 and c_n those ending in 2. Then $a_0 = 1$, $b_0 = c_0 = 0$, $a_1 = 3$ and $b_1 = c_1 = 1$. For $n \geq 2$, $a_n = 3a_{n-1} - b_{n-1} - c_{n-1}$ since we can add any of the three digits at the end of any sequence as long as we do not create 11 or 22. On the other hand, $b_n = a_{n-1} - b_{n-1}$ since we can add a 1 at the end of any sequence as long as we do not create 11. Similarly, $c_n = a_{n-1} - b_{n-1}$. Clearly, $b_n = c_n$ for all n. Hence $a_n = 3a_{n-1} - 2b_{n-1}$ and $b_n = a_{n-1} - b_{n-1}$. Substituting $2b_{n-1} = 3a_{n-1} - a_n$ into $2b_n = 2a_{n-1} - 2b_{n-1}$, we obtain $3a_n - a_{n+1} = 2a_{n-1} - 3a_{n-1} + a_n$. Simplifying and adjusting the indices, we have $a_n = 2a_{n-1} + a_{n-2}$. The characteristic equation is $x^2 - 2x - 1 = 0$ and the characteristic roots are $1 \pm \sqrt{2}$, so that $a_n = C_1(1+\sqrt{2})^n + C_2(1-\sqrt{2})^n$. Now $1 = a_0 = C_1 + C_2$ and $3 = a_1 = C_1(1+\sqrt{2}) + C_2(1-\sqrt{2})$. Using the former, the latter simplifies to $\sqrt{2} = C_1 - C_2$. Hence $C_1 = \frac{1+\sqrt{2}}{2}$ and $C_2 = \frac{1-\sqrt{2}}{2}$ so that $a_n = \frac{1}{2}((1+\sqrt{2})^{n+1} + (1-\sqrt{2})^{n+1})$.

Exercises

1. Solve the recurrence relation $a_n - 6a_{n-1} + 11a_{n-2} - 6a_{n-3} = 0$ with the initial conditions $a_0 = 1$, $a_1 = 4$ and $a_2 = 12$. Use the method of characteristic equations.

2. Solve the recurrence relation $a_n - 10a_{n-1} + 31a_{n-2} - 30a_{n-3} = 0$ with the initial conditions $a_0 = 0$, $a_1 = 4$ and $a_2 = 26$. Use the method of characteristic equations.

3. Solve the recurrence relation $a_n - 2a_{n-1} = n2^{n+1}$ with the initial condition $a_0 = 1$. Use the method of characteristic equations.

4. Solve the recurrence relation $a_n - a_{n-1} = 2\cos\frac{n\pi}{3}$ with the initial condition $a_0 = 1$. Use the method of characteristic equations.

5. A bug is at a vertex of an octagon. Each minute, it crawls to an adjacent vertex. In how many ways can it reach the opposite vertex for the first time after $2n$ minutes? Set up a recurrence relation and solve it by the method of characteristic equations.

6. A regular octahedron $ABCDEF$ is obtained by joining a square-based pyramid $E - ABCD$ to another square-based pyramid $F - ABCD$ across their common base. A bug begins at vertex B of this octahedron. In each minute, it moves from its current position to an adjacent vertex. Once it reaches vertex D, it ends there. What is the number of different paths the bug can take in order to finish at D after exactly n minutes? Set up a recurrence relation and solve it by the method of characteristic equations.

Section 2.5. The Method of Generating Functions

An alternative method for solving linear recurrence relations with constant coefficients makes use of generating functions introduced at the beginning of this chapter. This approach involves three steps. We illustrate with an example of a homogeneous recurrence relation.

Suppose $a_n - 2a_{n-1} - 5a_{n-2} + 6a_{n-3} = 0$, with $a_0 = 2$ and $a_1 = a_2 = -3$. Let $G(x)$ be the generating function for $\{a_n\}$. The first step is to obtain from the recurrence relation an equation from which we can determine $G(x)$.

This is called a **functional equation**. An algebraic equation of one variable contains numbers and a variable x, and we determine the numerical value of x which makes the equation true. Similarly, a functional equation of one variable contains expressions in a variable x and a function $G(x)$, and we determine the algebraic expression for $G(x)$ which makes the equation true.

In our example,

$$
\begin{array}{rrrrrrr}
G(x) &=& 2 & -3x & -3x^2 & +a_3x^3 & +\cdots \\
-2xG(x) &=& & -4x & +6x^2 & -2a_2x^3 & -\cdots \\
-5x^2G(x) &=& & & -10x^2 & -5a_1x^3 & -\cdots \\
6x^3G(x) &=& & & & 6a_0x^3 & +\cdots \\
\hline
(1-2x-5x^2+6x^3)G(x) &=& 2 & -7x & -7x^2 & &
\end{array}
$$

All subsequent terms on the right side are equal to 0 according to the recurrence relation.

The notation may be made more compact as follows. We multiply the recurrence relation by x^n, yielding

$$a_nx^n - 2xa_{n-1}x^{n-1} - 5x^2a_{n-2}x^{n-2} + 6x^3a_{n-3}x^{n-3} = 0.$$

Summing from $n = 3$ to infinity, we have

$$\sum_{n=3}^{\infty} a_nx^n - 2x\sum_{n=3}^{\infty} a_{n-1}x^{n-1} - 5x^2\sum_{n=3}^{\infty} a_{n-2}x^{n-2} + 6x^3\sum_{n=3}^{\infty} a_{n-3}x^{n-3} = 0.$$

The four terms on the left side are respectively equal to $G(x) - 2 + 3x + 3x^2$, $-2x(G(x) - 2 + 3x)$, $-5x^2(G(x) - 2)$ and $6x^3G(x)$. Simplification leads to the same functional equation as before.

The second step is to solve the functional equation obtained in the first step. This is usually straight-forward. In our example, we have

$$G(x) = \frac{2 - 7x - 7x^2}{1 - 2x - 5x^2 + 6x^3}.$$

The third step is to extract the sequence $\{a_n\}$ from the generating function $G(x)$ obtained in the second step. We have seen how this is done in Section 2.1.

In our example, note that $1 - 2x - 5x^2 + 6x^3 = (1-x)(1+2x)(1-3x)$. We apply the method of partial fractions. Let

$$\frac{2 - 7x - 7x^2}{1 - 2x - 5x^2 + 6x^3} = \frac{A}{1-x} + \frac{B}{1+2x} + \frac{C}{1-3x}.$$

Clearing fractions,

$$2 - 7x - 7x^2 = A(1+2x)(1-3x) + B(1-x)(1-3x) + C(1-x)(1+2x).$$

Setting $x = 1$, we have $-12 = -6A$ so that $A = 2$. Setting $x = -\frac{1}{2}$, we have $\frac{15}{4} = \frac{15}{4}B$ so that $B = 1$. Setting $x = \frac{1}{3}$, we have $\frac{-10}{9} = \frac{10}{9}C$ so that $C = -1$. Hence $G(x) = \frac{2}{1-x} + \frac{1}{1+2x} - \frac{1}{1-3x}$. The three terms on the right side generate respectively the sequences $\{2\}$, $\{(-2)^n\}$ and $\{-3^n\}$. It follows that $a_n = 2 + (-2)^n - 3^n$.

Note that the initial values are incorporated directly into the solution. Had we been asked to find the general solution of the recurrence relation without being given these values, then a_0, a_1, ... will appear in the coefficients, and the computation will be more involved.

Another characteristic of the generating function approach is that it treats homogeneous and non-homogeneous recurrence relations in the same way. Thus it is akin to the process of homogenization. We illustrate with an example.

Suppose $a_n - 3a_{n-1} - 4a_{n-2} = 3 \cdot 2^n - 6$ with $a_0 = 0$ and $a_1 = 1$. Let $G(x) = \sum_{n=0}^{\infty} a_n x^n$. We have

$$\sum_{n=2}^{\infty} a_n x^n - 3x \sum_{n=2}^{\infty} a_{n-1} x^{n-1} - 4x^2 \sum_{n=2}^{\infty} a_{n-2} x^{n-2} = 3 \sum_{n=2}^{\infty} 2^n x^n - 6 \sum_{n=2}^{\infty} x^n.$$

Hence $G(x) - x - 3xG(x) - 4x^2 G(x) = 3(\frac{1}{1-2x} - 1 - 2x) - 6(\frac{1}{1-x} - 1 - x)$. It follows that

$$G(x) = \frac{x + 3x^2 + 2x^3}{(1-x)(1+x)(1-2x)(1-4x)} = \frac{A}{1-x} + \frac{B}{1+x} + \frac{C}{1-2x} + \frac{D}{1-4x}.$$

Clearing fractions, we have

$$\begin{aligned}
x + 3x^2 + 2x^3 =\ & A(1+x)(1-2x)(1-4x) + B(1-x)(1-2x)(1-4x) \\
& + C(1-x)(1+x)(1-4x) + D(1-x)(1+x)(1-2x).
\end{aligned}$$

Setting $x = 1$, we have $6 = A(2)(-1)(-3)$ so that $A = 1$. Setting $x = -1$, we have $0 = B(2)(3)(5)$ so that $B = 0$. Setting $x = \frac{1}{2}$, we have $\frac{3}{2} = C(\frac{1}{2})(\frac{3}{2})(-1)$ so that $C = -2$. Setting $x = \frac{1}{4}$, we have $\frac{15}{32} = D(\frac{3}{4})(\frac{5}{4})(\frac{1}{2})$ so that $D = 1$. Now $\frac{1}{1-x}$ generates the sequence $\{1\}$, $\frac{-2}{1-2x}$ generates the sequence $\{-2^{n+1}\}$ while $\frac{1}{1-4x}$ generates the sequence $\{4^n\}$. It follows that $a_n = 1 - 2^{n+1} + 4^n$.

The method of generating functions can also be used to solve some non-linear recurrence relations. We illustrate with two examples.

Suppose $a_n = 1 \cdot 3^n + 2 \cdot 3^{n-1} + \cdots + n \cdot 3 + (n+1)$. Let $b_n = n+1$ and $c_n = 3^n$. Then $a_n = b_n c_0 + b_{n-1} c_1 + \cdots + b_0 c_n$. It follows that $G(x)$ is the product of the generating functions for $\{b_n\}$ and $\{c_n\}$, which are $\frac{1}{(1-x)^2}$ and $\frac{1}{1-3x}$ respectively. Hence

$$G(x) = \frac{1}{(1-3x)(1-x)^2} = \frac{A}{1-3x} + \frac{B}{1-x} + \frac{C}{(1-x)^2}.$$

Clearing fractions, $1 = A(1-x)^2 + B(1-3x)(1-x) + C(1-3x)$. Setting $x = \frac{1}{3}$, we have $1 = \frac{4}{9}A$ so that $A = \frac{9}{4}$. Setting $x = 1$, we have $1 = -2C$ so that $C = -\frac{1}{2}$. Setting $x = 0$, we have $1 = A + B + C$ so that $B = -\frac{3}{4}$. Hence

$$G(x) = \frac{9}{4} \sum_{n=0}^{\infty} 3^n x^n - \frac{3}{4} \sum_{n=0}^{\infty} x^n - \frac{1}{2} \sum_{n=0}^{\infty} (n+1)x^n.$$

It follows that $a_n = \frac{1}{4}(3^{n+2} - 2n - 5)$.

Suppose $3^n = a_0 \cdot 2^n + a_1 \cdot 2^{n-1} + \cdots + a_n$. Let $b_n = 2^n$ and $c_n = 3^n$. Then

$$c_n = a_n b_0 + a_{n-1} b_1 + \cdots + a_0 b_n.$$

It follows that $G(x)$ is the quotient of the generating functions for $\{c_n\}$ and $\{b_n\}$, which are $\frac{1}{1-3x}$ and $\frac{1}{1-2x}$ respectively. Hence $G(x) = \frac{1-2x}{1-3x} = 1 + \frac{x}{1-3x}$. Now the term 1 generates the isolated value $a_0 = 1$ while $\frac{x}{1-3x}$ generates the values $a_n = 3^{n-1}$ for $n \geq 1$.

We conclude with a problem involving a system of simultaneous recurrence relations.

(a) Determine the number of quaternary sequences of length n with an even number of 0s and an even number of 1s.

(b) Determine the number of quaternary sequences of length n with an odd number of 0s and an odd number of 1s.

(c) Determine the number of quaternary sequences of length n with an even number of 0s and an odd number of 1s.

(d) Determine the number of quaternary sequences of length n with an odd number of 0s and an even number of 1s.

We solve all four parts together. Let a_n be the number of quaternary sequences of length n with an even number of 0s and an even number of 1s. Then $a_0 = 1$. Let b_n be the number of quaternary sequences of length n with an odd number of 0s and an odd number of 1s. Then $b_0 = 0$. Let c_n be the number of quaternary sequences of length n with an even number of 0s and an odd number of 1s. Then $c_0 = 0$. Let d_n be the number of quaternary sequences of length n with an odd number of 0s and an even number of 1s. Then $d_0 = 0$.

We have

$$
\begin{aligned}
a_n &= 2a_{n-1} + c_{n-1} + d_{n-1}; \\
b_n &= 2b_{n-1} + c_{n-1} + d_{n-1}; \\
c_n &= 2c_{n-1} + a_{n-1} + b_{n-1}; \\
d_n &= 2d_{n-1} + a_{n-1} + b_{n-1}; \\
4^n &= a_n + b_n + c_n + d_n.
\end{aligned}
$$

Let $G(x)$, $H(x)$, $J(x)$ and $K(x)$ be the respective generating functions of $\{a_n\}$, $\{b_n\}$, $\{c_n\}$ and $\{d_n\}$. Then

$$
\begin{aligned}
G(x) - 1 &= 2xG(x) + xJ(x) + xK(x); & (1) \\
H(x) &= 2xH(x) + xJ(x) + xK(x); & (2) \\
J(x) &= 2xJ(x) + xG(x) + xH(x); & (3) \\
K(x) &= 2xK(x) + xG(x) + xH(x); & (4) \\
\frac{1}{1-4x} &= G(x) + H(x) + J(x) + K(x). & (5)
\end{aligned}
$$

Subtracting (2) from (1), we have $G(x) - H(x) - 1 = 2x(G(x) - H(x))$. This simplifies to $G(x) - H(x) = \frac{1}{1-2x}$. Subtracting x times (5) from (1), $G(x) - 1 - \frac{x}{1-4x} = x(G(x) - H(x)) = \frac{x}{1-2x}$. Hence $G(x) = 1 + \frac{x}{1-4x} + \frac{x}{1-2x}$, so that $H(x) = G(x) - \frac{1}{1-2x} = \frac{x}{1-4x} - \frac{x}{1-2x}$. Subtracting (4) from (3), we have $J(x) - K(x) = 2x(J(x) - K(x))$, which implies that $J(x) = K(x)$. From (5), $J(x) = K(x) = \frac{1}{2}(\frac{1}{1-4x} - G(x) - H(x)) = \frac{x}{1-4x}$.

It follows that the sequences generated are

$$
\begin{array}{llll}
a_0 = 1 & \text{and} & a_n = 4^{n-1} + 2^{n-1} & \text{for } n \geq 1; \\
b_0 = 0 & \text{and} & b_n = 4^{n-1} - 2^{n-1} & \text{for } n \geq 1; \\
c_0 = 0 & \text{and} & c_n = 4^{n-1} & \text{for } n \geq 1; \\
d_0 = 0 & \text{and} & d_n = 4^{n-1} & \text{for } n \geq 1.
\end{array}
$$

Examples

Example 2.5.1.
Solve the recurrence relation $a_n - 5a_{n-1} + 8a_{n-2} - 8a_{n-3} + 7a_{n-4} - 3a_{n-5} = 0$ with the initial conditions $a_0 = 3$, $a_1 = 4$, $a_2 = 9$, $a_3 = 28$ and $a_4 = 83$. Use the method of generating functions.

Solution:
Let $G(x) = \sum_{n=0}^{\infty} a_n x^n$. We have

$$
\begin{aligned}
0 &= \sum_{n=5}^{\infty} a_n x^n - 5x \sum_{n=5}^{\infty} a_{n-1} x^{n-1} + 8x^2 \sum_{n=5}^{\infty} a_{n-2} x^{n-2} \\
&\quad -8x^3 \sum_{n=5}^{\infty} a_{n-3} x^{n-3} + 7x^4 \sum_{n=5}^{\infty} a_{n-4} x^{n-4} - 3x^5 \sum_{n=5}^{\infty} a_{n-5} x^{n-5} \\
&= (G(x) - 3 - 4x - 9x^2 - 28x^3 - 83x^4) - 5x(G(x) - 3 - 4x - 9x^2 - 28x^3) \\
&\quad + 8x^2(G(x) - 3 - 4x - 9x^2) - 8x^3(G(x) - 3 - 4x) \\
&\quad + 7x^4(G(x) - 3) - 3x^5 G(x) \\
&= (1 - 5x + 8x^2 - 8x^3 + 7x^4 - 3x^5)G(x) - (3 - 11x + 13x^2 - 9x^3 + 4x^4).
\end{aligned}
$$

It follows that

$$
\begin{aligned}
G(x) &= \frac{3 - 11x + 13x^2 - 9x^3 + 4x^4}{1 - 5x + 8x^2 - 8x^3 + 7x^4 - 3x^5} \\
&= \frac{3 - 8x + 5x^2 - 4x^3}{1 - 4x + 4x^2 - 4x^3 + 3x^4} \\
&= \frac{A}{1 - 3x} + \frac{B}{1 - x} + \frac{C + Dx}{1 + x^2}.
\end{aligned}
$$

We have

$$
\begin{aligned}
&3 - 8x + 5x^2 - 4x^3 \\
&= A(1 - x)(1 + x^2) + B(1 - 3x)(1 + x^2) + (C + Dx)(1 - 3x)(1 - x).
\end{aligned}
$$

Setting $x = \frac{1}{3}$, we have $\frac{20}{27} = \frac{20}{27}A$ so that $A = 1$. Setting $x = 1$, we have $-4 = -4B$ so that $B = 1$. Setting $x = i$, we have $-2 - 4i = (C + Di)(-2 - 4i)$. Hence $C = 1$ and $D = 0$. We have $a_n = 3^n + 1 + \cos\frac{n\pi}{2}$. Compare with Example 2.4.1.

Example 2.5.2.
Solve the recurrence relation $a_n - a_{n-1} = 2^n + 2n$ with $a_0 = 3$. Use the method of generating functions.

Solution:
Let $G(x) = \sum_{n=0}^{\infty} a_n x^n$. We have

$$
\sum_{n=1}^{\infty} a_n x^n - x \sum_{n=1}^{\infty} a_{n-1} x^{n-1} = \sum_{n=1}^{\infty} 2^n x^n + 2 \sum_{n=1}^{\infty} (n+1)x^n - 2 \sum_{n=1}^{\infty} x^n.
$$

It follows that

$$G(x) - 3 - xG(x) = \left(\frac{1}{1-2x} - 1\right) + 2\left(\frac{1}{(1-x)^2} - 1\right) - 2\left(\frac{1}{1-x} - 1\right)$$

so that

$$G(x) = \frac{3 - 8x + 7x^2 - 4x^3}{(1-x)^3(1-2x)} = \frac{A}{1-x} + \frac{B}{(1-x)^2} + \frac{C}{(1-x)^3} + \frac{D}{1-2x}.$$

We have

$$3 - 8x + 7x^2 - 4x^3$$
$$= A(1-2x)(1-x)^2 + B(1-2x)(1-x) + C(1-2x) + D(1-x)^3.$$

Setting $x = 1$, we have $-2 = C(-1)$ so that $C = 2$. Setting $x = \frac{1}{2}$, we have $\frac{1}{4} = D(\frac{1}{8})$ so that $D = 2$. Comparing the coefficients of the cubic term, we have $-2A - D = -4$ so that $A = 1$. Comparing the constant terms, we have $A + B + C + D = 3$ so that $B = -2$. Hence

$$a_n = 1 - 2(n+1) + 2\binom{n+2}{2} + 2(2^n) = 1 + n + n^2 + 2^{n+1}.$$

Compare with Examples 2.3.1 and 2.4.2.

Example 2.5.3.
Determine the number of ternary sequences of length n with no two 1s adjacent and no two 2s adjacent. Set up a recurrence relation and solve it by the method of generating functions.

Solution:
Let a_n denote the number of such ternary sequences, b_n denote those ending in 1 and c_n those ending in 2. Then $a_0 = 1$, $b_0 = c_0 = 0$, $a_1 = 3$ and $b_1 = c_1 = 1$. For $n \geq 2$, $a_n = 3a_{n-1} - b_{n-1} - c_{n-1}$ since we can add any of the three digits at the end of any sequence as long as we do not create 11 or 22. On the other hand, $b_n = a_{n-1} - b_{n-1}$ since we can add a 1 at the end of any sequence as long as we do not create 11. Similarly, $c_n = a_{n-1} - b_{n-1}$. Clearly, $b_n = c_n$ for all n. Hence $a_n = 3a_{n-1} - 2b_{n-1}$ and $b_n = a_{n-1} - b_{n-1}$. Let $G(x)$ be the generating function for $\{a_n\}$ and $H(x)$ be that for $\{b_n\}$. Summing the recurrence relations from $n = 1$ to infinity, we have $G(x) - 1 = 3xG(x) - 2xH(x)$ and $H(x) = xG(x) - xH(x)$. From the latter equation, $H(x) = \frac{x}{1+x}G(x)$. Substituting into the former equation, we have $G(x)(1 - 3x + \frac{2x^2}{1+x}) = 1$ so that

$$G(x) = \frac{1+x}{1 - 2x - x^2} = \frac{A}{1 - (1+\sqrt{2})x} + \frac{B}{1 - (1-\sqrt{2})x}.$$

Clearing fractions, $1 + x = A(1 - (1-\sqrt{2})x) + B(1 - (1+\sqrt{2})x)$. Setting $x = -1 + \sqrt{2}$, we have $\sqrt{2} = (4 - 2\sqrt{2})A$ so that $A = \frac{1+\sqrt{2}}{2}$. Setting $x = -1 - \sqrt{2}$, we have $-\sqrt{2} = (4 + 2\sqrt{2})B$ so that $B = \frac{1-\sqrt{2}}{2}$. It follows that $a_n = \frac{1}{2}((1+\sqrt{2})^{n+1} + (1-\sqrt{2})^{n+1}$. Compare with Example 2.4.3.

Exercises

1. Solve the recurrence relation $a_n - 6a_{n-1} + 11a_{n-2} - 6a_{n-3} = 0$ with the initial conditions $a_0 = 1$, $a_1 = 4$ and $a_2 = 12$. Use the method of generating functions.

2. Solve the recurrence relation $a_n - 10a_{n-1} + 31a_{n-2} - 30a_{n-3} = 0$ with the initial conditions $a_0 = 0$, $a_1 = 4$ and $a_2 = 26$. Use the method of generating functions.

3. Solve the recurrence relation $a_n - 2a_{n-1} = n2^{n+1}$ with the initial condition $a_0 = 1$. Use the method of generating functions.

4. Solve the recurrence relation $a_n - a_{n-1} = 2\cos\frac{n\pi}{3}$ with the initial condition $a_0 = 1$. Use the method of generating functions.

5. A bug is at a vertex of an octagon. Each minute, it crawls to an adjacent vertex. In how many ways can it reach the opposite vertex for the first time after $2n$ minutes? Set up a recurrence relation and solve it by the method of generating functions.

6. A regular octahedron $ABCDEF$ is obtained by joining a square-based pyramid $E - ABCD$ to another square-based pyramid $F - ABCD$ across their common base. A bug begins at vertex B of this octahedron. In each minute, it moves from its current position to an adjacent vertex. Once it reaches vertex D, it ends there. What is the number of different paths the bug can take in order to finish at D after exactly n minutes? Set up a recurrence relation and solve it by the method of generating functions.

Section 2.6. Dirichlet Generating Functions

Given a sequence $\{a_1, a_2, a_3, \ldots\}$ of complex numbers (often, but not necessarily, integers), its **Dirichlet** generating function is $A(s) = \sum_{n=1}^{\infty} \dfrac{a_n}{n^s}$, where s is a complex number. Once again, we treat this as a formal power series, and do not worry about convergence.

We are primarily interested in sequences in which the terms are some arithmetic functions. We have come across two such functions in Chapter 1. The tau-function $\tau(n)$, defined on page 2, counts the number of positive divisors of the positive integer n, while the Euler phi-function $\phi(n)$, defined on page 5, counts the numbers of positive integers less than or equal to n and relatively prime to n. We now introduce two more.

The sigma-function $\sigma(n)$ counts the sum of the positive divisors of the positive integer n. Thus

n	$\sigma(n)$	n	$\sigma(n)$	n	$\sigma(n)$	n	$\sigma(n)$
1	1	5	6	9	13	13	14
2	3	6	12	10	18	14	24
3	4	7	8	11	12	15	24
4	7	8	15	12	28	16	31

The **Möbius** mu-function, $\mu(n)$, is defined for all positive integers n by

$$\mu(n) = \begin{cases} 1 & \text{if } n = 1, \\ (-1)^t & \text{if } n \text{ is the product of } t \text{ distinct primes,} \\ 0 & \text{if } n \text{ is divisible by the square of any prime.} \end{cases}$$

Thus

n	$\mu(n)$	n	$\mu(n)$	n	$\mu(n)$	n	$\mu(n)$
1	1	5	-1	9	0	13	-1
2	-1	6	1	10	1	14	1
3	-1	7	-1	11	-1	15	1
4	0	8	0	12	0	16	0

The Möbius mu-function satisfies the formula

$$\sum_{d \mid n} \mu(d) = \begin{cases} 1 \text{ if } n = 1, \\ 0 \text{ if } n > 1. \end{cases}$$

By definition, $\sum_{d \mid 1} \mu(d) = 1$ while

$$\sum_{d \mid n} \mu(p^k) = \mu(1) + \mu(p) + \mu(p^2) + \cdots + \mu(p^k) = 1 - 1 = 0$$

for any prime p and any positive integer k. Hence $\sum_{d \mid n} \mu(d) = 0$ for $n > 1$.

Dirichlet's Product Formula.

Suppose $f(n)$ and $g(n)$ are two functions defined for all positive integers n. Then

$$\sum_{n=1}^{\infty} \frac{f(n)}{n^s} \sum_{m=1}^{\infty} \frac{g(m)}{m^s} = \sum_{r=1}^{\infty} \frac{1}{r^s} \sum_{n|r} f(n) g\left(\frac{r}{n}\right).$$

Proof:

Let $mn = r$, so that $m|r$, $n|r$, $\frac{r}{m} = n$ and $\frac{r}{n} = m$. Then,

$$F(s)G(s) = \sum_{n=1}^{\infty} \frac{f(n)}{n^s} \sum_{m=1}^{\infty} \frac{g(m)}{m^s} = \sum_{n=1}^{\infty} \sum_{m=1}^{\infty} \frac{f(n)g(m)}{(nm)^s} = \sum_{r=1}^{\infty} \frac{1}{r^s} \sum_{n|r} f(n) g\left(\frac{r}{n}\right).$$

The Dirichlet generating function for the sequence $\{a_n\}$ where $a_n = 1$ for all n is called the **Riemann zeta-function**, $\zeta(s) = \sum_{n=1}^{\infty} \frac{1}{n^s}$. It is of paramount importance in analytic number theory.

We now use Dirichlet's Product Formula to determine the Dirichlet generating function for $\{\mu(n)\}$ in terms of the Riemann zeta-function. Using

$$\sum_{d|n} \mu(d) = \begin{cases} 1 \text{ if } n = 1, \\ 0 \text{ if } n > 1, \end{cases}$$

we have

$$\sum_{n=1}^{\infty} \frac{\mu(n)}{n^s} \sum_{m=1}^{\infty} \frac{1}{m^s} = \sum_{r=1}^{\infty} \frac{1}{r^s} \sum_{n|r} \mu(n) = 1.$$

Since $\sum_{m=1}^{\infty} \frac{1}{m^s} = \zeta(s)$, this yields $\sum_{n=1}^{\infty} \frac{\mu(n)}{n^s} = \frac{1}{\zeta(s)}$.

Möbius' Inversion Formula.

Let $F(n) = \sum_{d|n} f(d)$ where $f(n)$ is any function defined for all positive integers n. Then

$$f(n) = \sum_{d|n} \mu(d) F\left(\frac{n}{d}\right) = \sum_{d|n} \mu\left(\frac{n}{d}\right) F(d).$$

Proof:

From Dirichlet's Product Formula,

$$\begin{aligned}
\zeta(s) \sum_{m=1}^{\infty} \frac{f(m)}{m^s} &= \sum_{n=1}^{\infty} \frac{1}{n^s} \sum_{m=1}^{\infty} \frac{f(m)}{m^s} \\
&= \sum_{r=1}^{\infty} \frac{1}{r^s} \sum_{m|r} f(m) \\
&= \sum_{r=1}^{\infty} \frac{F(r)}{r^s}.
\end{aligned}$$

It follows that

$$\sum_{m=1}^{\infty} \frac{1}{m^s} f(m) = \frac{1}{\zeta(s)} \sum_{m=1}^{\infty} \frac{F(m)}{m^s}$$

$$= \sum_{n=1}^{\infty} \frac{\mu(n)}{n^s} \sum_{m=1}^{\infty} \frac{F(m)}{m^s}$$

$$= \sum_{r=1}^{\infty} \frac{1}{r^s} \sum_{n|r} \mu(n) F\left(\frac{r}{n}\right)$$

$$= \sum_{r=1}^{\infty} \frac{1}{r^s} \sum_{m|r} \mu\left(\frac{r}{m}\right) F(m).$$

We shall take for granted that if two Dirichlet series yield identical functions for all s, then their coefficients must be identical. Hence the desired conclusion follows.

Euler's Product Formula:

The Riemann zeta-function satisfies $\dfrac{1}{\zeta(s)} = \prod \left(1 - \dfrac{1}{p^s}\right)$ where the product is extended over all prime numbers.

First Proof:

Using an approach along the line of the Sieve of Eratosthenes, we have

$$\zeta(s) = 1 + \frac{1}{2^s} + \frac{1}{3^s} + \frac{1}{4^s} + \frac{1}{5^s} + \frac{1}{6^s} + \frac{1}{7^s} + \frac{1}{8^s} + \frac{1}{9^s} + \cdots;$$

$$\frac{1}{2^s}\zeta(s) = \frac{1}{2^s} + \frac{1}{4^s} + \frac{1}{6^s} + \frac{1}{8^s} + \cdots;$$

$$\left(1 - \frac{1}{2^s}\right)\zeta(s) = 1 + \frac{1}{3^s} + \frac{1}{5^s} + \frac{1}{7^s} + \frac{1}{9^s} + \cdots;$$

$$\frac{1}{3^s}\left(1 - \frac{1}{2^s}\right)\zeta(s) = \frac{1}{3^s} + \frac{1}{9^s} + \cdots;$$

$$\left(1 - \frac{1}{3^s}\right)\left(1 - \frac{1}{2^s}\right)\zeta(s) = 1 + \frac{1}{5^s} + \frac{1}{7^s} + \cdots.$$

Let the prime numbers be $2 = p_1 < p_2 < p_3 < \cdots$. Continuing with this pattern, we have

$$\zeta(s)\left(1 - \frac{1}{2^s}\right)\left(1 - \frac{1}{3^s}\right)\left(1 - \frac{1}{5^s}\right)\cdots\left(1 - \frac{1}{p_t^s}\right) = 1 + \sum \frac{1}{n^s},$$

where the summation extends over those integers $n > 1$ which have no prime divisors less than or equal to p_t. This sum goes to 0 as t goes to infinity since it is the tail of a convergent series. Thus

$$\zeta(s)\prod\left(1 - \frac{1}{p^s}\right) = 1 \quad \text{and} \quad \frac{1}{\zeta(s)} = \prod\left(1 - \frac{1}{p^s}\right),$$

where the product is extended over all prime numbers.

Second Proof:
Using $\frac{1}{1-x} = 1 + x + x^2 + x^3 + \cdots$, we have

$$\prod_{i=1}^{t}\left(1 - \frac{1}{p_i^s}\right)^{-1} = \prod_{i=1}^{t}\left(1 + \frac{1}{p_i^s} + \frac{1}{p_i^{2s}} + \frac{1}{p_i^{3s}} + \cdots\right)$$

$$= 1 + \frac{1}{2^s} + \frac{1}{3^s} + \cdots + \frac{1}{p_t^s} + \sum\frac{1}{n^s}.$$

The last summation is over those integers $n > p_t$ which have only prime divisors less than or equal to p_t. This sum goes to 0 as t goes to infinity since it is the tail of a convergent series. Thus

$$\prod\left(1 - \frac{1}{p^s}\right)^{-1} = \sum_{n=1}^{\infty}\frac{1}{n^s} = \zeta(s) \text{ and } \frac{1}{\zeta(s)} = \prod\left(1 - \frac{1}{p^s}\right),$$

where the products are extended over all prime numbers.

Remark:
Both of these proofs were found by L. Euler in the 18th century. We now present another proof, due to **Solomon Golomb**.

Third Proof:
If we define a probability distribution on the positive integers where $pr_s(n) \approx \frac{1}{n^s}$, the normalization must be to divide all the values $\frac{1}{n^s}$ by $\sum_{n=1}^{\infty}\frac{1}{n^s} = \zeta(s)$.
Thus $pr_s(n) = \frac{n^{-s}}{\zeta(s)}$. In this distribution, the probability that a "random" integer is a multiple of a given integer m is

$$\frac{1}{\zeta(s)}\sum_{k=1}^{\infty}\frac{1}{(km)^s} = \frac{m^{-s}}{\zeta(s)}\sum_{k=1}^{\infty}\frac{1}{k^s} = \frac{1}{m^s}.$$

The probability that a "random" integer is not divisible by a given integer m is $1 - \frac{1}{m^s}$. The probability that a "random" integer is not divisible by any of a set of primes $\{p_1, p_2, \ldots, p_t\}$ is $(1 - \frac{1}{p_1^s})(1 - \frac{1}{p_2^s})\cdots(1 - \frac{1}{p_t^s})$. Thus the probability that a positive integer has no prime factors at all is $\prod\left(1 - \frac{1}{p^s}\right)$ where the product is extended over all prime numbers. However, the only such positive integer is 1, where $pr_s(1) = \frac{1}{\zeta(s)}$.

By Euler's Product Formula, we have

$$\sum_{n=1}^{\infty}\frac{\mu(n)}{n^s} = \frac{1}{1^s} - \frac{1}{2^s} - \frac{1}{3^s} - \frac{1}{5^s} + \frac{1}{2^s3^s} - \frac{1}{7^s} + \frac{1}{2^s5^s} - \cdots$$

$$= (1 - \frac{1}{2^s})(1 - \frac{1}{3^s})(1 - \frac{1}{5^s})\cdots$$

$$= \frac{1}{\zeta(s)},$$

a result we have proved earlier.

We conclude this section by treating some topics related to arithmetic functions.

If $\sigma(n) = 2n$, n is called a **perfect** number. In other words, n is the sum of its positive divisors excluding n itself. For example. 6=1+2+3 and 28=1+2+4+7+14 are perfect numbers.

Perfect numbers are intimately related to primes of the form $2^m - 1$ which are called **Mersenne** primes. (Primes of the form $2^m + 1$, on the other hand, are called **Fermat** primes.) We claim that if $2^m - 1$ is a Mersenne prime, then m itself is prime. Suppose to the contrary that m is composite. Then $m = ab$ for some integers a and b where $1 < a < m$, so that $1 < 2^a - 1 < 2^m - 1$. However, $2^m - 1$ is divisible by $2^a - 1$. This contradiction justifies the claim.

Euclid's Theorem.

If $n = 2^{m-1}(2^m - 1)$ and $2^m - 1$ is a Mersenne prime, then n is a perfect number.

Proof:

Since $p = 2^m - 1$ is prime, the positive divisors of n are those listed in the chart below.

	1	2	2^2	2^3	$\cdots$	2^{m-1}
1	1	2	2^2	2^3	$\cdots$	2^{m-1}
p	p	$2p$	$2^2 p$	$2^3 p$	$\cdots$	$2^{m-1} p$

Hence $\sigma(n) = (1 + 2 + 2^2 + \cdots + 2^{m-1})(1 + p) = (2^m - 1)2^m = 2n$.

It is not known if the converse of Euclid's Theorem is true. A partial result is the following.

Euler's Theorem.

If n is an even perfect number, then for some m, $n = 2^{m-1}(2^m - 1)$, where $2^m - 1$ is prime.

Proof:

Let n be an even perfect number. Then $n = 2^{m-1}k$ where $m \geq 2$ is an integer and k is odd. Then $2^m k = 2n = \sigma(n) = \sigma(2^{m-1})\sigma(k) = (2^m - 1)\sigma(k)$. It follows that $2^m - 1 = \frac{k}{\sigma(k)-k}$. Now $\sigma(k) - k$ is a positive divisor of k less than k, and at the same time it is the sum of all positive divisors of k less than k. This is only possible if $\sigma(k) - k = 1$. It follows that $k = 2^m - 1$ is a prime, and indeed, $n = 2^{m-1}(2^m - 1)$.

Examples

Example 2.6.1.
Determine the Dirichlet generating function for $\{\tau(n)\}$ in terms of the Riemann zeta-function.

Solution:
By Dirichlet's Product Formula, $\displaystyle\sum_{n=1}^{\infty} \frac{1}{n^s} \sum_{m=1}^{\infty} \frac{1}{m^s} = \sum_{n=1}^{\infty} \frac{1}{r^s} \sum_{n|r} 1 = \sum_{r=1}^{\infty} \frac{\tau(r)}{r^s}$.

It follows that $\displaystyle\sum_{n=1}^{\infty} \frac{\tau(n)}{n^s} = \zeta^2(s)$.

Example 2.6.2.

(a) Prove that $\displaystyle\sum_{d|n} \phi(d) = n$.

(b) Prove that $\displaystyle\sum_{d|n} \mu(d)\frac{n}{d} = \phi(n)$.

Solution:

(a) For any prime p and any positive integer k, we have

$$
\begin{aligned}
\sum_{d|n} \phi(p^k) &= \phi(1) + \phi(p) + \phi(p^2) + \cdots + \phi(p^k) \\
&= 1 + (p-1) + (p^2 - p) + \cdots + (p^k - p^{k-1}) \\
&= p^k.
\end{aligned}
$$

Hence $\displaystyle\sum_{d|n} \phi(d) = n$.

(b) The result follows from (a) and Möbius' Inversion Formula.

Example 2.6.3.
Use the formula

$$
\sum_{d|n} \mu(d) = \begin{cases} 1 \text{ if } n = 1, \\ 0 \text{ if } n > 1. \end{cases}
$$

to give an alternative proof of Möbius' Inversion Formula and its converse.

Solution:

Assume first that $F(n) = \sum_{d|n} f(d)$. Then

$$
\begin{aligned}
\sum_{d|n} \mu(d) F\left(\frac{n}{d}\right) &= \sum_{d|n} \mu(d) \sum_{t|\frac{n}{d}} f(t) \\
&= \sum_{t|d} f(t) \sum_{d|\frac{n}{t}} \mu(d) \\
&= f(n).
\end{aligned}
$$

Conversely, assume that $f(n) = \sum_{d|n} \mu(d) F\left(\frac{n}{d}\right)$. Then

$$
\begin{aligned}
\sum_{d|n} f(d) &= \sum_{d|n} \sum_{t|d} \mu(t) F\left(\frac{d}{t}\right) \\
&= \sum_{t|n} F(t) \sum_{d|\frac{n}{t}} \mu(d) \\
&= F(n).
\end{aligned}
$$

Remark:

The above proof of Möbius' Inversion Formula involves a change in the order of summation. We illustate with $n = 6$. In the chart below, $\sum_{d|n} \mu(d) \sum_{t|\frac{n}{d}} f(t)$ sums the entries row by row, while $\sum_{t|d} f(t) \sum_{d|\frac{n}{t}} \mu(d)$ sums them column by column.

$$
\begin{array}{llll}
\mu(1)f(1) & +\mu(1)f(2) & +\mu(1)f(3) & +\mu(1)f(6) \\
+\mu(2)f(1) & & +\mu(2)f(3) & \\
+\mu(3)f(1) & +\mu(3)f(2) & & \\
+\mu(6)f(1) & & &
\end{array}
$$

Exercises

1. Determine the Dirichlet generating function for $\{\phi(n)\}$ in terms of the Riemann zeta-function.

2. Determine the Dirichlet generating function for $\{\sigma(n)\}$ in terms of the Riemann zeta-function.

3. Evaluate $\displaystyle\sum_{d|n} \mu(d)\tau\left(\frac{n}{d}\right)$.

4. Evaluate $\displaystyle\sum_{d|n} \mu(d)\sigma\left(\frac{n}{d}\right)$.

5. Prove that $\tau(n)$ is odd if and only if n is a square.

6. (a) Prove that if $2^m + 1$ is a Fermat prime, then m itself must be a power of 2.

   (b) Prove that if $\phi(n)$ is a power of 2, then n is the product of a power of 2 and distinct Fermat primes.

Practice Questions

Generating Functions

1. What is the generating function for the sequence $\{4^n - 3^n\}$?

2. What is the generating function for the sequence $\{2^{n+1} - (\frac{1}{2})^{n+1}\}$?

3. What is the generating function for the sequence $\{n - 2\}$?

4. What sequence is generated by the function $\frac{3}{(1-2x)(1-5x)}$?

5. What sequence is generated by the function $\frac{5x}{(1-3x)(2-x)}$?

6. What sequence is generated by the function $\frac{1}{(1-3x)^2}$?

Setting up Recurrence Relations

7. Let a_n denote the number of sequences of 1s, 2s and 3s with sum n. We have $a_0 = 1$, $a_1 = 1$ and $a_2 = 3$. Find a recurrence relation for a_n.

8. Let a_n denote the number of sequences of 1s, 3s and 5s with sum n. We have $a_0 = 1$, $a_1 = 1$, $a_2 = 1$, $a_3 = 2$ and $a_4 = 3$. Find a recurrence relation for a_n.

9. Let a_n denote the number of ways of giving away 1, 2 or 3 books each day for n days, such that the number of days on which 1 book is given away is even. We have $a_1 = 2$. Find a recurrence relation for a_n.

10. Let a_n denote the number of ways of giving away 1, 2 or 3 books each day for n days, such that the number of days on which 1 book is given away is odd. We have $a_1 = 1$. Find a recurrence relation for a_n.

11. Let a_n be the number of regions into which the plane is divided by $n \geq 2$ lines such that every two lines intersect at a unique point except that two of the lines are parallel. We have $a_2 = 3$. Find a recurrence relation for a_n.

12. Let a_n be the number of regions into which the plane is divided by $n \geq 3$ lines such that every two lines intersect at a unique point except that three of these points coincide. We have $a_2 = 3$. Find a recurrence relation for a_n.

Homogeneous Recurrence Relations

Solve the following homogeneous recurrence relations:

13. $a_n - 9a_{n-1} + 20an - 2 = 0$ with $a_0 = -3$ and $a_1 = -10$;

14. $a_n - 5a_{n-1} + 6a_{n-2} = 0$ with $a_0 = 1$ and $a_1 = -2$;

15. $a_n - 6a_{n-1} + 11a_{n-2} - 6a_{n-3} = 0$ with $a_0 = 2$, $a_1 = 5$ and $a_2 = 13$;

16. $a_n - a_{n-1} - 9a_{n-2} + 9a_{n-3} = 0$ with $a_0 = 0$, $a_1 = 12$ and $a_2 = 24$;

17. $a_n - 5a_{n-1} + 8a_{n-2} - 4a_{n-3} = 0$ with $a_0 = 2$, $a_1 = 3$ and $a_2 = 7$;

18. $a_n - 3a_{n-2} + 2a_{n-3} = 0$ with $a_0 = 9$, $a_1 = 0$ and $a_2 = 0$.

Non-homogeneous Recurrence Relations

Solve the following non-homogeneous recurrence relations:

19. $a_n - a_{n-1} = 3^n$ with $a_0 = 1$.

20. $a_n - a_{n-1} = \frac{1}{n(n+1)}$ with $a_0 = 1$;

21. $a_n - 4a_{n-1} - 5a_{n-2} = -9 \cdot 2^{n-1}$ with $a_0 = 2$ and $a_1 = 10$;

22. $a_n + 2a_{n-} - 3a_{n-2} = 4 \cdot 3^n$ with $a_0 = 1$ and $a_1 = 19$;

23. $a_n - 4a_{n-1} + 4_{n-2} = 9(-1)^n$ with $a_0 = 0$ and $a_1 = 1$;

24. $a_n - 3a_{n-1} + 2a_{n-2} = 2^n$ with $a_0 = 0$ and $a_1 = 1$.

Answers

(1) $\frac{1}{(1-3x)(1-4x)}$ (2) $\frac{3}{(1-2x)(2-x)}$ (3) $\frac{-2+3x}{(1-x)^2}$ (4) $\{5^{n+1} - 2^{n+1}\}$

(5) $\{3^n - (\frac{1}{2})^n\}$ (6) $\{(n+1)3^n\}$ (7) $a_n = a_{n-1} + a_{n-2} + a_{n-3}$

(8) $a_n = a_{n-1} + a_{n-3} + a_{n-5}$ (9) $a_n = a_{n-1} + 3^{n-1}$ (10) $a_n = a_{n-1} + 3^{n-1}$

(11) $a_n = a_{n-1} + n$ (12) $a_n = a_{n-1} + n$ (13) $a_n = 2 \cdot 5^n - 5 \cdot 4^n$

(14) $a_n = 5 \cdot 2^n - 4 \cdot 3^n$ (15) $a_n = 3^n + 2^n$ (16) $a_n = 4 \cdot 3^n - (-3)^n - 3$

(17) $a_n = n2^n - 2^n + 3$ (18) $a_n = 8 - 6n + (-2)^n$ (19) $a_n = \frac{1}{2}(3^{n+1} - 1)$

(20) $a_n = \frac{2n+1}{n+1}$ (21) $a_n = 5^n - (-1)^n + 2^{n+1}$ (22) $a_n = 1 + (-3)^{n+1} + 3^{n+1}$

(23) $a_n = n2^n - 2^n + (-1)^n$ (24) $a_n = n2^n - 2^n + 1$

Chapter Three

Permutations

Section 3.1. Permutations with or without Repetitions

Up to now, we have been considering combination problems where the order in which the objects are selected does not matter. We turn now to the more difficult permutation problems where the order in which the objects are arranged in a row does matter.

By the Multiplication Principle, the number of arrangements of k elements from a set of size n is given by $n(n-1)\cdots(n-k+1) = \frac{n!}{(n-k)!}$. This allows us to reinterpret the formula $\binom{n}{k} = \frac{n!}{k!(n-k)!}$ in the following manner. The k elements in a subset can be rearranged among themselves in $k!$ ways. Hence each such subset is counted $k!$ times among the arrangements. Dividing by $k!$ yields the number of subsets of size k.

Seven children are to be seated in a row. The number of different seating arrangements is 7!=5040. Suppose two of them must be next to each other. We can treat them as a single person. Then there are 6! ways of seating the children. This must be multiplied by 2 since they can be rearranged between themselves in 2 ways. The total is therefore $2 \times 6! = 1440$.

Suppose instead these two children must not be next to each other. The simplest way to find the number of different seating arrangements is to subtract 1440 from 5040 and obtain 3600. Alternatively, arrange the other five children in 5!=120 ways. Now there are six places where the two children may be inserted, at either end of the row or between two other children. Since they must choose different places, the first one has 6 choices and the second 5. Hence the total number of different seating arrangements is $120 \times 6 \times 5 = 3600$.

Suppose instead the seven children are to be seated at round table, but with no other restrictions. Moreover, seating arrangements in the same cyclic order are not considered different. We may seat an arbitrary child in an arbitrary seat. The remaining seats are now transformed into a row of six by the occupied seat, and they can be filled in 6!=720 ways. In general, the number of **cyclic** permutations of n elements is $(n-1)!$.

How many ways are there of seating n couples in a row so that no couple is together? Let S denote the set of all seating arrangements and, for $1 \leq i \leq n$, let A_i be the subset in which the i-th couple is together. Then $|S| = (2n)!$. By tying couples together as in the introductory problem, we have $|A_i| = 2(2n-1)!$, $|A_i \cap A_j| = 2^2(2n-2)!$, and so on.

© The Author(s), under exclusive license to Springer Nature Switzerland AG 2021
S. W. Golomb, A. Liu, *Solomon Golomb's Course on Undergraduate Combinatorics*,
https://doi.org/10.1007/978-3-030-72228-9_4

By the Principle of Inclusion-Exclusion,

$$|\overline{A_1} \cap \overline{A_2} \cap \cdots \cap \overline{A_n}| = (2n)! - \binom{n}{1}2(2n-1)! + \binom{n}{2}2^2(2n-2)! - \cdots + (-1)^n 2^n n!.$$

Suppose the same element can be used any number of times in a permutation. Then we have n choices for each position, so that the total number of such arrangements is n^k. Thus the number of binary sequences of length k is 2^k, the number of ternary sequences of length k is 3^k, and so on.

Suppose the element i, $1 \le i \le n$, has multiplicity m_i in a multi-set, and we wish to arrange all of its elements in a row. If we consider all the elements distinct, the number of permutations is clearly $(m_1 + m_2 + \cdots + m_n)!$. Consider all permutations in which the m_i copies of i occupy the same m_i positions. Since they are indistinguishable, and they can be rearranged among themselves in $m_i!$ ways, these m_i permutations merge into a single one. It follows that the total number of permutations is

$$\frac{(m_1 + m_2 + \cdots + m_n)!}{m_1! m_2! \cdots m_n!}.$$

Thus the total number of ways of rearranging the letters in MICROSCOPIC is $\dfrac{11!}{1!2!3!1!2!1!1!}$.

Certain divisibility problems may be solved by interpreting them as permutation problems with limited repetitions. Suppose we wish to prove that $(n^2)!$ is divisible by $(n!)^n$. Consider n^2 objects consisting of n of each of n kinds. The total number of ways of arranging them is $\frac{(n^2)!}{n!n!\cdots n!} = \frac{(n^2)!}{(n!)^n}$. Since this must be an integer, the desired conclusion follows.

We can in fact prove that $(n^2)!$ is divisible by $(n!)^{n+1}$. Define two of the permutations above to be equivalent if they are obtainable from each other by renaming the objects. In other words, the same positions occupied by objects of one type in the first permutation are also occupied by objects of one type in the second permutation, though not necessarily the same type. Since there are $n!$ permutations in each equivalence class, the number of permutations, namely $\frac{(n^2)!}{(n!)^n}$, is divisible by $n!$. In other words, $(n^2)!$ is divisible by $(n!)^{n+1}$.

We now return to the problem of forming letter strings that can be spelled with the letters of MICROSCOPIC. Suppose we only want letter strings of length 4. We consider four cases.

Case 1. Three identical letters.

The letters can be chosen in 6 ways since the identical letters can be chosen in 1 way and the other letter in 6 ways. The chosen letters can be arranged in 4 ways as there are $\binom{4}{3}$ ways of placing the identical letters. This yields a total of 24 ways.

Case 2. Two distinct pairs of identical letters.

The letters can be chosen in $\binom{3}{2} = 3$ ways. The chosen letters can be arranged in 6 ways as there are $\binom{4}{2} = 6$ ways of placing one pair of identical letters. This yields a total of 18 ways. **Case 3.** One pair of identical letters. The letters can be chosen in 45 ways since the identical letters can be chosen in 3 ways and the other letters in $\binom{6}{2} = 15$ ways. The chosen letters can be arranged in 12 ways since the two letters not in the identical pair can be placed in 4×3 ways. This yields a total of 540 ways.

Case 4. No identical letters.

The letters can be chosen in $\binom{7}{4} = 35$ ways. The chosen letters can be arranged in $4! = 24$ ways. This yields a total of 840 ways.

The grand total is 24+18+540+840=1422 ways.

Examples

Example 3.1.1.
Determine the number of ways of putting five flags on three distinct poles, if

(a) the flags are distinct;

(b) the flags are identical;

(c) the flags are identical and each pole must have at least one flag;

(d) the flags are distinct and each pole must have at least one flag.

Solution:

(a) Arrange the flags in any order, which is irrelevant, and put them on the poles one at a time. The first one can be placed in 3 ways, on any of the empty poles. The next one can be placed in 4 ways, on either of the empty poles or on the same pole with the first flag, either above or below it. Similarly, the remaining flags may be placed in 5, 6 and 7 ways respectively. Hence the total is $3 \times 4 \times 5 \times 6 \times 7 = 2520$.

(b) The total is $2520 \div 5! = 21$ since we no longer distinguish among the flags.

(c) The total is 6 because the possible partitions are 3+1+1, 1+3+1, 1+1+3, 2+2+1, 2+1+2 and 1+2+2.

(d) The total is $6 \times 5! = 720$ since we now distinguish among the flags.

Example 3.1.2.
Prove that $(n!)!$ is divisible by $(n!)^{(n-1)!}$.

Solution:
Suppose we have $n!$ objects consisting of n of each of $(n-1)!$ kinds. The total number of ways of arranging them is $\frac{(n!)!}{n!n!\cdots n!} = \frac{(n!)!}{(n!)^{(n-1)!}}$. Since this must be an integer, the desired conclusion follows.

Example 3.1.3.
Find the number of letter strings of length 4 that can be spelled with the letters of MELANESIANS.

Solution:
We consider three cases.
Case 1. Two distinct pairs of identical letters.
The letters can be chosen in $\binom{4}{2} = 6$ ways. The chosen letters can be arranged in 6 ways as there are $\binom{4}{2} = 6$ ways of placing one pair of identical letters. This yields a total of 36 ways.

Case 2. One pair of identical letters.

The letters can be chosen in 60 ways since the identical letters can be chosen in 4 ways and the other letters in $\binom{6}{2} = 15$ ways. The chosen letters can be arranged in 12 ways since the two letters not in the identical pair can be placed in 4×3 ways. This yields a total of 720 ways.

Case 3. No identical letters.

The letters can be chosen in $\binom{7}{4} = 35$ ways. The chosen letters can be arranged in $4! = 24$ ways. This yields a total of 840 ways.

The grand total is $36+720+840=1596$ ways.

Exercises

1. Determine the number of ways of putting two red flags, two blue flags and one white flag on three distinct poles, each of which must have at least one flag.

2. Determine the number of ways of putting three red flags, one blue flag and one white flag on three distinct poles, each of which must have at least one flag.

3. Prove that $(2n)!$ is divisible by 2^n.

4. Prove that $(3n)!$ is divisible by $2^n 3^n$.

5. Find the number of letter strings of length 4 that can be spelled with the letters of MATHEMATICS.

6. Find the number of letter strings of length 4 that can be spelled with the letters of MISSISSIPPI.

Section 3.2. Exponential Generating Functions

The **exponential generating function** for the sequence $\{a_n\}$ is defined as

$$E(x) = a_0 + a_1 x + a_2 \frac{x^2}{2!} + a_3 \frac{x^3}{3!} + \cdots.$$

For example, e^x generates $\{1, 1, 1, \ldots\}$. e^{-x} generates $\{1, -1, 1, -1, \ldots\}$, $\frac{1}{2}(e^x + e^{-x})$ generates $\{1, 0, 1, 0, \ldots\}$ and $\frac{1}{2}(e^x - e^{-x})$ generates $\{0, 1, 0, 1, \ldots\}$.

The reason why we use the functions $\frac{x^n}{n!}$ instead of just x^n is to give us a tool that can handle permutation problems. The ordinary generating function is ideal for handling combination problems. However, when order is taken into consideration, the factorial function cannot be far behind.

Nevertheless, we need to justify why the exponential generating function is useful in dealing with order. Let us illustrate with an example, forming all possible "words" from the letters of "ADA". Vacuously, there is 1 word of length 0. Trivially there are 2 words of length one, namely, A and D. There are three words of length 2, namely, AA, AD and DA, Finally there are 3 words of length 3, namely, AAD, ADA and DAA.

We now turn to the approach using exponential generating functions. For D, it is $1 + x$, and for A it is $1 + x + \frac{x^2}{2!}$. The composite exponential generating function for the problem is

$$
\begin{aligned}
(1+x)\left(1 + x + \frac{x^2}{2!}\right) &= 1 + 2x + \left(1 + \frac{1}{2!}\right)x^2 + \frac{1}{2!}x^3 \\
&= 1 + 2x + \left(2! + \frac{2!}{2!}\right)\frac{x^2}{2!} + \frac{3!}{2!}\frac{x^3}{3!} \\
&= 1 + 2x + 3\frac{x^2}{2!} + 3\frac{x^3}{3!}.
\end{aligned}
$$

The result certainly agrees with our empirical data. Let us look at the term $\frac{x^3}{3!}$ more closely. We have 3 letters which can be permuted in $3!$ ways. However, 2 of the letters are identical and can be permuted internally in $2!$ ways. Hence the overall number of permutations is $\frac{3!}{2!} = 3$. Note that the $2!$ comes from the exponential generating function $1 + x + \frac{x^2}{2!}$ for A, anticipating that if both copies of A are taken, then we have to divide out by $2!$. The $3!$ is built into the composite exponential generating function, indicating that 3 letters are being permuted without regard to their nature.

Similar observations can be made about the term $\frac{x^2}{2!}$, which is the sum of two terms. When we take 1 copy of each of A and D, there are $2!$ ways of permuting them, but when we take 2 copies of A, the number of permutations is $\frac{2!}{2!}$. So everything works out according to plan.

For a more elaborate illustration, form all possible "words" of length up to 4 from the letters of "MICROSCOPIC". The exponential generating function for each of M, R, S and P is $1 + x$, that for each of I and O is $1 + x + \frac{1}{2}x^2$, and that of C is $1 + x + \frac{1}{2}x^2 + \frac{1}{6}x^3$.

The overall exponential generating function is

$$(1+x)^4 \left(1 + x + \frac{1}{2}x^2\right)^2 \left(1 + x + \frac{1}{2}x^2 + \frac{1}{6}x^3\right)$$

$$= (1 + 4x + 6x^2 + 4x^3 + x^4)\left(1 + 2x + 2x^2 + x^3 + \frac{1}{4}x^4\right)$$

$$\left(1 + x + \frac{1}{2}x^2 + \frac{1}{6}x^3\right)$$

$$= (1 + 4x + 6x^2 + 4x^3 + x^4)\left(1 + 3x + \frac{9}{2}x^2 + \frac{25}{6}x^3 + \frac{31}{12}x^4 + \cdots\right)$$

$$= 1 + 7x + \frac{45}{2}x^2 + \frac{265}{6}x^3 + \frac{237}{6}x^4 + \cdots$$

$$= 1 + 7x + 45\frac{x^2}{2!} + 265\frac{x^3}{3!} + 1422\frac{x^4}{4!} + \cdots.$$

The respective numbers are 7, 45, 265 and 1422.

The number of permutations of n elements from a set of m with no repetition is $\frac{m!}{(m-n)!}$ for $0 \leq n \leq m$, and 0 when $n > m$. The exponential generating function for this sequence is

$$\sum_{n=0}^{m} \frac{m!}{(m-n)!} \frac{x^n}{n!} = \sum_{n=0}^{m} \binom{m}{n} x^n = (1+x)^m.$$

The number of permutations of n elements from a set of m with unlimited repetition is m^n. The exponential generating function for this sequence is

$$\sum_{n=0}^{\infty} m^n \frac{x^n}{n!} = \sum_{n=0}^{\infty} \frac{(mx)^n}{n!} = e^{mx}.$$

In particular, the exponential generating function for the numbers of quaternary sequences of different lengths is e^{4x}.

We now revisit a solved case of the Distribution Problem introduced in Section 0.1. In Case (1), we are distributing n distinct objects into k distinct cells, with empty cells permitted. For each cell, it may contain any of the objects. Thus the exponential generating function for this cell is $1 + x + \frac{x^2}{2!} + \frac{x^3}{3!} + \cdots = e^x$. It follows that the overall exponential generating function is $e^{kx} = 1 + kx + k^2\frac{x^2}{2!} + k^3\frac{x^3}{3!} + \cdots$, indeed generating exponentially the sequence $\{k^n\}$.

In our final illustration, consider quaternary sequences of length n. Let a_n denote the number of those with an even number of 0s and an even number of 1s, b_n those with an odd number of 0s and an odd number of 1s, c_n those with an even number of 0s and an odd number of 1s, and d_n those with an odd number of 0s and an even number of 1s.

The exponential generating function for a digit which appears an even number of times is $\frac{1}{2}(e^x + e^{-x})$. The exponential generating function for a digit which appears an odd number of times is $\frac{1}{2}(e^x - e^{-x})$. The exponential generating function for a digit which appears an arbitrary number of times is e^x.

It follows that the exponential generating function for the sequence $\{a_n\}$ is $\frac{1}{4}(e^{4x} + 2e^{2x} + 1)$. Hence the sequence generated is $a_0 = \frac{1}{4}(1+2+1) = 1$ and $a_n = 4^{n-1} + 2^{n-1}$ for $n \geq 1$.

The exponential generating function for the sequence $\{b_n\}$ is given by $\frac{1}{4}(e^{4x} - 2e^{2x} + 1)$. Hence the sequence generated is $b_0 = \frac{1}{4}(1 - 2 + 1) = 0$ and $b_n = 4^{n-1} - 2^{n-1}$ for $n \geq 1$.

The exponential generating function for the sequences $\{c_n\}$ and $\{d_n\}$ is $\frac{1}{4}(e^{4x} - 1)$. Hence the sequences generated are $c_0 = d_0 = \frac{1}{4}(1 - 1) = 0$ and $c_n = d_n = 4^{n-1}$ for $n \geq 1$.

Examples

Example 3.2.1.
Find the numbers of letter strings of length 1, 2, 3 and 4 that can be spelled with the letters of MELANESIANS.

Solution:
The exponential generating function for each of M, L and I is $1 + x$, and that for each of E, A, N and S is $1 + x + \frac{1}{2}x^2$. Thus the overall exponential generating function is

$$(1+x)^3 \left(1 + x + \frac{1}{2}x^2\right)^4$$

$$= (1 + 3x + 3x^2 + x^3)\left(1 + 2x + 2x^2 + x^3 + \frac{1}{4}x^4\right)^2$$

$$= (1 + 3x + 3x^2 + x^3)\left(1 + 4x + 8x^2 + 10x^3 + \frac{17}{2}x^4 + \cdots\right)$$

$$= 1 + 7x + 23x^2 + 47x^3 + \frac{133}{2}x^4 + \cdots$$

$$= 1 + 7x + 46\frac{x^2}{2!} + 282\frac{x^3}{3!} + 1596\frac{x^4}{4!} + \cdots.$$

The respective numbers are 7, 46, 282 and 1596. The last value agrees with the answer in Example 3.1.3.

Example 3.2.2.
Determine the number of ternary sequences of length n

(a) with an even number of 0s;

(b) with an odd number of 0s.

Solution:

(a) The exponential generating function for the digit 0 is $\frac{1}{2}(e^x + e^{-x})$. The exponential generating function for the digit 1 or 2 is e^x. The overall exponential generating function is

$$\frac{1}{2}(e^x + e^{-x})e^{2x}$$

$$= \frac{1}{2}(e^{3x} + e^x)$$

$$= \frac{1}{2}\left(\sum_{n=0}^{\infty} 3^n \frac{x^n}{n!} + \sum_{n=0}^{\infty} \frac{x^n}{n!}\right).$$

Hence the sequence generated is $a_n = \frac{1}{2}(3^n + 1)$.

(b) The exponential generating function for the digit 0 is $\frac{1}{2}(e^x - e^{-x})$. The exponential generating function for the digit 1 or 2 is e^x. Hence the overall exponential generating function is

$$\frac{1}{2}(e^x - e^{-x})e^{2x} = \frac{1}{2}(e^{3x} - e^x)$$

$$= \frac{1}{2}\left(\sum_{n=0}^{\infty} 3^n \frac{x^n}{n!} - \sum_{n=0}^{\infty} \frac{x^n}{n!}\right).$$

Hence the sequence generated is $a_n = \frac{1}{2}(3^n - 1)$.

Compare with Example 2.3.3.

Example 3.2.3.

(a) Find the number of quaternary sequences of length n with at least one 0 and at least one 1.

(b) Find the number of quaternary sequences of length n with at least one 0 and at most one 1.

(c) Find the number of quaternary sequences of length n with at most one 0 and at most one 1.

Solution:

(a) The exponential generating function is $(e^x - 1)^2 e^{2x} = e^{4x} - 2e^{3x} + e^{2x}$. The sequence generated is $4^n - 2 \cdot 3^n + 2^n$.

(b) The exponential generating function is

$$(1 + x)(e^x - 1)e^{2x}$$

$$= e^{3x} - e^{2x} + xe^{3x} - xe^{2x}$$

$$= \sum_{n=0}^{\infty} 3^n \frac{x^n}{n!} - \sum_{n=0}^{\infty} 2^n \frac{x^n}{n!} + \sum_{n=0}^{\infty} 3^n \frac{x^{n+1}}{n!} - \sum_{n=0}^{\infty} 2^n \frac{x^{n+1}}{n!}$$

$$= \sum_{n=0}^{\infty} 3^n \frac{x^n}{n!} - \sum_{n=0}^{\infty} 2^n \frac{x^n}{n!} + \sum_{n=1}^{\infty} n3^{n-1} \frac{x^n}{n!} - \sum_{n=1}^{\infty} n2^{n-1} \frac{x^n}{n!}.$$

The sequence generated is $3^n - 2^n + n3^{n-1} - n2^{n-1}$.

(c) The exponential generating function is

$$(1 + x)^2 e^{2x} = e^{2x} + 2xe^{2x} + x^2 e^{2x}$$

$$= \sum_{n=0}^{\infty} 2^n \frac{x^n}{n!} + 2 \sum_{n=0}^{\infty} 2^n \frac{x^{n+1}}{n!} + \sum_{n=0}^{\infty} 2^n \frac{x^{n+2}}{n!}$$

$$= \sum_{n=0}^{\infty} 2^n \frac{x^n}{n!} + 2 \sum_{n=1}^{\infty} n2^{n-1} \frac{x^n}{n!} + \sum_{n=2}^{\infty} n(n-1)2^{n-2} \frac{x^n}{n!}.$$

The sequence generated is $2^n + n2^n + n(n-1)2^{n-2}$.

Exercises

1. Find the numbers of letter strings of length 1, 2, 3 and 4 that can be spelled with the letters of MATHEMATICS. Use the method of exponential generating functions.

2. Find the numbers of letter strings of length 1, 2, 3 and 4 that can be spelled with the letters of MISSISSIPPI. Use the method of exponential generating functions.

3. Determine the number of quaternary sequences of length n with an even number of 0s and at least one 3. Use the method of exponential generating functions.

4. Determine the number of quaternary sequences of length n with an even number of 0s and at most one 3. Use the method of exponential generating functions.

5. Determine the number of quaternary sequences of length n with an even number of 0s and at least two 1s. Use the method of exponential generating functions.

6. Determine the number of quaternary sequences of length n with an odd number of 0s and at most two 1's. Use the method of exponential generating functions.

Section 3.3. Derangements

If the i-th term of a permutation of 1, 2, $\ldots$, n is equal to i, we say that i is a **fixed point** of the permutation. The number of permutations with exactly k fixed points is denoted by $D_n(k)$. The permutations with no fixed points are called **derangements**, and $D_n(0) = D_n$ are called the **derangement numbers**. The following table shows some values of $D_n(k)$ for small n and k.

	$k =$	0	1	2	3	4	5
	1	0	1	0	0	0	0
	2	1	0	1	0	0	0
$n =$	3	2	3	0	1	0	0
	4	9	8	6	0	1	0
	5	44	45	20	10	0	1

Two immediate observations are $D_n(k) = 0$ if $k > n$ and $D_n(k) = 1$ if $k = n$.

Our initial focus is on the special case $k = 0$. We wish to find an expression for D_n. Denote by S the set of all permutations of 1, 2, $\ldots$, n. For $1 \le i \le n$, let A_i be the set of all permutations of which i is a fixed point. By the Principle of Inclusion-Exclusion,

$$
\begin{aligned}
D_n &= |\overline{A_1} \cap \overline{A_2} \cap \cdots \cap \overline{A_n}| \\
&= S - \Sigma_n |A_i| + \Sigma_n |A_i \cap A_j| - \cdots \\
&= n! - \binom{n}{1}(n-1)! + \binom{n}{2}(n-2)! - \cdots \\
&= n!\left(1 - \frac{1}{1!} + \frac{1}{2!} - \cdots + (-1)^n \frac{1}{n!}\right).
\end{aligned}
$$

Let us compute D_n in a different way, by finding a recurrence relation. Since 1 is not a fixed point, we must have k in the first position where $2 \le k \le n$. If 1 is forbidden to be in the k-th position, the numbers other than k can be deranged in D_{n-1} ways. Otherwise, 1 is in position k, and the remaining numbers can be deranged in D_{n-2} ways. It follows that $D_n = (n-1)(D_{n-1} + D_{n-2})$.

We can derive a second recurrence relation by rewriting the first recurrence relation in the form $D_n - nD_{n-1} = -(D_{n-1} - (n-1)D_{n-2})$. Iteration yields

$$
\begin{aligned}
D_n - nD_{n-1} &= -(D_{n-1} - (n-1)D_{n-2}), \\
D_{n-1} - (n-1)D_{n-2} &= -(D_{n-2} - (n-2)D_{n-3}), \\
D_{n-2} - (n-2)D_{n-3} &= -(D_{n-3} - (n-3)D_{n-4}), \\
\cdots &= \cdots, \\
D_2 - 2D_1 &= -(D_1 - D_0).
\end{aligned}
$$

Multiplication yields $D_n - nD_{n-1} = (-1)^n$.

We now solve this recurrence relation as follows. We have

$$
\begin{aligned}
D_n &= nD_{n-1} + (-1)^n, \\
nD_{n-1} &= n(n-1)D_{n-2} + n(-1)^{n-1}, \\
n(n-1)D_{n-2} &= n(n-1)(n-2)D_{n-3} + n(n-1)(-1)^{n-2}, \\
\cdots &= \cdots, \\
n(n-1)\cdots 2D_1 &= n!D_0 + n(n-1)\cdots 2(-1).
\end{aligned}
$$

Addition yields

$$
\begin{aligned}
D_n &= n!D_0 + n(n-1)\cdots 2(-1) + \cdots \\
&\quad + n(n-1)(-1)^{n-2} + n(-1)^{n-1} + (-1)^n \\
&= n!\left(1 - \frac{1}{1!} + \frac{1}{2!} - \cdots + (-1)^n\frac{1}{n!}\right).
\end{aligned}
$$

Another approach makes use of exponential generating functions. Let $G(x) = \sum_{n=0}^{\infty} D_n \frac{x^n}{n!}$. We have

$$
D_n - nD_{n-1} = (-1)^n,
$$

$$
\sum_{n=1}^{\infty} D_n \frac{x^n}{n!} - x\sum_{n=1}^{\infty} D_{n-1}\frac{x^{n-1}}{(n-1)!} = \sum_{n=1}^{\infty}(-1)^n\frac{x^n}{n!},
$$

$$
G(x) - 1 - xG(x) = e^{-x} - 1.
$$

Hence $G(x) = \frac{e^{-x}}{1-x}$. Considered as an ordinary generating function, e^{-x} generates the sequence $\{1, -\frac{1}{1!}, \frac{1}{2!}, \ldots, (-1)^n\frac{1}{n!}, \ldots\}$. Recall that $\frac{1}{1-x}$ is the summation operator. Hence the sequence generated by $G(x)$ as an ordinary generating function is $\{1 - \frac{1}{1!} + \frac{1}{2!} - \ldots + (-1)^n\frac{1}{n!}\}$. It follows that the sequence generated by $G(x)$ as an exponential generating function is $\{D_n\}$ where

$$
D_n = n!\left(1 - \frac{1}{1!} + \frac{1}{2!} - \ldots + (-1)^n\frac{1}{n!}\right).
$$

We now turn our attention to the function $D_n(k)$. That $D_n(n-1) = 0$ is classic folklore. A less obvious relation is $D_n(0) - D_n(1) = (-1)^n$. Recall that $D_n - nD_{n-1} = (-1)^n$. Our result will follow if $D_n(1) = nD_{n-1}(0)$. In fact, $D_n(k) = \binom{n}{k}D_{n-k}(0)$ for $0 \le k \le n$. This is because there are $\binom{n}{k}$ ways of choosing the fixed points, and $D_{n-k}(0)$ ways of deranging the remaining points. It follows that

$$
D_n(k) = \frac{n!}{k!}\left(1 - \frac{1}{1!} + \frac{1}{2!} - \cdots + (-1)^{n-k}\frac{1}{(n-k)!}\right).
$$

We can prove two recurrence relations for $D_n(k)$, analogous to the earlier ones for D_n.

We claim that

$$D_n(k) = D_{n-1}(k-1) + (n-1)(D_{n-2}(k) + D_{n-1}(k) - D_{n-2}(k-1)).$$

Count the number of permutations with exactly k fixed points according to which point is in the first place. If it is 1, then it is a fixed point, and we have $k-1$ others. The number of such permutations is $D_{n-1}(k-1)$. Suppose it is 2. We consider two cases. First, suppose 1 is in the second place. Then neither 1 nor 2 is a fixed point, and we have to get the k fixed points from the remaining $n-2$. Hence there are $D_{n-2}(k)$ such permutations. Second, suppose 1 is not in the second place. We will treat it as though it is 2. Then there are $D_{n-1}(k)$ ways of getting k fixed points in the last $n-1$ places. However, in $D_{n-2}(k-1)$ of these, the fake 2 is counted as a fixed point. Hence the number of desired permutations when 2 is in the first place is given by $D_{n-2}(k) + D_{n-1}(k) - D_{n-2}(k-1)$. The same argument applies if the point in the first place is k for $2 \le k \le n$.

We now prove that

$$D_n(k) - nD_{n-1}(k) = (-1)^{n-k}\binom{n}{k}.$$

We define $f(n,k) = (D_n(k) - nD_{n-1}(k))(-1)^{n-k}$. For $k = 0$, we have $f(n,0) = (D_n - nD_{n-1})(-1)^n = 1$, and $f(n,n) = D_n(n) - nD_{n-1}(n) = 1$ for $k = n$. Now the recurrence relation in the preceding paragraph may be rewritten as

$$\begin{aligned}
D_n(k) - nD_{n-1}(k) &= -(D_{n-1}(k) - (n-1)D_{n-2}(k)) \\
&\quad +(D_{n-1}(k-1) - (n-1)D_{n-2}(k-1)).
\end{aligned}$$

This is equivalent to $f(n,k) = f(n-1,k) + f(n-1,k-1)$. It follows that $f(n,k) = \binom{n}{k}$, a hidden connection to the Pascal Triangle.

Examples

Example 3.3.1.

During a function at a posh club, n gentlemen checked their hats as they came in. When they left, their hats were returned to them at random. What is the probability that none of them got back his own hat?

Solution:

The number of ways of returning the hats is clearly $n!$. The number of ways in which no gentleman got back his own hat is D_n. Hence the desired probability is $1 - \frac{1}{1!} + \frac{1}{2!} - \cdots + (-1)^n \frac{1}{n!} \approx \frac{1}{e}$.

Example 3.3.2.

Prove that $kD_n(k) = nD_{n-1}(k-1)$.

Solution:

We count the pairs (f, π) where f is a fixed point of the permutation π of $1, 2, \ldots, n$ which has exactly k fixed points. Since there are $D_n(k)$ such permutations, and k fixed points for each, the desired number of pairs is $kD_n(k)$. On the other hand, there are n possible points. It can be a fixed point of a permutation with $k-1$ other fixed points, and the number of such permutations is $D_{n-1}(k-1)$. Hence the desired number of pairs is also given by $nD_{n-1}(k-1)$.

Example 3.3.3.

The integers $1, 2, \ldots, n$ are arranged in a circle. Let R_n denote the number of arrangements in which i is not followed immediately clockwise by $i+1$ for $1 \le i < n$ and n is not followed immediately clockwise by 1.

  (a) Find an expression for R_n.

  (b) Prove that $R_n + R_{n+1} = D_n$ for $n \ge 2$.

Solution:

  (a) The total number of cyclic permutations of $1, 2, \ldots, n$ is $(n-1)!$. Whenever a forbidden configuration occurs, we have one less point to permute. By the Principle of Inclusion-Exclusion,

$$R_n = (n-1)! - \binom{n}{1}(n-2)! + \binom{n}{2}(n-3)! - \cdots$$
$$+ (-1)^{n-1}\binom{n}{n-1}0! + (-1)^n.$$

(b) We have

$$
\begin{aligned}
R_n + R_{n+1} &= \sum_{i=0}^{n-1}(-1)^i \binom{n}{i}(n-i+1)! + (-1)^n \\
&+ \sum_{i=0}^{n}(-1)^i \binom{n+1}{i}(n-i)! + (-1)^{n+1} \\
&= \sum_{i=1}^{n}(-1)^{i-1} \binom{n}{i-1}(n-i)! + n! \\
&+ \sum_{i=1}^{n}(-1)^i \binom{n+1}{i}(n-i)! \\
&= n! + \sum_{i=1}^{n}(-1)^n \left(\binom{n+1}{i} - \binom{n}{i-1} \right)(n-i)! \\
&= n! + \sum_{i=1}^{n}(-1)^i \binom{n}{i}(n-i)! \\
&= D_n.
\end{aligned}
$$

Exercises

1. In how many ways can 1, 2, 3, 4, 5, 6, 7, 8 and 9 be permuted such that every odd integer is not in its natural place?

2. Two professors are giving oral examinations to the same four students in the same hour, each lasting a quarter of an hour. In how many different ways can a conflict-free examination schedule be set?

3. Prove that $\binom{n}{0}D_0 + \binom{n}{1}D_1 + \cdots + \binom{n}{n}D_n = n!$.

4. Prove that $D_n(1) + 2D_n(2) + \cdots + nD_n(n) = n!$.

5. Consider the permutations of 1, 2, ..., n in which k is not followed immediately by $k+1$ for $1 \leq k \leq n-1$. Prove that their number is given by $D_n + D_{n-1}$.

6. Consider the permutations of 1, 2, ..., n in which k is not followed immediately by $k+1$ for $1 \leq k \leq n-1$, and n is not followed immediately by 1. Prove that their number is given by nD_{n-1}.

Section 3.4. Rook Polynomials

How many permutations of 1, 2, 3 and 4 are there such that 1 is not in the first or second position, 2 is not in the second or third position, and 3 is not in the third or fourth position? It is not hard to see that there are only 4 such permutations, namely, (2,3,1,4), (2,3,4,1), (3,4,1,2) and (4,3,1,2).

However, when the size of the problem increases, this brute force method would not work well. We now use a more systematic approach. Let S be the set of all permutations without restriction, A be the set of those with 1 in the first position, B be the set of those with 1 in the second position, C be the set of those with 2 in the second position, D be the set of those with 2 in the third position, E be the set of those with 3 in the third position, and F be the set of those with 3 in the fourth position. We seek the value of $|\overline{A} \cap \overline{B} \cap \overline{C} \cap \overline{D} \cap \overline{E} \cap \overline{F}|$.

We have $|S| = 4!$ and $|A| = |B| = |C| = |D| = |E| = |F| = 3!$. The size of the intersection of two of A, B, C, D, E and F is either 0 or 2!, depending on whether the conditions which define the two sets are conflicting or not. Here, $A \cap B$, $B \cap C$, $C \cap D$, $D \cap E$ and $E \cap F$ are the only empty intersections. Each of the other 10 has size 2!. The only non-empty triple intersections are $A \cap C \cap E$, $A \cap C \cap F$, $A \cap D \cap F$ and $B \cap D \cap F$, and each has size 1!. Finally. $A \cap B \cap C \cap D$ is empty. By the Principle of Inclusion-Exclusion, $|\overline{A} \cap \overline{B} \cap \overline{C} \cap \overline{D} \cap \overline{E} \cap \overline{F}| = 4! - 6 \times 3! + 10 \times 2! - 4 \times 1! = 4$.

We have not succeeded in finding a completely systematic approach because the number of non-empty intersections of two or more sets is still determined by brute force. Let us draw a diagram of the forbidden positions as shown below.

Here, the rows represent the numbers 1, 2, 3 and 4 and the columns represent the positions. A marked cell indicates a forbidden position for a particular element. Our problem is equivalent to finding the number of ways of placing a number of non-attacking rooks on the cells of a chessboard defined by the forbidden positions. Here, a chessboard is defined as any finite subset of the cells of an infinite chessboard. Two rooks attack each other if they lie on the same row or column, even if not all of the intervening cells belong to the chessboard.

Clearly, the number of ways of placing one non-attacking rook on a chessboard is equal to the number of cells in the chessboard. The diagram below shows the 10 ways of placing two non-attacking rooks and 4 ways of placing three non-attacking rooks on the chessboard above. The maximum number of rooks that can be placed is at most the number of rows or the number of columns, whichever is less.

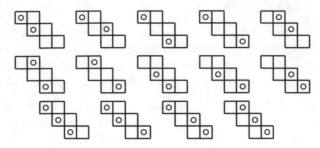

Denote by B a chessboard, and by r_k the number of ways of placing k non-attacking rooks on B. We can record the composite result in the form of a polynomial $R(B) = 1 + r_1 x + r_2 x^2 + r_3 x^3 + \cdots$. This is called the **rook polynomial** of B. For the chessboard B under consideration, we have $R(B) = 1 + 6x + 10x^2 + 4x^3$.

Technically, we should have used the notation $R_B(x)$ since the rook polynomial is a function of x. However, since x is just a formal variable, we wish to emphasize the dependency of the rook polynomial on the chessboard.

Note that the rook polynomial of a chessboard is unchanged if the chessboard is rotated or reflected. It also remains unchanged if the rows and columns of the chessboard are permuted. Thus there are 15 distinct chessboards of size up to 4. They are shown in the diagram below, along with their rook polynomials.

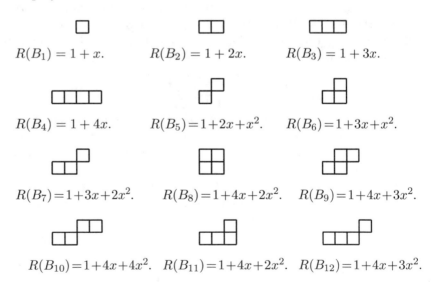

$R(B_1) = 1 + x.$ $R(B_2) = 1 + 2x.$ $R(B_3) = 1 + 3x.$

$R(B_4) = 1 + 4x.$ $R(B_5) = 1 + 2x + x^2.$ $R(B_6) = 1 + 3x + x^2.$

$R(B_7) = 1 + 3x + 2x^2.$ $R(B_8) = 1 + 4x + 2x^2.$ $R(B_9) = 1 + 4x + 3x^2.$

$R(B_{10}) = 1 + 4x + 4x^2.$ $R(B_{11}) = 1 + 4x + 2x^2.$ $R(B_{12}) = 1 + 4x + 3x^2.$

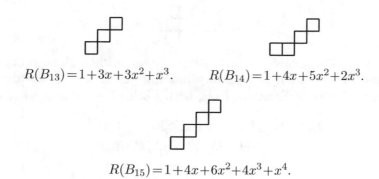

$$R(B_{13})=1+3x+3x^2+x^3. \qquad R(B_{14})=1+4x+5x^2+2x^3.$$

$$R(B_{15})=1+4x+6x^2+4x^3+x^4.$$

Note that $R(B_8) = R(B_{11})$ and $R(B_9) = R(B_{12})$, so that different chessboards may have the same rook polynomial. Some of the rook polynomials can be factored.

$$
\begin{aligned}
R(B_5) &= R(B_1)^2. \\
R(B_7) &= R(B_1)R(B_2). \\
R(B_{10}) &= R(B_2)^2. \\
R(B_{12}) &= R(B_1)R(B_3). \\
R(B_{13}) &= R(B_1)^3. \\
R(B_{14}) &= R(B_1)^2 R(B_2). \\
R(B_{15}) &= R(B_1)^4.
\end{aligned}
$$

These observations generalize to the following result.

Multiplication Rule.
If $B = B' \cup B''$ and no cell in B' is in the same row or column as any cell in B'', then we have $R(B) = R(B')R(B'')$.

Proof:
This follows from the Multiplication Principle because the rook placement in B' is independent of the rook placement in B''.

To compute rook polynomials systematically, we need one other result.

Inclusion-Exclusion Rule.
Let s be a cell in B. Let $B - s$ be the board obtained from B by deleting s, and let B/s be the board obtained from B by deleting all squares in the same row or column as s. Then we have $R(B) = R(B - s) + xR(B/s)$.

Proof:
The number of ways of placing k non-attacking rooks on B without putting one on s is equal to the number of ways of placing k non-attacking rooks on $B - s$. The number of ways of placing k non-attacking rooks on B with one on s is equal to the number of ways of placing $k - 1$ non-attacking rooks on B/s. The desired result follows from the Addition Principle since these two types of placements on B are exclusive and exhaustive.

As an illustration, we shall determine again the rook polynomial of the chessboard B in our introductory problem. Using the Multiplication Rule and the Inclusion-Exclusion Rule, with s being the cell marked $\bullet$, we have

$$
\begin{aligned}
R(B) &= R(B-s) + xR(B/s) \\
&= R(B_2)R(B_6) + xR(B_1)R(B_2) \\
&= (1+2x)(1+3x+x^2) + x(1+x)(1+2x) \\
&= 1 + 6x + 10x^2 + 4x^3.
\end{aligned}
$$

Examples

Example 3.4.1.
Determine the rook polynomial of the chessboard shown in the diagram below.

Solution:
Permuting the rows and columns of the given chessboard, we can convert it to the one shown below. Its rook polynomial is

$$R(B_6)^2 = (1 + 3x + x^2)^2 = 1 + 6x + 11x^2 + 6x^3 + x^4.$$

Example 3.4.2.
Determine the number of permutations of 1, 2, 3, 4 and 5 such that 1 is not in the first or second position, 2 is not in the second or third position, 3 is not in the third or fourth position, 4 is not in the fourth or fifth position, and 5 not in the first or fifth position.

Solution:
We first compute the rook polynomial of the chessboard B of forbidden positions.

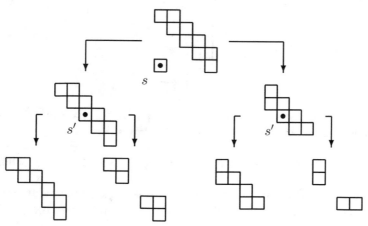

We have

$$
\begin{aligned}
R(B) &= R(B-s) + xR(B/s) \\
     &= R((B-s)-s') + xR((B-s)/s') \\
     &\quad + xR((B/s)-s') + x^2 R((B/s)/s') \\
     &= R(B_9)^2 + xR(B_6)^2 + xR(B_6)^2 + x^2 R(B_2)^2 \\
     &= (1+4x+3x^2)^2 + 2x(1+3x+x^2)^2 + x^2(1+2x)^2 \\
     &= 1 + 10x + 35x^2 + 50x^3 + 25x^4 + 2x^5.
\end{aligned}
$$

We now return to the problem of forbidden positions. By the Principle of Inclusion-Exclusion, the number of desired permutations is

$$
5! - 10 \times 4! + 35 \times 3! - 50 \times 2! + 25 \times 1! - 2 \times 0! = 13.
$$

Example 3.4.3.

Determine the number of ways of arranging two As, two Bs and two Cs in a row, with neither A in the first two positions, neither B in the middle two positions and neither C in the last two positions.

Solution:

We call the two As 1 and 2, the two Bs 3 and 4 and the two Cs 5 and 6, and determine the number of desired permutations of 1, 2, 3, 4, 5 and 6. The rook polynomial of the chessboard B of forbidden positions is

$$
\begin{aligned}
R(B) &= R(B_8)^3 \\
     &= (1+4x+2x^2)^3 \\
     &= 1 + 12x + 54x^2 + 112x^3 + 108x^4 + 48x^5 + 8x^6.
\end{aligned}
$$

We now return to the problem of forbidden positions. By the Principle of Inclusion-Exclusion, the number of desired permutations is

$$
6! - 12 \times 5! + 54 \times 4! - 112 \times 3! + 108 \times 2! - 48 \times 1! + 8 \times 0! = 80.
$$

To account for the fact that the two As are indistinguishable, as are the two Bs and the two Cs, we divide by 8 to obtain the final answer 10.

Exercises

1. Determine the rook polynomial of the chessboard shown in the diagram below.

2. Determine the rook polynomial of the chessboard shown in the diagram below.

3. Determine the number of permutations of 1, 2, 3 and 4 with 2 not in position 3 or 4, 3 not in position 3, and 4 not in positions 1, 2 or 3. Use the method of rook polynomials.

4. Determine the number of permutations of 1, 2, 3 and 4 with 1 not in position 4, 2 not in position 1 or 4, and 3 not in positions 1, 2 or 3. Use the method of rook polynomials.

5. Two distinct dice are rolled six times. The combinations (2,1), (2,2), (2,3), (3,2), (3,3) and (4,4) do not appear. In how many different ways can all faces of both dice show?

6. Two distinct dice are rolled six times. The combinations (1,1), (1,2), (2,2), (2,3), (2,4) and (3,2) do not appear. In how many different ways can all faces of both dice show?

Section 3.5. Cycles

So far, we have treated permutations as static configurations. For example, $\{2, 5, 1, 8, 6, 3, 7, 4\}$ is regarded simply a rearrangement of the elements $\{1, 2, 3, 4, 5, 6, 7, 8\}$. However, we can also consider permutations as dynamic transformations. We form an array by putting the permutation in the first row, and the elements in the second row are in numerical order. This yields

$$\begin{pmatrix} 2 & 5 & 1 & 8 & 6 & 3 & 7 & 4 \\ 1 & 2 & 3 & 4 & 5 & 6 & 7 & 8 \end{pmatrix}.$$

We now rearrange the columns so that the elements in the top row are in numerical order. We obtain

$$\begin{pmatrix} 1 & 2 & 3 & 4 & 5 & 6 & 7 & 8 \\ 3 & 1 & 6 & 8 & 2 & 5 & 7 & 4 \end{pmatrix}.$$

This sends the element 1 to position 3, the element 2 to position 1, the element 3 to position 6, and so on, bringing us back to $\{2, 5, 1, 8, 6, 3, 7, 4\}$. We call $\{3, 1, 6, 8, 2, 5, 7, 4\}$ its **inverse** permutation.

As another illustration, consider the permutation $\{5,3,7,6,8,4,2,1\}$. We have

$$\begin{pmatrix} 5 & 3 & 7 & 6 & 8 & 4 & 2 & 1 \\ 1 & 2 & 3 & 4 & 5 & 6 & 7 & 8 \end{pmatrix}$$

and

$$\begin{pmatrix} 1 & 2 & 3 & 4 & 5 & 6 & 7 & 8 \\ 8 & 7 & 2 & 6 & 1 & 4 & 3 & 5 \end{pmatrix}.$$

When we perform two permutations in succession, the end result is another permutation called the **product** of the two permutations. For instance, the product of the permutation $\{2,5,1,8,6,3,7,4\}$ followed by the permutation $\{5,3,7,6,8,4,2,1\}$ is

$$\begin{pmatrix} 1 & 2 & 3 & 4 & 5 & 6 & 7 & 8 \\ 3 & 1 & 6 & 8 & 2 & 5 & 7 & 4 \\ 2 & 8 & 4 & 5 & 7 & 1 & 3 & 6 \end{pmatrix}.$$

The final permutation is $\{6,1,7,3,4,8,5,2\}$.

There is another way to represent permutations. In the permutation $\{2,5,1,8,6,3,7,4\}$, 1 goes to 3, 3 goes to 6, 6 goes to 5, 5 goes to 2 and 2 goes to 1. This completes a cycle (1,3,6,5,2) with five elements. Performing this cycle alone yields the permutation $\{2,5,1,4,6,3,7,8\}$. We start another cycle with the first element not in preceding cycles, namely 4, and obtain a cycle (4,8) with two elements. Continuing the transformation by performing this cycle, we have $\{2,5,1,8,6,3,7,4\}$. Finally, we also have a cycle (7) with only one element.

The permutation is then represented as a product of disjoint **cycles**, namely, $(1,3,6,5,2)(4,8)(7)$. Similarly, the permutation $\{5,3,7,6,8,4,2,1\}$ is represented by $(1,8,5)(2,7,3)(4,6)$. It is easy to see that such a factorization is unique, up to the order of the cycles. This order does not matter since the cycles are disjoint.

In the terminology of Section 3.3, the lone element in a cycle of length 1 is a fixed point of the permutation. A cycle of length 2 is called a **transposition**. Note that every permutation can be represented as a product of transpositions, which are not necessarily disjoint.

This follows from the fact that every cycle can also be so represented. Indeed,

$$(a_1, a_2, \ldots, a_n) = (a_1, a_2)(a_1, a_3) \cdots (a_1, a_n),$$

with the understanding that (a_i, a_j) means switching the elements in the a_i-th and the a_j-th positions, rather than switching the elements a_i and a_j. Because the cycles are not disjoint, the order in which the action of the transpositions are taken is important. We illustrate with

$$(1, 3, 6, 5, 2) = (1, 3)(1, 6)(1, 5)(1, 2)$$

as follows.

Starting position	$\{1,2,3,4,5,6,7,8\}$
Action of $(1,3)$	$\{3,2,1,4,5,6,7,8\}$
Action of $(1,6)$	$\{6,2,1,4,5,3,7,8\}$
Action of $(1,5)$	$\{5,2,1,4,6,3,7,8\}$
Action of $(1,2)$	$\{2,5,1,4,6,3,7,8\}$

The end result is exactly the permutation produced by the cycle $(1,3,6,5,2)$.

Note that this time, the factorization into transpositions is not unique. The cycle $(1,3,6,5,2)$ may also be factorized as

$$(1, 2)(2, 3)(3, 5)(5, 6)(2, 3)(3, 5)(5, 6)(3, 5).$$

Even the number of transpositions in the factorization is not constant. However, there is an invariant, in that the parity of this number is unchanged.

To see this, consider the product of differences $X(n) = \prod_{1 \leq i < j \leq n} (j - i)$. When we perform a permutation $p = \{a_1, a_2, \ldots, a_n\}$ of $\{1, 2, \ldots, n\}$ on $X(n)$, it becomes $X_p(n) = \prod_{1 \leq i < j \leq n} (a_j - a_i)$. Note that both $X(n)$ and $X_p(n)$ consist of $\binom{n}{2}$ factors. Moreover, the factors of $X(n)$ can be put in a one to one correspondence with the factors of $X_p(n)$ so that the two in each pair are differences of the same two numbers, except possibly in reverse orders. It follows that $X_p(n) = \pm X(n)$.

Consider the action of the permutation $p = \{2, 5, 1, 8, 6, 3, 7, 4\}$ on $X(8)$. Then $X_p(8)$ is a product of 28 factors, namely,

$$
\begin{aligned}
X_p(8) \;=\; & (5-2)(1-5)(1-2)(8-1)(8-5)(8-2)(6-8) \\
& (6-1)(6-5)(6-2)(3-6)(3-8)(3-1)(3-5) \\
& (3-2)(7-3)(7-6)(7-8)(7-1)(7-5)(7-2) \\
& (4-7)(4-3)(4-6)(4-8)(4-1)(4-5)(4-2).
\end{aligned}
$$

The differences $1-5$, $1-2$, $6-8$, $3-6$, $3-8$, $3-5$, $7-8$, $4-7$, $4-6$, $4-8$ and $4-5$ have reversed order. Since there are 11 of them, $X_p(8) = -X(8)$.

A permutation p of $\{1, 2, \ldots, n\}$ is said to be an **even** permutation if $X_p(n) = X(n)$, and an **odd** permutation if $X_p(n) = -X(n)$. We claim that an even permutation is a product of an even number of transpositions, and that an odd permutation is a product of an odd number of transpositions.

Consider the action of a transposition (a_i, a_j), $1 \le i < j \le n$, on $X_p(n)$. This switches a_i with a_j. Differences involving neither a_i nor a_j are unchanged. The difference $a_j - a_i$ has reversed its sign. Consider now the two differences involving one of a_i and a_j and another element a_k. If $k < i$, then $a_k - a_i$ and $a_k - a_j$ just trade places. The same applies if $k > j$. Suppose $i < k < j$. Then both $a_k - a_i$ and $a_j - a_k$ have reversed their signs. Hence $X_{(i,j)}(n) = -X(n)$.

Since each transposition reverses the sign of $X(n)$, the final result is $X(n)$ if the number of transpositions is even, and the final result is $-X(n)$ otherwise. This justifies our claim. It follows easily that the product of two even permutations is another even permutation, the product of two odd permutations is an even permutation, and the product of an even permutation and an odd permutation is an odd permutation.

It is easy to prove that for $n \ge 2$, exactly half of the $n!$ permutations of $\{1, 2, \ldots, n\}$ are even permutations. Just form the permutations into pairs which differ only in that the first two elements have been switched. In each pair, exactly one of them is an even permutation.

The concepts of even and odd permutations play an essential role in linear algebra, particularly in the evaluation of the determinant. We give here a combinatorial application.

The famous 14-15 puzzle consists of 15 tiles arranged in a 4×4 board as shown in the diagram below, with an empty space at the bottom right corner. A tile adjacent to the empty space may slide onto it. The objective is to arrive at a configuration exactly the same as the starting position except for the interchange of positions of the 14 and 15 tiles.

1	2	3	4
5	6	7	8
9	10	11	12
13	14	15	

We now prove that this task is impossible. Color the board black and white in checkerboard fashion, as shown in the diagram above. We imagine that there is a 16 tile on the empty space. Each slide then becomes a permutation of the positions {1,2,3,4,5,6,7,8,9,10,11,12,13,14,15,16}, in the form of a transposition involving the 16 tile and the sliding tile. There are twenty-four permissible transpositions, (1,2), (2,3), (3,4), (5,6), (6,7), (7,8), (9,10), (10,11), (11,12), (13,14), (14,15), (15,16), (1,5), (5,9), (9,13), (2,6), (6,10), (10,14), (3,7), (7,11), (11,15), (4,8), (8,12) and (12,16). Recall that the numbers here represent the spaces and not the tiles.

Each move of the 16 tile takes it to an adjacent space which is of the opposite color. Since this tile eventually returns to its starting space, it must have made an even number of moves. The eventual permutation, as a product of an even number of transpositions, must be an even permutation. However, since the target permutation may be obtained from the initial permutation by performing just (14,15), it is an odd permutation, and therefore cannot be attained.

In the next problem, a king had 6 strongboxes constructed to store his most valuable possessions. Each strongbox had a unique key. In considering what were his most valuable possessions, the king came to the conclusion that whoever got hold of the keys could help themselves to the contents of the strongboxes. It followed that his most valuable possessions were the keys themselves. So he tossed them at random into the strongboxes, a key in each strongbox. Then he slammed them shut. Afterwards, he realized that without the appropriate key, the only way to open a strongbox is to smash it. The King ordered 1 of the strongboxes to be smashed, and to use keys retrieved in the process to open the remaining 5 strongboxes without smashing another one. What was the probability of success?

We may assume that the King smashes box number 1. With any retrieved key, as long as it is not that of box number 1, he can open another box. What he wants is for the key to box number 1 to show up in the last box he opens. The following is an example of a successful scenario.

Boxes	1	2	3	4	5	6
Keys	2	5	4	6	3	1

This permutation forms a single cycle (1,2,5,3,4,6). The following is an example of an unsuccessful scenario.

Boxes	1	2	3	4	5	6
Keys	2	5	1	4	6	3

This time, the permutation consists of two separate cycles, (1,2,5,6,3) and (4). The number of permutations consisting of a single cycle is 5! because we can start with any number, and then there are respectively 5, 4, 3, 2 and 1 choices for the remaining numbers. Since there are 6! permutations overall, the probability of success is $\frac{5!}{6!} = \frac{1}{6}$.

Among the $n!$ permutations on $\{1, 2, \ldots, n\}$, it is of interest to determine the number $\ell(k, n)$ of those whose longest cycle is of length k. For instance, $\ell(1, 4) = 1$ since the only such permutation is (1)(2)(3)(4). Counting the permutations (1)(2)(3,4), (1,2)(3)(4), (1,2)(3,4), (1)(3)(2,4), (1,3)(2)(4), (1,3)(2,4), (1)(4)(2,3), (1,4)(2)(3) and (1,4)(2,3), we have $\ell(2, 4) = 9$. Next, $\ell(3, 4) = 8$, counting (1)(2,3,4), (1)(2,4,3), (1,2,3)(4), (1,3,2)(4), (1,2,4)(3), (1,4,2)(3), (1,3,4)(2) and (1,4,3)(2). Finally, we have $\ell(4, 4) = 6$, counting (1,2,3,4), (1,2,4,3), (1,3,2,4), (1,3,4,2), (1,4,2,3) and (1,4,3,2). We remark that $\ell(1, 4) + \ell(2, 4) + \ell(3, 4) + \ell(4, 4) = 24$. In general, $\ell(1, n) = 1$,

$$\ell(n, n) = (n-1)! \text{ and } \sum_{k=1}^{n} \ell(k, n) = n!.$$

In a permutation on $\{1, 2, \ldots, n\}$, the **expected length** of the longest cycle is defined to be $E(n) = \dfrac{1}{n!} \sum_{k=1}^{n} k\ell(k, n)$, and the **relative length** of the longest cycle is defined to be $\dfrac{E(n)}{n}$, which converges to $0.62432965\ldots$ as n approaches infinity. This number is known as **Golomb's constant**.

Examples

Example 3.5.1.
Consider the permutations $\{2,5,1,8,6,3,7,4\}$ and $\{5,3,7,6,8,4,2,1\}$. Verify that their product is $\{6,1,7,3,4,8,5,2\}$ by decomposing the cycle representation of the second permutation into transpositions.

Solution:
The factorization of the second permutation is
$$(1,8,5)(2,7,3)(4,6) = (1,8)(1,5)(2,7)(2,3)(4,6).$$

The computation below leads to the desired permutation.

Starting position	$\{2,5,1,8,6,3,7,4\}$
Action of $(1,8)$	$\{4,5,1,8,6,3,7,2\}$
Action of $(1,5)$	$\{6,5,1,8,4,3,7,2\}$
Action of $(2,7)$	$\{6,7,1,8,4,3,5,2\}$
Action of $(2,3)$	$\{6,1,7,8,4,3,5,2\}$
Action of $(4,6)$	$\{6,1,7,3,4,8,5,2\}$

Example 3.5.2.
In the strongbox problem, what if the King smashed at random one box first, and a second one if necessary?

Solution:
The King will be successful if the permutation of the keys consists of at most two cycles. With six numbers, there are eleven kinds of cycle structures.

6	$\binom{6}{6}$	$5!$	120
5+1	$\binom{6}{5}$	$4!$	144
4+2	$\binom{6}{4}$	$3!$	90
3+3	$\binom{6}{3} \div 2!$	$2! \times 2!$	40
4+1+1	$\binom{6}{4}$	$3!$	90
3+2+1	$\binom{6}{3}\binom{3}{2}$	$2!$	120
2+2+2	$\binom{6}{2}\binom{4}{2} \div 3!$	1	15
3+1+1+1	$\binom{6}{3}$	$2!$	40
2+2+1+1	$\binom{6}{2}\binom{4}{2} \div 2!$	1	45
2+1+1+1+1	$\binom{6}{2}$	1	15
1+1+1+1+1+1	$\binom{6}{0}$	1	1

The second column counts the number of ways of separating the six numbers into the respective cycles, and the third column counts the number of ways of forming these cycles. The fourth column, which is the total number of cycles of the specific type, is the product of the preceding two columns. Actually, we only need the first four rows. We work out everything just to verify that the total number of permutations is indeed $6!=720$. The total number of permutations which lead to successful scenarios is $120+144+90+40 =394$, and the probability of success is $\frac{394}{720} = \frac{197}{360}$.

Example 3.5.3.

Six prisoners are getting a reprieve. They will be shown into a room one at a time. The room contains six numbered boxes, each with the name of a different one of them, distributed completely at random. Each prisoner is trying to find the number of the box containing his own name. He may ask to look inside a box. If he does not find his name, he may ask to look inside a second box, and a third one if necessary. However, that is as far as he may go. The prisoners all go free if each finds the number of the box containing his own name, but will remain in prison if at least one fails. The prisoners meet to discuss their strategy before they are shown into that room. After that, no further communication is allowed among them. Without any strategy, the probability of success for each prisoner is $\frac{3}{6} = \frac{1}{2}$, so that the overall probability of success is $(\frac{1}{2})^6 = \frac{1}{64}$. However, there is a strategy by which they can improve their chances to over $\frac{1}{3}$. How can they do that?

Solution:

The prisoners assign numbers to themselves, from 1 to 6. When shown into the room, each asks to look inside the box bearing his number. If he does not find his own name, he asks to look inside the box bearing the number of the prisoner whose name is inside the box just examined. He repeats this step if he still does not find his own name. As an illustration, consider the permutation

$$\begin{pmatrix} 1 & 2 & 3 & 4 & 5 & 6 \\ 4 & 3 & 2 & 6 & 1 & 5 \end{pmatrix}.$$

The first row contains the box numbers and the second row the numbers of the prisoners whose names are in the respective boxes. The cycle structure is $(1465)(23)$. So prisoner number 2 and prisoner number 3 will find their own names in the second box they examine, while the others will fail in their task. So success or failure depends on whether the longest cycle has length 3 or less, or 4 or more. From the chart in Example 3.5.2, the overall probability of success is $\frac{40+120+15+40+45+15+1}{720} = \frac{23}{60}$.

Exercises

1. Compute the product of the permutations

$$\{3, 6, 2, 5, 7, 1, 8, 4\} \text{ and } \{3, 1, 2, 8, 7, 4, 5, 6\}$$

   (a) directly;

   (b) by decomposing the cycle representation of the second permutation into transpositions.

2. Compute the product of the permutations

$$\{8, 6, 2, 5, 1, 7, 3, 4\} \text{ and } \{2, 1, 6, 3, 5, 7, 4, 8\}$$

   (a) directly;

   (b) by decomposing the cycle representation of the second permutation into transpositions.

3. In the strongbox Problem, suppose the King smashed 2 of them at random. What was the probability that the king could open the remaining 4 strongboxes without smashing another one?

4. In the strongbox Problem, suppose that when the King tossed the keys at random into the strongboxes, it was not necessarily true that exactly one key went into each strongbox. After smashing 2 of them at random, what was the probability that the king could open the remaining 4 strongboxes without smashing another one?

5. Prove that $\frac{E(n-1)}{n-1} > \frac{E(n)}{n}$ for $n \geq 2$.

6. Prove that $\frac{E(n)}{n+1} < \frac{E(n+1)}{n+2}$ for $n \geq 2$.

Section 3.6. Comma-free Dictionaries

Let n and k be positive integers. An (n, k)-**dictionary** is the complete set of words of length k using an alphabet with n letters. For example, a $(2,4)$-dictionary with alphabet $\{1,2\}$ consists of the words 1111, 1112, 1121, 1122, 1211, 1212, 1221, 1222, 2111, 2112, 2121, 2122, 2211, 2212, 2221 and 2222. There are n choices for each of the k letters. By the Multiplication Principle, the size of an (n, k)-dictionary is n^k.

A word in an (n, k)-dictionary is said to have period d for some positive divisor d of k if the word consists of the same block of d letters repeated $\frac{k}{d}$ times, and d is the smallest positive integer with this property. A word of period k in an (n, k)-dictionary is said to be **primitive**, and the number of such words is denoted by $P_k(n)$.

In our introductory example, 1111 and 2222 have period 1, 1212 and 2121 have period 2 while the remaining words have period 4 and are primitive. Hence $P_4(2) = 12$.

Note that a word of period d corresponds to a primitive word in the (n, d)-dictionary. Since every word is of period d for some positive divisor d of k, we have $\sum_{d|k} P_d(n) = n^k$. By Möbius' Inversion Formula which we have encountered in Section 2.6, $P_k(n) = \sum_{d|k} \mu(d) n^{\frac{k}{d}}$. We indeed have

$$P_4(2) = \mu(1)2^4 + \mu(2)2^2 + \mu(4)2^1 = 16 - 4 = 12.$$

Alternatively, this formula may be derived from the Principle of Inclusion-Exclusion.

A subset of an (n, k)-dictionary constitutes a **comma-free** dictionary if whenever $a_1 a_2 \ldots a_k$ and $b_1 b_2 \ldots b_k$ are words in the subset, then the word $a_{i+1} a_{i+2} \ldots a_k b_1, b_2, \ldots b_i$ does not belong to the subset for $1 \leq i \leq n - 1$. The maximum number of words in a comma-free (n, k)-dictionary is denoted by $W_k(n)$.

It should be clear that non-primitive words cannot be included in a comma-free dictionary. The letters of each primitive k-letter word may be permuted cyclically in k ways, and at most one of these k words may be included. It follows that $W_k(n) \leq \frac{1}{k} P_k(n)$.

In particular, $W_4(2) \leq \frac{1}{4} P_4(2) = 3$. We now construct a comma-free $(2,4)$-dictionary with 3 words. The 16 words in our introductory example may be divided into four groups of four.
Group I: 1111, 2222, 1212, 2121;
Group II: 1112. 1121, 1211, 2111;
Group III: 1122, 1221, 2211, 2112;
Group IV: 1222, 2122, 2212, 2221.

The first group consists of the non-primitive words which must be excluded. Each of the subsequent groups consists of cyclic permutations of the same word, and at most one from each group may be included. Finally, 1211, 1221 and 1222 constitute a comma-free dictionary. Hence $W_4(2) = 3$.

For all odd k, the upper bound $W_k(n) \leq \dfrac{1}{k} \sum_{d|k} \mu(d) n^{\frac{k}{d}}$ is sharp. This was known at one point as the **Golomb-Gordon-Welch Conjecture**, later proved by **W. L. Eastman** and **S. Even**, and independently by **R. Scholtz**.

The result is trivial for $k = 1$. For odd $k \geq 3$, we shall construct maximal comma-free (n, k)-dictionaries by **Scholtz's Algorithm**. We use as alphabet the set $S_0 = \{1, 2, \ldots, n\}$. In each iteration, we drop the first among the shortest words of odd length and use it as a prefix on all other words, as often as possible. We denote by S_j the infinite dictionary after the j-th iteration, and let $P_j(x) = \sum_{i=1}^{\infty} n_{i,j} x^i$, where $n_{i,j}$ is the number of words of length i in S_j.

We first illustrate with $n = 2$.

Initialization.
We have $S_0 = \{1, 2\}$ and $P_0(x) = 2x$.

First Iteration. Prefix=1.
S_1 consists of

$$2 \quad 12 \quad 112 \quad 1112 \quad 11112 \quad 111112 \quad \ldots$$

We have $P_1(x) = (P_0(x) - x) \sum_{h=0}^{\infty} x^h = x + x^2 + x^3 + x^4 + x^5 + x^6 + \cdots$.

Second Iteration. Prefix=2.
S_2 consists of

$$
\begin{array}{llllll}
12 & 112 & 1112 & 11112 & 111112 & \ldots \\
 & 212 & 2112 & 21112 & 211112 & \ldots \\
 & & 2212 & 22112 & 221112 & \ldots \\
 & & & 22212 & 222112 & \ldots \\
 & & & & 222212 & \ldots \\
 & & & & & \ldots
\end{array}
$$

We have $P_2(x) = (P_1(x) - x) \sum_{h=0}^{\infty} x^h = x^2 + 2x^3 + 3x^4 + 4x^5 + 5x^6 + \cdots$.

A $(2,3)$-dictionary consists of $\{112, 212\}$.

Third Iteration: Prefix=112.

S_3 consists of

12		1112	11112	111112	...
	212	2112	21112	211112	...
		2212	22112	221112	...
			22212	222112	...
			11212	222212	...
				112212	...
					...

We have $P_3(x) = (P_2(x) - x^3) \sum_{h=0}^{\infty} x^{3h} = x^2 + x^3 + 3x^4 + 5x^5 + 6x^6 + \cdots$.

Fourth Iteration: Prefix=212.

S_4 consists of

12	1112	11112	111112	...
	2112	21112	211112	...
	2212	22112	221112	...
		22212	222112	...
		11212	222212	...
			112212	...
	21212			...
				...

We have $P_4(x) = (P_3(x) - x^3) \sum_{h=0}^{\infty} x^{3h} = x^2 + 3x^4 + 6x^5 + 6x^6 + \cdots$.

A $(2,5)$-dictionary consists of $\{11112, 21112, 22112, 22212, 11212, 21212\}$.

In general, we have $P_j(x) = (P_{j-1}(x) - x^{\ell_j}) \sum_{h=0}^{\infty} x^{h\ell_j}$, where ℓ_j is the length of the prefix used in the j-th iteration. The factor before the summation represents the removal of the prefix from S_{j-1} while the summation represents the application of the prefix h times, h ranging from 0 to ∞.

Note that

$$
\begin{aligned}
1 - P_j(x) &= 1 - (P_{j-1} - x^{\ell_j}) \frac{1}{1 - x^{\ell_j}} \\
&= \frac{1 - x^{\ell_j} - P_{j-1}(x) + x^{\ell_j}}{1 - x^{\ell_j}} \\
&= \frac{1 - P_{j-1}(x)}{1 - x^{\ell_j}}.
\end{aligned}
$$

It follows that in our illustration, $1 - P_1(x) = \frac{1-2x}{1-x}$, $1 - P_2(x) = \frac{1-2x}{(1-x)^2}$, $P_3(x) = \frac{1-2x}{(1-x)^2(1-x^3)}$ and $1 - P_4(x) = \frac{1-2x}{(1-x)^2(1-x^3)^2}$. In general, each iteration introduces an additional factor in the denominator on the right side. This implies that the order in which the iterations are carried out does not affect the number of words of each length in the resulting dictionary.

Here is a quick illustration with $n = 3$ and $k = 3$. We have

$$
\begin{aligned}
S_1 &= \{1, 2, 3\}, \\
S_2 &= \{2, 3, 12, 13, 112, 113, \ldots\}, \\
S_3 &= \{3, 12.13.23, 112, 113, 223, 212, 213, \ldots\}, \\
S_4 &= \{12.13.23, 112, 113, 212, 213, 223, 312, 313, 323, \ldots\}.
\end{aligned}
$$

A (3,3)-dictionary consists of the 8 words of length 3 in S_4.

We claim that Scholtz's Algorithm always yields maximal dictionaries for odd lengths. This can be justified if we can show that for every string of the alphabet of odd length, there is exactly one word which results from Scholtz's Algorithm if and only if the string is primitive.

We illustrate with the string 21331322212, and consider it as wrapped around a circle. We put a comma after each letter and obtain

$$(2, 1, 3, 3, 1, 3, 2, 2, 2, 1, 2,).$$

The first prefix we use is 1. So we erase all commas after 1s and obtain (2,13,3,13,2,2,2,12,). The next prefix is 2, and we have (213,3,13,22212,). Note that the final comma is not erased since it comes after 12 rather than just 2. The next prefix is 3, yielding (213,313,22212,). The next prefix is 213, and we have (213313,22212,). The last prefix here is 22212, and the final result is (213313,22212), and the unique word from the cyclic permutations of the given string is 22212213313.

We call this procedure the *word scan* and it can be applied in general. Suppose at least two commas remain. Since the word has odd length, at least one block of letters between two adjacent commas has odd length and can be used as the next prefix. This means that the word scan has not yet terminated. It will terminate when at most one comma remains. If a single comma remains, then we have the unique word from the cyclic class. If no commas remain, then the last batch must disappear together. This happens if and only if the cyclic class is non-primitive. Thus our claim is justified.

Finally, we prove that Scholtz's Algorithm always yields comma-free dictionaries for odd lengths. Let A and B be two words in the dictionary, and let W be a word of the same length such that its initial segment W_1 coincides with the final segment of A and its final segment W_2 coincides with the initial segment of B. We claim that W cannot be a word in this dictionary.

We first consider the case where W_1 has even length. We illustrate with $A = 3131212$, $B = 3222112$ and $W = 1212322$. Applying the word scan to W, we have in succession $(1,2,1,2,3,2,2,)$, $(12,12,3,2,2,)$, $(12,12,3,22)$, $(12,12,322)$ and $(1212,322)$. We see that the last comma is at the end of $W_1 = 1212$, rather than at the end of $W_2 = 322$.

This will be the case in general. If not, the comma at the end of W_1 will disappear at some point during the word scan. In order for it to do so, it must be at the end of a block P of letters of odd length. Since W_1 has even length, there must be another comma before P. However, this comma, at the end of a block Q of odd length as part of A, disappears before the comma after P. If Q has been truncated, the residual part must also have odd length, and the comma after the truncated Q will disappear even sooner. It follows that W is not a word in the dictionary.

We now consider the case when W_2 has even length. We illustrate with the same A and B but with $W = 2123222$. Applying the word scan to W, we have in succession $(2,1,2,3,2,2,2,)$, $(2,12,3,2,2,2,)$, $(212,3,222)$ and $(212,3222)$. Once again, the last comma is at the end of $W_1 = 212$, rather than at the end of $W_2 = 3222$.

This will be the case in general. When the word scan is applied to B, the surviving comma is of course at its end. When the comma after W_2 disappears, it must be at the end of a block P of odd length since W_2 has even length. P lies within B, and therefore within W_2. Let Q be the part of W_2 excluding P. When the word scan is applied to W, we claim that the comma after P still disappears before the comma after Q. Note that Q is also of odd length. To delay the disappearance of the comma after Q, the block containing Q must extend to part of W_1. Call that R, and note that it has even length. So the part R of W_1 excluding S has odd length since so does W_1. The comma after R disappears after the comma after S. However, this will also be the case when the word scan is applied to A, and this is impossible since the comma after R is at the end of A. This justifies our claim, and completes the proof that Scholtz's Algorithm does produce maximal comma-free dictionaries.

It is known that the upper bound $W_k(n) \leq \dfrac{1}{k} \sum_{d|k} \mu(d) n^{\frac{k}{d}}$ cannot be attained for any even k, provided that n is sufficiently large. The conjecture is that $n > 3$ is sufficiently large. We shall focus only on $k = 2$.

Consider a comma-free $(n, 2)$-dictionary D. Let A be the set of letters which are found at the beginning of some words in D but never at the end of any word in D. Let B be the set of letters which are found at the beginning of some words in D, and also at the end of some words in D. Let C be the set of letters which are found at the end of some words in D but never at the beginning of any word in D.

Note that the words in D must either start with a letter in A and end with a letter in B, start with a letter in A and end with a letter in C, or start with a letter in B and end with a letter in C. It cannot start with a letter x in B and end with another letter y in B, because there must be some word in D which ends with x and some word in D which starts with y.

Let $|A| = \alpha$, $|B| = \beta$ and $|C| = \gamma$. Then $\alpha + \beta + \gamma = n$ and we have $|D| = \alpha\beta + \alpha\gamma + \beta\gamma$. To maximize $|D|$, we should choose α, β and γ to be as nearly equal as possible. For $n = 3t$, we choose $\alpha = \beta = \gamma = t$ so that $D = 3t^2$. For $n = 3t + 1$, we choose $\alpha = \gamma = t$ and $\beta = t + 1$ so that $|D| = t(3t + 2)$. For $n = 3t + 2$, we choose $\alpha = \gamma = t + 1$ and $\beta = t$ so that $|D| = (t+1)(3t+1)$. We can also summarize all three cases in a single formula $|D| = \lfloor \frac{n^2}{3} \rfloor$.

For $n = 2$, it is only known that the upper bound is attained for $k = 2$, 4, 6, 8 and 10. For $n = 3$, it is only known that the upper bound is attained for $k = 2$ and 4.

Examples

Example 3.6.1.
Construct a maximal comma-free (2,7)-dictionary.

Solution:
From the illustration in the text, we have after four iterations one 2-letter word 12, six 5-letter words 11112, 21112, 22112, 22212, 11212 and 21212, along with twelve 7-letter words

$$
\begin{array}{lll}
1111112, & 2111112 & 2211112 \\
2221112 & 2222112 & 2222212 \\
1121112 & 1122112 & 1122212 \\
2121112 & 2122212 & 2122212.
\end{array}
$$

In the next six iterations, each of the 5-letter words is used as a prefix, yielding six more 7-letter words

$$
\begin{array}{lll}
1111212 & 2111212 & 2211212 \\
2221212 & 1121212 & 2121212.
\end{array}
$$

These eighteen words constitute a maximal comma-free (2,7)-dictionary.

Example 3.6.2.
Construct a maximal comma-free (6,3)-dictionary.

Solution:
Initialization: We have $S_0 = \{1, 2, 3, 4, 5, 6\}$.
First Iteration. Prefix=1.
$S_1 = S_0 - \{1\} \cup \{12, 13, 14, 15, 16, 112, 113, 114, 115, 116, \ldots\}$.
Second Iteration. Prefix=2.
$S_2 = S_1 - \{2\} \cup \{23, 24, 25, 26, 212, 213, 214, 215, 216, \ldots\}$.
Third Iteration. Prefix=3.
$S_3 = S_2 - \{3\}$
$\cup \{34, 35, 36, 312, 313, 314, 315, 316, 323, 324, 325, 326, 334, 335, 336\}$.
Fourth Iteration. Prefix=4.
$S_4 = S_3 - \{4\}$
$\cup \{45, 46, 412, 413, 414, 415, 416, 423, 424, 425, 426, 434, 435, 436, 445, 446\}$.
Fifth Iteration. Prefix=5.
$S_5 = S_4 - \{5\}$
$\cup \{56, 512, 513, 514, 515, 516, 523, 524, 525, 526, 534, 535, 536, 545, 546, 556\}$.
Sixth Iteration. Prefix=6.
$S_6 = S_5 - \{6\}$
$\cup \{612, 613, 614, 615, 616, 623, 624, 625, 626, 634, 635, 636, 645, 646, 656\}$.
There are $5 + 9 + 12 + 14 + 15 + 15 = 70 = \frac{6^3 - 6}{3}$ words of length 3 in S_6.
They constitute a maximal comma-free (6,3)-dictionary.

Example 3.6.3.

Construct a maximal comma-free (5,2)-dictionary.

Solution:

Take $A = \{1, 2\}$, $B = \{3\}$ and $C = \{4, 5\}$. Then

$$D = \{13, 23, 14, 24, 15, 25, 34, 35\}.$$

The size of this dictionary is 8, which is less than the upper bound $\frac{5^2 - 5}{2} = 10$.

Exercises

1. Construct a maximal comma-free (4,3)-dictionary.

2. Construct a maximal comma-free (5,3)-dictionary.

3. Construct a maximal comma-free (3,5)-dictionary.

4. Construct a maximal comma-free (2,9)-dictionary.

5. Construct a maximal comma-free (6,2)-dictionary.

6. Construct a maximal comma-free (7,2)-dictionary.

Practice Questions

Permutations

1. How many different four-digit numbers divisible by 4 can be written using the digits 1, 2, 3 and 4 if each digit can be used only once?

2. How many different four-digit numbers divisible by 4 can be written using the digits 1, 2, 3 and 4 if each digit can be used any number of times?

3. Find the number of integers from 0 through 999999 which do not contain two equal digits adjacent to each other?

4. Find the number of integers from 0 through 999999 which do not contain two equal digits separated by exactly one digit in between?

5. In how many ways can eight identical rooks be placed on a chessboard so that no two are in the same row or the same column?

6. For a meeting, 64 chairs are arranged in an 8×8 formation. Eight delegates are the first group to arrive. In how many ways can they be seated so that no two are in the same row or the same column?

Exponential Generating Functions

7. What is the exponential generating function for the sequence $\{2^{2n+1}\}$?

8. What is the exponential generating function for the sequence $\{n(n-1)\}$?

9. What is the exponential generating function for the sequence $\{n!\}$?

10. What sequence is generated exponentially by the function e^{-3x}?

11. What sequence is generated exponentially by the function xe^{2x}?

12. What sequence is generated exponentially by the function $\frac{1}{(1-x)^2}$?

Ternary Sequences

Find the number of ternary sequences with length n with the following properties:

13. It contains an even number of 0s and an even number of 1s.

14. It contains an even number of 0s and an odd number of 1s.

15. It contains an odd number of 0s and an odd number of 1s.

16. It contains at least one 0 and at least one 1.

17. It contains at least one 0 and at most one 1.

18. It contains at most one 0 and at most one 1.

Cycles

19. Find the inverse of the permutation $\{3,1,5,8,7,6,4,2\}$.

20. Consider the permutations $\{3,1,5,8,7,6,4,2\}$ and $\{6,2,1,3,8,5,4,7\}$. Find their product.

21. Decompose the permutations $\{3,1,5,8,7,6,4,2\}$ into disjoint cycles.

22. Decompose the permutations $\{6,2,1,3,8,5,4,7\}$ into transpositions.

23. What permutation decomposes into $(17)(13)(26)(14)(18)$?

24. What permutation decomposes into $(13)(18)(26)(84)(37)$?

Answers

(1) 6 (2) 64 (3) 597871 (4) 664300 (5) 40320 (6) 1625702400
(7) $2e^{4x}$ (8) $x^2 e^x$ (9) $\frac{1}{1-x}$ (10) $\{(-1)^n 2^n\}$ (11) $\{n2^{n-1}\}$ (12) $\{(n+1)!\}$
(13) $\frac{1}{4}(3^n + 2 + 2(-1)^n)$ (14) $\frac{1}{4}(3^n + (-1)^n)$ (15) $\frac{1}{4}(3^n - 2 + 2(-1)^n)$
(16) $3^n - 2^{n+1} + 1$ (17) $n2^{n-1} + 2^n - n - 1$ (18) $n^2 + n + 1$
(19) $\{2,8,1,7,3,6,5,4\}$ (20) $\{1,6,8,7,4,5,3,2\}$ (21) $(1357482)(6)$
(22) $(16)(15)(18)(17)(14)(13)$ (23) $\{7,6,4,8,5,2,3,1\}$ (24) $\{7,6,4,8,5,2,3,1\}$

Chapter Four

Special Numbers

Section 4.1. Fibonacci Numbers

The famous Fibonacci numbers $\{F_n\}$ are defined by the recurrence relation $F_n = F_{n-1} + F_{n-2}$ with the initial conditions $F_0 = 0$ and $F_1 = 1$. The first few numbers are 0, 1, 1, 2, 3, 5, 8, 13, 21, 34, 55, 89, 144 and 233.

Fibonacci, son of Bonaccio, was the Italian mathematician Leonardo of Pisa who lived in the twelfth and thirteenth centuries. His monumental work *Liber Abaci*, "Book of Calculation", summarized the arithmetic and algebra that were known at the time.

On one of the pages of this huge volume is the following problem: We start off with one pair of baby rabbits. In each month, a pair of adult rabbits will give birth to a pair of baby rabbits, and in each month, a pair of baby rabbits will grow into a pair of adult rabbits. How many pairs of rabbits will there be in one year, assuming that none of them has vanished?

Let $\{a_n\}$ be the number of pairs of adult rabbits in month n, and b_n be the number of pairs of baby rabbits in month n. We have $a_0 = 0$ and $b_0 = 1$. Since each pair of adult rabbits gives birth to a pair of baby rabbits in each month, $b_n = a_{n-1}$. Since each pair of baby rabbits grows into a pair of adult rabbits, $a_n = a_{n-1} + b_{n-1} = a_{n-1} + a_{n-2}$. It follows that $\{a_n\} = \{F_n\}$, $\{b_n\} = \{F_{n-1}\}$ and $\{a_n + b_n\} = F_{n+1}$. In particular, $a_{12} + b_{12} = F_{13} = 233$.

That was the only mention of rabbits in *Liber Abaci*. If Fibonacci ever finds out that he has come down in history as Dr. Rabbit, he will surely turn in his grave (meaning turn his grave in) and come back to settle the score!

Since the Fibonacci numbers satisfy a homogeneous linear recurrence relation with constant coefficients, we have standard methods for finding an explicit formula for its terms.

We first apply the method of characteristic equations. The equation here is $x^2 - x - 1 = 0$ and the characteristic roots are $x = \frac{1 \pm \sqrt{5}}{2}$. Hence

$$F_n = C_1 \left(\frac{1 + \sqrt{5}}{2} \right)^n + C_2 \left(\frac{1 - \sqrt{5}}{2} \right)^n.$$

We have $0 = F_0 = C_1 + C_2$ and $1 = F_1 = C_1(\frac{1+\sqrt{5}}{2}) + C_2(\frac{1-\sqrt{5}}{2})$. The latter simplifies to $C_1 - C_2 = \frac{2}{\sqrt{5}}$. Hence $C_1 = \frac{1}{\sqrt{5}}$ and $C_2 = -\frac{1}{\sqrt{5}}$, so that

$$F_n = \frac{1}{\sqrt{5}} \left(\left(\frac{1 + \sqrt{5}}{2} \right)^n - \left(\frac{1 - \sqrt{5}}{2} \right)^n \right).$$

© The Author(s), under exclusive license to Springer Nature Switzerland AG 2021
S. W. Golomb, A. Liu, *Solomon Golomb's Course on Undergraduate Combinatorics*,
https://doi.org/10.1007/978-3-030-72228-9_5

We now apply the method of generating functions. Let

$$F(x) = F_0 + F_1 x + F_2 x^2 + \cdots.$$

We have

$$\sum_{n=2}^{\infty} F_n x^n = x \sum_{n=2}^{\infty} F_{n-1} x^{n-1} + x^2 \sum_{n=2}^{\infty} F_{n-2} x^{n-2}.$$

Hence $F(x) - x = x F(x) + x^2 F(x)$ so that

$$x = (1 - x - x^2) F(x) = \left(1 - \frac{1 + \sqrt{5}}{2} x\right) \left(1 - \frac{1 - \sqrt{5}}{2} x\right) F(x).$$

It follows that

$$F(x) = \frac{A}{1 - \frac{1+\sqrt{5}}{2} x} + \frac{B}{1 - \frac{1-\sqrt{5}}{2} x}.$$

Clearing fractions, we have $x = A(1 - \frac{1-\sqrt{5}}{2} x) + B(1 - \frac{1+\sqrt{5}}{2} x)$. Setting $x = -\frac{1-\sqrt{5}}{2}$, we have $-\frac{1-\sqrt{5}}{2} = A(1 + (\frac{1-\sqrt{5}}{2})^2)$ so that $A = \frac{1}{\sqrt{5}}$. Setting $x = -\frac{1+\sqrt{5}}{2}$, we have $-\frac{1+\sqrt{5}}{2} = B(1 + (\frac{1+\sqrt{5}}{2})^2)$ so that $B = -\frac{1}{\sqrt{5}}$. It follows that

$$F_n = \frac{1}{\sqrt{5}} \left(\left(\frac{1 + \sqrt{5}}{2}\right)^n - \left(\frac{1 - \sqrt{5}}{2}\right)^n \right).$$

The term $(\frac{1-\sqrt{5}}{2})^n$ has alternating signs and an absolute value less than 1. For sufficiently large value of n, F_n is the integer closest to $(\frac{1+\sqrt{5}}{2})^n$. The limiting value of the ratio $\dfrac{F_{n+1}}{F_n}$ is equal to $\frac{1+\sqrt{5}}{2}$, which is known as the **Golden Ratio** and denoted by τ (unrelated to $\tau(n)$). It follows that an alternative representation of the Fibonacci numbers is

$$F_n = \frac{\tau^n - (1 - \tau)^n}{\sqrt{5}}.$$

We may also write $F(x) = \frac{x}{1 - (x + x^2)} = x + x(x + x^2) + x(x + x^2)^2 + \cdots$. The coefficient of x^n is $\binom{n-1}{0}$ in $x(x + x^2)^{n-1}$, $\binom{n-2}{1}$ in $x(x + x^2)^{n-2}$, $\binom{n-3}{2}$ in $x(x + x^2)^{n-3}$, and so on. Hence for $n \geq 1$,

$$F_n = \binom{n-1}{0} + \binom{n-2}{1} + \binom{n-3}{2} + \cdots + \binom{0}{n-1}.$$

Of course, the last non-zero term is $\binom{\frac{n}{2}}{\frac{n-2}{2}}$ if n is even, and $\binom{\frac{n-1}{2}}{\frac{n-1}{2}}$ if n is odd.

We can verify by mathematical induction that this formula does give the Fibonacci numbers. We have $\binom{0}{0} = 1 = F_1$ and $\binom{1}{0} + \binom{0}{1} = 1 = F_2$. Suppose the result holds up to $n - 1$. For $n \geq 3$,

$$\sum_{k=0}^{n-1} \binom{n-k-1}{k} = \binom{n-1}{0} + \sum_{k=1}^{n-2} \binom{n-k-1}{k}$$

$$= \binom{n-2}{0} + \sum_{k=1}^{n-2} \left(\binom{n-k-2}{k} + \binom{n-k-2}{k-1} \right)$$

$$= \sum_{k=0}^{n-2} \binom{n-k-2}{k} + \sum_{k=0}^{n-3} \binom{n-k-3}{k}$$

$$= F_{n-1} + F_{n-2}$$

$$= F_n.$$

This shows that the Fibonacci numbers are hiding inside Pascal's Triangle. To make them stand out more clearly, we draw parallel lines through Pascal's Triangle, as shown in the diagram below. Then the Fibonacci numbers are sums of the numbers on these lines.

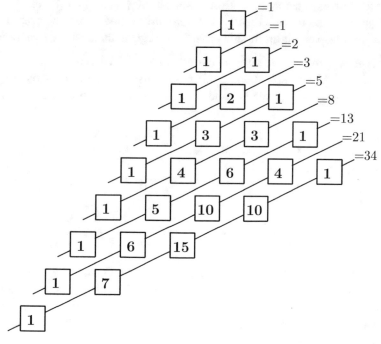

We now consider a problem which appears to have no bearing on the relation between the Fibonacci numbers and Pascal's Triangle. From the set $\{1, 2, \ldots, n\}$, we wish to choose a subset of size k not containing consecutive integers. We claim that the number of such subsets is $\binom{n-k+1}{k}$.

We illustrate with the case $n = 7$ and $k = 3$. Since the numbers involved are relatively small, we can write out all such subsets, and there are indeed $\binom{7-3+1}{3} = 10$ of them, shown in separate columns in the chart below.

1	1	1	1	1	1	2	2	2	3
3	3	3	4	4	5	4	4	5	5
5	6	7	6	7	7	6	7	7	7

Suppose we subtract 1 from each number in the second row, and subtract 2 from each number in the third row. Since each column does not contain consecutive integers initially, it now contains distinct integers. The largest integer in the new chart is $7 - 3 + 1 = 5$.

1	1	1	1	1	1	2	2	2	3
2	2	2	3	3	4	3	3	4	4
3	4	5	4	5	5	4	5	5	5

Each column in the new chart is a subset of $\{1,2,3,4,5\}$ of size 3. Moreover, all such subsets appear in this chart since we can apply the reverse operation of expansion to obtain a subset of $\{1,2,3,4,5,6,7\}$ of size 3 not containing consecutive integers. Thus we have a one-to-one correspondence, which justifies the claim in this case. The argument is easily generalized.

We now present an alternative argument. We represent each subset of $\{1,2,3,4,5,6,7\}$ of size 3 containing no consecutive integers by a vertical column consisting of three Os and four Xs, with no two Os adjacent.

O	O	O	O	O	O	X	X	X	X
X	X	X	X	X	X	O	O	O	X
O	O	O	X	X	X	X	X	X	O
X	X	X	O	O	X	O	O	X	X
O	X	X	X	X	O	X	X	O	O
X	O	X	O	X	X	O	X	X	X
X	X	O	X	O	O	X	O	O	O

One of the four Xs must go between the first O and the second O, and another of them between the second O and the third O. Let us suppress them. Now we have an arbitrary binary sequence consisting of three Os and two Xs. Hence their number is $\binom{5}{3}$. Again, the argument can be generalized easily.

Suppose we fix n and take all possible values of k. Continuing to take $n = 7$, we have $\binom{7-0+1}{0} = 1$ subset for $k = 0$, namely, the empty set. For $k = 1$, we have $\binom{7-1+1}{1} = 7$ subsets, namely, all the singletons. For $k = 2$, we have $\binom{7-2+1}{2} = 15$ subsets. For $k = 3$, we have the 10 subsets found above. For $k = 4$, we have $\binom{7-4+1}{4} = 1$ subset, namely, $\{1,3,5,7\}$. Since there are no more subsets for $k \geq 5$, the total is $F_8 = 1 + 7 + 15 + 10 + 1 = 34$. The five summands form a straight line in Pascal's Triangle, as we have pointed out earlier.

We now show directly that for $n \geq 2$, F_n is the number of subsets of $\{1, 2, \ldots, n-2\}$ which do not contain two consecutive numbers. We call such a subset a *good* subset. For $n = 2$, the overall set is $\emptyset$. Hence $F_2 = 1$ holds since $\emptyset$ is good. For $n = 3$, the overall set is $\{1\}$. Hence $F_3 = 2$ also holds since both $\emptyset$ and $\{1\}$ are good. Suppose the result holds up to $n - 1$. For $n \geq 4$, a good subset either contains or does not contain $n - 2$. If it does not, then it is one of the good subsets of $\{1, 2, \ldots, n-3\}$. If it does, then it is the union of $\{n-2\}$ and a good subset of $\{1, 2, \ldots, n-4\}$. It follows that $F_n = F_{n-1} + F_{n-2}$ indeed.

Examples

Example 4.1.1.

The sequences $\{a_n\}$ and $\{b_n\}$ satisfy the recurrence relations

$$a_n = a_{n-1} + b_{n-1} \text{ and } b_n = a_n + b_{n-1},$$

with the initial conditions $a_0 = 0$ and $b_0 = 1$. Use the method of generating functions to determine these two sequences.

Solution:

Let $A(x)$ and $B(x)$ be the respective generating functions. We have

$$\sum_{n=1}^{\infty} a_n x^n = x \sum_{n=1}^{\infty} a_{n-1} x^{n-1} + x \sum_{n=1}^{\infty} b_{n-1} x^{n-1}$$

and

$$\sum_{n=1}^{\infty} b_n x^n = \sum_{n=1}^{\infty} a_n x^n + x \sum_{n=1}^{\infty} b_{n-1} x^{n-1}.$$

It follows that $A(x) = xA(x) + xB(x)$ and $B(x) - 1 = A(x) + xB(x)$. Eliminating $B(x)$, we have $\frac{(1-x)A(x)}{x} = B(x) = \frac{A(x)+1}{1-x}$ so that

$$A(x) = \frac{x}{1 - 3x + x^2} = \frac{C_1}{1 - \frac{3+\sqrt{5}}{2}x} + \frac{C_2}{1 - \frac{3-\sqrt{5}}{2}x}.$$

Clearing fractions, we have $x = C_1(1 - \frac{3-\sqrt{5}}{2}x) + C_2(1 - \frac{3+\sqrt{5}}{2}x)$. Comparing the constant terms, we have $0 = C_1 + C_2$. Comparing the linear terms, we have $1 = C_1(-\frac{3-\sqrt{5}}{2}) + C_2(-\frac{3+\sqrt{5}}{2})$. This simplifies to $C_1 - C_2 = \frac{2}{\sqrt{5}}$. Hence $C_1 = \frac{1}{\sqrt{5}}$ and $C_2 = -\frac{1}{\sqrt{5}}$. Using $(\frac{1\pm\sqrt{5}}{2})^2 = \frac{3\pm\sqrt{5}}{2}$, we have

$$
\begin{aligned}
a_n &= \frac{1}{\sqrt{5}}\left(\frac{3+\sqrt{5}}{2}\right)^n - \frac{1}{\sqrt{5}}\left(\frac{3-\sqrt{5}}{2}\right)^n \\
&= \frac{1}{\sqrt{5}}\left(\left(\frac{1+\sqrt{5}}{2}\right)^{2n} - \left(\frac{1-\sqrt{5}}{2}\right)^{2n}\right) = F_{2n}.
\end{aligned}
$$

On the other hand,

$$B(x) = \frac{(1-x)A(x)}{x} = \frac{1-x}{1 - 3x - x^2} = \frac{C_3}{1 - \frac{3+\sqrt{5}}{2}x} + \frac{C_4}{1 - \frac{3-\sqrt{5}}{2}x}.$$

Clearing fractions, we have $1 - x = C_3(1 - \frac{3-\sqrt{5}}{2}x) + C_4(1 - \frac{3+\sqrt{5}}{2}x)$. Comparing the constant terms, we have $1 = C_3 + C_4$. Comparing the linear terms, we have $-1 = C_3(-\frac{3-\sqrt{5}}{2}) + C_4(-\frac{3+\sqrt{5}}{2})$. This simplifies to $C_3 - C_4 = \frac{1}{\sqrt{5}}$.

Hence $C_3 = \frac{1+\sqrt{5}}{2\sqrt{5}}$ and $C_4 = -\frac{1-\sqrt{5}}{2\sqrt{5}}$. Using the fact that $(\frac{1\pm\sqrt{5}}{2})^2 = \frac{3\pm\sqrt{5}}{2}$, we have

$$
\begin{aligned}
b_n &= \frac{1+\sqrt{5}}{2\sqrt{5}}\left(\frac{3+\sqrt{5}}{2}\right)^n - \frac{1-\sqrt{5}}{2\sqrt{5}}\left(\frac{3-\sqrt{5}}{2}\right)^n \\
&= \frac{1}{\sqrt{5}}\left(\left(\frac{1+\sqrt{5}}{2}\right)^{2n+1} - \left(\frac{1-\sqrt{5}}{2}\right)^{2n+1}\right) = F_{2n+1}.
\end{aligned}
$$

Remark:
The two subsequences of the Fibonacci numbers, consisting respectively of the terms in even and odd positions, have some remarkable properties. Denote $\{1, 2, 5, 13, 34, \ldots\}$ by A and $\{1, 3, 8, 21, 55, \ldots\}$ by B. Then the total number of ways a number can be expressed as a difference of two terms in A or two terms in B is at most 1. The sequences A and B have only one common element, namely 1. For any positive integers a and b, the sequences $A + a = \{1 + a, 2 + a, 5 + a, 13 + a, 34 + a, \ldots\}$ and $B + b = \{1 + b, 3 + b, 8 + b, 21 + b, 55 + b \ldots\}$ have at most one common element.

Example 4.1.2.
Along the Grand Trunk Road in the northwestern part of the Indian subcontinent were the cities Lahore, Umballa, Delhi, Alighur and Benares in that order. Kim, the title character of a novel by Rudyard Kipling, started from Lahore and headed for Benares. It would take him one day to go from any city to the next one. At any intermediate city, except at Umballa during the first night, he would decide either to continue in the same direction or to turn back the next day. His journey would come to an end if he reached Benares or returned to Lahore, but continue otherwise. If he changed directions n times altogether, how many different forms could his path take?

Solution:
Let k_n denote the number of possible forms for Kim's path if he changed directions n times. We claim that $k_n = k_{n-1} + k_{n-2}$. Consider first the case where n is even. This means that Kim would go onto Benares. Hence the last change in directions must occur in Delhi or Umballa. If it occurred in Delhi, this means that he must have returned from Alighur, and had gone there earlier from Delhi. If we shorten the last segment "Delhi — Alighur — Delhi" to just "Delhi", we will get a path counted in k_{n-2}. On the other hand, if the last change in directions occurred in Umballa, this means that he could have gone back to Lahore had he not done so. If we replace the segment "Delhi — Alighur — Benares" at the very end by "Lahore", we will get a path counted in k_{n-1}. The case where n is odd can be handled in an analogous manner. Just interchange "Lahore" with "Benares" and "Umballa" with "Alighur". This justifies the claim. From the initial conditions, $k_n = F_{n+2}$.

Example 4.1.3.

An OE-sequence is an increasing sequence of positive integers such that all terms in odd positions are odd and all terms in even positions are even. For instance, there are 8 OE-sequences in which the largest term is at most 4, namely, $\emptyset$, $\{1\}$, $\{3\}$, $\{1,2\}$, $\{1,4\}$, $\{3,4\}$, $\{1,2,3\}$ and $\{1,2,3,4\}$. How many OE-sequences are there in which the largest term is equal to n?

Solution:

Let a_n be the number of OE-sequences in which the largest term is equal to n. Note that we have $a_1 = a_2 = 1$ for the sequences $\{1\}$ and $\{1,2\}$. We claim that for all $n \geq 3$, $a_n = a_{n-1} + a_{n-2}$. Consider any OE-sequence counted in a_n. If the second largest term is $n - 1$, we can drop the n and obtain an OE-sequence counted in a_{n-1}. If the second largest term is not $n - 1$, it cannot be $n - 2$ either. We can change the n into $n - 2$ and obtain an OE-sequence counted in a_{n-2}. Since the process is reversible, the claim is justified. From the initial conditions, $a_n = F_n$.

Exercises

1. Let $a_n = \sum_{k=0}^{n} \binom{n+k}{2k}$.

   (a) Prove that $a_n - 3a_{n-1} + a_{n-2} = 0$.

   (b) Using (a), prove that $a_n = F_{2n}$.

   (c) Without using (a), prove that $a_n = F_{2n}$.

2. Let $b_n = \sum_{k=0}^{n} \binom{n+k+1}{2k+1}$.

   (a) Prove that $b_n - 3b_{n-1} + b_{n-2} = 0$.

   (b) Using (a), prove that $b_n = F_{2n+1}$.

   (c) Without using (a), prove that $b_n = F_{2n+1}$.

3. In how many ways can a $1 \times n$ chessboard be covered by a combination of 1×1 squares and 1×2 rectangles?

4. In how many ways can a $2 \times n$ chessboard be covered by a combination of 1×2 and 2×1 rectangles?

5. A bug begins at vertex B of a regular pentagon $ABCDE$. In each minute, it moves from its current position to an adjacent vertex. Once it reaches vertex E, it ends there. Determine the number of different paths the bug can take in order to reach E after exactly n minutes.

6. A bug starts at the bottom left hexagon in the diagram below. It can move from one hexagon to an adjacent hexagon provided that the center of the second hexagon is to the right of the first one. What is the number of different paths for the bug to reach the top right hexagon, if there are n hexagons in each of the two rows?

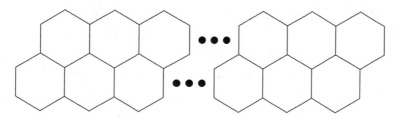

Section 4.2. Fibonacci Identities

The Fibonacci numbers have so many fascinating properties that an entire mathematical journal, *The Fibonacci Quarterly*, is devoted to this and related topics. Here, we highlight some of the simpler Fibonacci identities. We begin with

$$F_m F_n + F_{m-1} F_{n-1} = F_{m+n-1}. \tag{6}$$

We prove this by induction on n. For $n = 1$, $F_m = F_m$ is true for all $m \geq 1$. Suppose $F_m F_n + F_{m-1} F_{n-1} = F_{m+n-1}$ for some $n \geq 1$ and all $m \geq 1$. Then

$$
\begin{aligned}
F_{m+n} &= F_{(m+1)+n-1} \\
&= F_{m+1} F_n + F_{(m+1)-1} F_{n-1} \\
&= (F_m + F_{m-1}) F_n + F_m (F_{n+1} - F_n) \\
&= F_m F_n + F_{m-1} F_n + F_m F_{n+1} - F_m F_n \\
&= F_m F_{n+1} + F_{m-1} F_{(n+1)-1}.
\end{aligned}
$$

By setting $m = n$ in (1), we have

$$F_n^2 + F_{n-1}^2 = F_{2n-1}. \tag{7}$$

On the other hand, we have

$$F_n^2 - F_{n-1}^2 = F_{n+1} F_{n-2}. \tag{8}$$

This follows from $F_n^2 - F_{n-1}^2 = (F_n + F_{n-1})(F_n - F_{n-1}) = F_{n+1} F_{n-2}$.

Here are some similar identities. We have

$$F_{n+3}^2 + F_n^2 = 2F_{n+1}^2 + 2F_{n+2}^2. \tag{9}$$

This is because

$$
\begin{aligned}
F_{n+3}^2 + F_n^2 &= (F_{n+2} + F_{n+1})^2 + F_n^2 \\
&= F_{n+2}^2 + 2F_{n+2}F_{n+1} + F_{n+1}^2 + F_n^2 \\
&= (F_{n+1} + F_n)^2 + 2F_{n+2}(F_{n+2} - F_n) + F_{n+1}^2 + F_n^2 \\
&= 2F_{n+1}^2 + 2F_{n+1}F_n + 2F_n^2 + 2F_{n+2}^2 - 2F_{n+2}F_n \\
&= 2F_{n+2}^2 + 2F_{n+1}^2 + 2F_n(F_{n+1} + F_n - F_{n+2}) \\
&= 2F_{n+2}^2 + 2F_{n+1}^2.
\end{aligned}
$$

On the other hand,

$$F_{n+2}^2 - F_n^2 = F_{2n+2}. \tag{10}$$

Using (1) with $m = n + 1$, we have

$$
\begin{aligned}
F_{n+2}^2 - F_n^2 &= (F_{n+1} + F_n)^2 - F_n^2 \\
&= F_{n+1}^2 + 2F_{n+1}F_n \\
&= F_{n+1}(F_{n+1} + F_n) + F_{n+1}F_n \\
&= F_{n+2}F_{n+1} + F_{n+1}F_n \\
&= F_{2n+2}.
\end{aligned}
$$

We consider one more identity, namely,

$$
F_{n+1}F_{n+2} - F_{n-1}F_n = F_{2n+1}. \tag{11}
$$

Using (2), we have

$$
\begin{aligned}
F_{n+1}F_{n+2} - F_{n-1}F_n &= (F_n + F_{n-1})(F_{n+1} + F_n) - F_{n-1}F_n \\
&= F_n^2 + F_{n+1}F_{n-1} + F_nF_{n+1} \\
&= F_n^2 + F_{n+1}(F_{n-1} + F_n) \\
&= F_n^2 + F_{n+1}^2 \\
&= F_{2n+1}.
\end{aligned}
$$

We now turn to some Fibonacci identities involving summations. First, we have

$$
F_0 + F_1 + F_2 + \cdots + F_n = F_{n+2} - 1. \tag{12}
$$

The left side is equal to $F_2 + (F_3 - F_1) + \cdots + (F_{n+1} - F_{n-1})$, which is a telescopic sum equal to $F_n + F_{n+1} - 1 = F_{n+2} - 1$. Next, we have for $n \geq 1$,

$$
F_0 - F_1 + F_2 - \cdots + (-1)^n F_n = (-1)^n F_{n-1} - 1. \tag{13}
$$

This time, the left side is equal to $-(F_2 - F_0) + (F_3 - F_1) - \cdots + (-1)^n F_n$, which is another telescopic sum equal to $(-1)^n F_{n-1} - 1$. Finally, we have

$$
F_0^2 + F_1^2 + F_2^2 + \cdots + F_n^2 = F_n F_{n+1}. \tag{14}
$$

This is because

$$
\begin{aligned}
F_n F_{n+1} &= F_n(F_n + F_{n-1}) \\
&= F_n^2 + F_{n-1}F_n \\
&= F_n^2 + F_{n-1}(F_{n-1} + F_{n-2}) \\
&= \cdots \\
&= F_n^2 + \cdots + F_1^2 + F_1 F_0 \\
&= F_n^2 + \cdots + F_1^2 + F_0^2.
\end{aligned}
$$

We conclude by examining some arithmetical properties of the Fibonacci numbers. When divided by 2, the remainders of the Fibonacci numbers form a cycle $(0,1,1)$. Hence F_n is divisible by 2 if and only if n is divisible by 3. When the divider is 3, the cycle is $(0,1,1,2,0,2,2,1)$, so that F_n is divisible by 3 if and only if n is divisible by 4. When the divider is 5, the cycle is $(0,1,1,2,3,0,3,3,1,4,0,4,4,3,2,0,2,2,4,1)$, so that F_n is divisible by 5 if and only if n is divisible by 5. When the divider is 7, the cycle is $(0,1,1,2,3,5,6,0,6,6,5,4,2,6,1)$, so that F_n is divisible by 7 if and only if n is divisible by 8. Other similar results can be obtained in the same way.

We now prove that if n divides m, then F_n divides F_m. Let $m = kn$. We use induction on k. For $k = 1$, F_n clearly divides F_n. Suppose the result holds for some $k \geq 1$. Then F_n divides F_{kn}. Consider the next case where $m = (k+1)n$. By (1), we have $F_m = F_{kn+(n+1)-1} = F_{kn}F_{n+1} + F_{kn-1}F_n$. Since F_n divides the factor F_{kn} in the first term and the factor F_n in the second term, F_n divides F_m.

The Fibonacci numbers are strongly tied to the concept of greatest common divisor. Consecutive Fibonacci numbers F_n and F_{n-1} are relatively prime to each other. When determining their greatest common divisor by the Euclidean Algorithm, we discover that it is equal to the greatest common divisor of F_{n-1} and F_{n-2}, which is in turn equal to that of F_{n-2} and F_{n-3}, and so on. In the last step, we have 1 as the greatest common divisor of F_1 and F_0.

Let d be the greatest common divisor of m and n. We claim that the greatest common divisor of F_m and F_n is given by F_d. We may assume that $m \geq n$. By the Division Algorithm, $m = qn + r$ for unique integers q and r. By (1), we have $F_m = F_{qn}F_{r+1} + F_{qn-1}F_r$. Now n divides qn. Hence F_n divides F_{qn} so that the greatest common divisor of F_n and F_m is the same as the greatest common divisor of F_n and $F_{qn-1}F_r$. Now F_{qn} and F_{qn-1} are relatively prime. Since F_n is a divisor of the latter, it is relatively prime to the former. It follows that the greatest common divisor of F_n and F_m is equal to the greatest common divisor of F_n and F_r. Thus this computation corresponds step by step to the Euclidean Algorithm by which we determine that d is the greatest common divisor of n and m. This justified the claim.

Examples

Example 4.2.1.

(a) Prove that $F_{2n+1}^2 - F_{2n-1}^2 = F_{4n}$.

(b) Prove that $F_{2n+1}F_{2n+2} - F_{2n-1}F_{2n} = F_{4n+1}$.

(c) Prove that $F_{2n+1}F_{2n+3} - F_{2n-1}F_{2n+1} = F_{4n+2}$.

(d) Prove that $F_{2n+2}F_{2n+3} - F_{2n}F_{2n+1} = F_{4n+3}$.

Solution:

(a) We have $F_{(2n-1)+2}^2 - F_{2n-1}^2 = F_{2(2n-1)+2} = F_{4n}$ by (5).

(b) We have $F_{(2n)+1}F_{(2n)+2} - F_{(2n)-1}F_{(2n)} = F_{2(2n)+1} = F_{4n+1}$ by (6).

(c) By (1), we have

$$
\begin{aligned}
& F_{2n+1}(F_{2n+2} + F_{2n+1}) - (F_{2n+1} - F_{2n})F_{2n+1} \\
={} & F_{(2n+1)}F_{(2n+2)} + F_{(2n+1)-1}F_{(2n+2)-1} \\
={} & F_{(2n+1)+(2n+2)-1} \\
={} & F_{4n+2}.
\end{aligned}
$$

(d) We have $F_{(2n+1)+1}F_{(2n+1)+2} - F_{(2n+1)-1}F_{(2n+1)} = F_{2(2n+1)+1} = F_{4n+3}$ by (6).

Example 4.2.2.

(a) Prove that $F_0 + F_2 + F_4 + \cdots + F_{2n} = F_{2n+1} - 1$.

(b) Prove that $F_1 + F_3 + F_5 + \cdots + F_{2n-1} = F_{2n}$.

Solution:

(a) The sum is equal to

$$(F_3 - F_1) + (F_5 - F_3) + \cdots + (F_{2n+1} - F_{2n-1}) = F_{2n+1} - 1.$$

(b) The sum is equal to

$$(F_2 - F_0) + (F_4 - F_2) + (F_6 - F_4) + \cdots + (F_{2n} - F_{2n-2}) = F_{2n}.$$

Example 4.2.3.

Prove that $F_n + F_{n+1} + \cdots + F_{n+9}$ is divisible by 11 for $n \geq 0$.

Solution:

Let $F_n = x$ and $F_{n+1} = y$. The computation below shows that the featured sum is equal to $55x + 88y$, which is clearly divisible by 11.

k	F_{n+k}	$\sum\limits_{i=0}^{k} F_{n+i}$
0	x	x
1	y	$x + y$
2	$x + y$	$2x + 2y$
3	$x + 2y$	$3x + 4y$
4	$2x + 3y$	$5x + 7y$
5	$3x + 5y$	$8x + 12y$
6	$5x + 8y$	$13x + 20y$
7	$8x + 13y$	$21x + 33y$
8	$13x + 21y$	$34x + 54y$
9	$21x + 34y$	$55x + 88y$

Exercises

1. Prove that $F_n F_{n+1} - F_{n-1} F_{n+2} = (-1)^{n-1}$.

2. Prove that $F_n^2 - F_{n+1} F_{n-1} = (-1)^{n-1}$.

3. (a) Prove that $F_0 + F_3 + F_6 + \cdots + F_{3n} = \frac{1}{2}(F_{3n+2} - 1)$.

   (b) Prove that $F_1 + F_4 + F_7 + \cdots + F_{3n+1} = \frac{1}{2} F_{3n+3}$.

   (c) Prove that $F_2 + F_5 + F_8 + \cdots + F_{3n+2} = \frac{1}{2}(F_{3n+4} - 1)$.

4. (a) Prove that $F_0 + F_4 + F_8 + \cdots + F_{4n} = F_{2n+1}^2 - 1$.

   (b) Prove that $F_1 + F_5 + F_9 + \cdots + F_{4n+1} = F_{2n+1} F_{2n+2}$.

   (c) Prove that $F_2 + F_6 + F_{10} + \cdots + F_{4n+2} = F_{2n+1} F_{2n+3} - 1$.

   (d) Prove that $F_3 + F_7 + F_{11} + \cdots + F_{4n+3} = F_{2n+2} F_{2n+3}$.

5. Prove Hogatt's Theorem which states that every positive integer is a sum of distinct Fibonacci numbers.

6. (a) Prove that for $n \geq 1$, $F_{n+5} > 10 F_n$.

   (b) Prove that for $n \geq 1$ and $t \geq 1$, $F_{n+5t} > 10^t F_n$.

   (c) Prove Lamé's Theorem, which states that the number of divisions necessary to find the greatest common divisor of two positive integers by means of the Euclidean Algorithm is not greater than 5 times the number of digits in the base-10 representation of the smaller of the two numbers.

   (d) Prove that the number 5 in Lamé's Theorem cannot be replaced by the number 4.

Remark:
The point of this result is that in using the Euclidean Algorithm to find the greatest common divisor of two numbers, the greatest number of steps required, relative to the sizes of the numbers, is when the two numbers are consecutive Fibonacci numbers.

Section 4.3. Catalan Numbers

A little girl lives at the origin $(0,0)$ of the coordinate plane, and goes to a school at the point (n, n), where n is some positive integer. She is not supposed to dilly-dally on the way, but to go there in as direct a way as possible. Thus she will be taking n unit steps towards the east, and n unit steps towards the north, in some order. However, she is warned not to go into the north-west part of the town. More specifically, she is not allowed to visit points (x, y) where $y > x$. How many different paths can she follow to go from home to school?

The number of such paths from $(0,0)$ to (n, n) is called the n-th **Catalan** number, after the nineteenth century Belgian mathematician Eugene Catalan. In the following chart, we generate the number of different paths which lead from $(0,0)$ to a specific point according to the above rules. A forbidden point is simply marked X. The Catalan numbers are in boldface.

X	X	X	X	X	X	**132**
X	X	X	X	X	**42**	132
X	X	X	X	14	42	90
X	X	X	**5**	14	28	48
X	X	**2**	5	9	14	20
X	**1**	2	3	4	5	6
1	1	1	1	1	1	1

The recurrence relation satisfied by the Catalan numbers $\{C_n\}$ is not a simple one. We have $C_n = C_0 C_{n-1} + C_1 C_{n-2} + \cdots + C_{n-2} C_1 + C_{n-1} C_0$. This is arrived at as follows. The term $C_{k-1} C_{n-k}$ counts the number of paths which visit the line $y = x$ for the first time at (k, k), not counting $(0,0)$. Such a path is divided into two independent parts, that before arriving at (k, k) and that after arriving at (k, k).

The number of paths in the second part is simply C_{n-k}. The first part is a path from $(1,0)$ to $(k, k-1)$, without visiting any point (x, y) where $y \geq x$. Defining $z = x - 1$, our path now starts at $(0,0)$ and goes to $(k-1, k-1)$, without visiting any points (z, y) where $y > z$. Hence the number of such paths is C_{k-1}.

To solve this convoluted recurrence relation, we use the method of generating functions. Let $C(x) = C_0 + C_1 x + C_2 x^2 + C_3 x^3 + \cdots$. We have

$$
\begin{aligned}
C^2(x) &= (C_0 + C_1 x + C_2 x^2 + C_3 x^3 + \cdots)(C_0 + C_1 x + C_2 x^2 + C_3 x^3 + \cdots) \\
&= C_0 C_0 + (C_0 C_1 + C_1 C_0)x + (C_0 C_2 + C_1 C_1 + C_2 C_0)x^2 \\
&\quad + (C_0 C_3 + C_1 C_2 + C_2 C_1 + C_3 C_0)x^3 + \cdots \\
&= C_1 + C_2 x + C_3 x^2 + C_4 x^3 + \cdots \\
&= \frac{C(x) - 1}{x}.
\end{aligned}
$$

It follows that $xC^2(x) - C(x) + 1 = 0$. By the Quadratic Formula, $C(x) = \frac{1 \pm \sqrt{1-4x}}{2x}$. This may be rewritten as $2xC(x) = 1 \pm \sqrt{1-4x}$. Setting $x = 0$, we have $0 = 1 \pm 1$, so that we must take the minus sign.

Now

$$
\begin{aligned}
2xC(x) &= 1 - (1 - 4x)^{\frac{1}{2}} \\
&= 1 - \sum_{n=0}^{\infty} \binom{\frac{1}{2}}{n} (-4x)^n \\
&= \sum_{n=1}^{\infty} \frac{2}{n} \binom{2n-2}{n-1} x^n.
\end{aligned}
$$

It follows that $C_n = \frac{1}{n+1}\binom{2n}{n}$.

Like the Fibonacci numbers, the Catalan numbers are the answers to a surprisingly large number of problems. Here is one of them. How many binary sequences consisting of n 0s and n 1s are there if the running total of 0s never falls behind the running total of 1s? There is an easy one-to-one correspondence between a "good" sequence and a "good" path from the problem above. If we denote a move to the east by 0 and a move to the north by 1, we have a binary sequence consisting of n 0s and n 1s. That the running total of 0s never falls behind the running total of 1s means not straying into the north-west part of the town. It follows that the number of "good" sequences is equal to the number of "good" paths, namely, C_n.

Suppose a cinema charges 5 dollars for a movie, and the till is initially empty. In the line-up are n customers each with a 5 dollar bill, and another n customers each with a 10 dollar bill, where n is a positive integer. In how many line-ups will no customer have to wait to get change? (Customers with the same amount of money are considered indistinguishable.) It is easy to see that we have the same problem if customers with 5 dollar bills are identified with 0s and customers with 10 dollar bills are identified with 1s.

The relatively simple form of the Catalan numbers leads us to believe that there may be a simpler way to arrive at the answer. Let us return to our initial problem. Suppose the little girl is expected to prepare a plan of how she is going to the school, that is, write down a sequence of n 0s for the moves to the East, and n 1s for the moves to the North. These plans are not pre-screened, and as a result, she may find herself in the north-west part of the town. She is instructed to follow her plan, as soon as she is over the line $y = x$, but treat the 0s as moves to the North and the 1s as moves to the East. She will invariably end up in the police station where help is available. Where is the police station?

Suppose the little girl finds herself in the north-west part of the town for the first time after having made $2k+1$ moves. (The number of moves to get to the line $y = x$ is of course even, so that the number of moves to cross this line for the first time must be odd.) Then she has made k moves to the East and $k+1$ moves to the North. According to her plan, she would have $n-k$ more moves to the East and $n-k-1$ more moves to the North, but now these are interchanged. Hence she will end up at the point $(n-1, n+1)$.

There is a one-to-one correspondence between "bad" paths from $(0,0)$ to (n,n) and arbitrary paths from $(0,0)$ to $(n-1, n+1)$. The number of the latter is given by $\binom{2n}{n-1}$. Hence the number of "good" paths from $(0,0)$ to (n,n) is given by $C_n = \binom{2n}{n} - \binom{2n}{n-1} = \frac{1}{n+1}\binom{2n}{n}$ as before.

We now turn to a problem in combinatorial geometry. Let a_n be the number of ways of triangulating a convex $(n+1)$-gon by diagonals which intersect only at the vertices. We consider a line segment as a convex 2-gon and take $a_1 = 1$. It is easy to see that $a_2 = 1$ and $a_3 = 2$. The following diagram illustrates $a_4 = 5$.

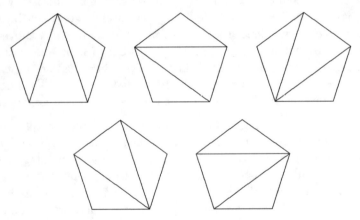

In a convex $(n+1)$-gon $A_1 A_2 \ldots A_{n+1}$, the side $A_1 A_{n+1}$ must be a side of one of the triangles, say $A_1 A_k A_{n+1}$, $2 \le k \le n$. This triangle partitions the remaining part of the polygon into a convex k-gon and a convex $(n-k+2)$-gon. It follows that $a_n = a_1 a_{n-1} + a_2 a_{n-2} + \cdots + a_{n-2}a_2 + a_{n-1}a_1$. Hence the sequence $\{a_n\}$ satisfies the same recurrence relation as the Catalan numbers. From the initial conditions, we have $a_n = C_{n-1}$.

The next problem does not at first seem to have anything to do with the Catalan numbers. We have a product of n variables $x_1 x_2 \cdots x_n$. Since multiplication is a binary operation, it must be performed $n-1$ times to evaluate this product. The sequence of operations may be indicated by putting in brackets the two factors in each of the $n-1$ operations. For instance, when $n=4$, we have five different expressions, namely, $(x_1 x_2)(x_3 x_4)$, $x_1((x_2 x_3)x_4)$, $((x_1 x_2)x_3)x_4$, $x_1(x_2(x_3 x_4))$ and $(x_1(x_2 x_3))x_4$.

While multiplication is associative, exponentiation is most certainly not. For instance, the expression $2^{2^{3^2}}$ may be interpreted in five different ways, yielding five different values. These are $((2^2)^3)^2 = 2^{12}$, $(2^{(2^3)})^2 = 2^{16}$, $(2^2)^{(3^2)} = 2^{18}$, $2^{(2^3)^2} = 2^{64}$ and $2^{(2^{(3^2)})} = 2^{512}$.

As a digression, note that the English language is non-associative. An ambiguous statement such as the "biggest book sale" has two non-equivalent interpretations in that the "biggest (book sale)" is not the same as the "(biggest book) sale". Try to insert brackets in the ambiguous statement a "pretty little girls' school" in five different ways, yielding the following interpretations:

(1) a very small school for girls;
(2) a beautiful school for small girls;
(3) a school for beautiful small girls;
(4) a beautiful small school for girls;
(5) a school for very small girls.

Let a_n be the number of such expressions. Let b_n be the number of expressions in which the variables $x_1, x_2, \ldots, x_n$ can appear in any order. Clearly, $b_n = n! a_n$. Consider any expression in b_{n-1}. Now x_n may be combined with either factor in any of the $n-2$ multiplications, and it can appear before or after that factor. Moreover, x_n may appear at the very front or the very back of the expression. Hence

$$b_n = (4(n-2) + 2)b_{n-1} = (4n-6)b_{n-1}.$$

It follows that

$$a_n = \frac{b_n}{n!} = \frac{(4n-6)b_{n-1}}{n!} = \frac{4n-6}{n} \cdot \frac{b_{n-1}}{(n-1)!} = \frac{4n-6}{n}a_{n-1}.$$

Iteration yields

$$\begin{aligned}
a_n &= \frac{4n-6}{n} \cdot \frac{4n-10}{n-1}a_{n-2} \\
&= \cdots \\
&= \frac{4n-6}{n} \cdot \frac{4n-10}{n-1} \cdots \frac{2}{2}a_1 \\
&= \frac{2^{n-1}(2n-3)(2n-5)\cdots 3 \cdot 1}{n!} \\
&= \frac{2^{n-1}(2n-2)(2n-3)\cdots 2 \cdot 1}{2^{n-1}(n-1)! n!} \\
&= \frac{1}{n}\binom{2n-2}{n-1}.
\end{aligned}$$

Once again, we have $a_n = C_{n-1}$.

We now establish a one-to-one correspondence between the number of triangulations of polygons and the number of ways of bracketing products. Consider a particular triangulation. Label the edges $A_1A_2, A_2A_3,$ $\dots, A_nA_{n+1}$ with the variables $x_1, x_2, \dots, x_n$. For any triangle in which two of the edges are labeled, label the third edge with the product of those two expressions, retaining any parentheses. The label on the edge A_1A_{n+1} will be a parenthesized expression of the sum of the n variables. Different triangulations lead to different parenthesized expressions. As an illustration, the triangulation in the diagram below yields the parenthesized expression $(x_1x_2)(x_3x_4)$.

x_2 x_3

x_1x_2 x_3x_4

x_1 x_4

$(x_1x_2)(x_3x_4)$

Conversely, given a parenthesized expression of the product of the n variables, label the edges $A_1A_2, A_2A_3, \dots, A_nA_{n+1}$ with the variables $x_1, x_2,$ $\dots, x_n$. For any two labeled edges with a common endpoint, complete the triangle if the product of these two partial expressions is computed in the parenthesized expression. Label this new edge with the product. Continuing this process yields a triangulation of the $(n+1)$-gon, and different expressions lead to different triangulation. As an illustration, the parenthesized expression $(x_1(x_2x_3))x_4$ yields the following triangulation.

x_2x_3 $x_1(x_2x_3)$

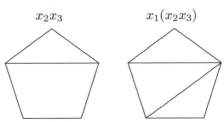

Consider a convex n-gon $A_1A_2 \dots A_n$. Choose any of its n vertices, say A_1. Any of A_1A_k, $3 \le k \le n-1$, can appear as one of the $n-3$ diagonals in one of the a_{n-1} triangulations. This diagonal partitions the remaining part of the polygon into a convex k-gon and a convex $(n-k+2)$-gon. It follows that

$$2(n-3)a_{n-1} = n \sum_{k=3}^{n-1} a_{k-1}a_{n-k+1}.$$

Note that the factor 2 accounts for each diagonal being considered from both endpoints, and the factor n accounts for the fact that any of the n vertices may be under consideration. Recall that

$$a_n = a_1 a_{n-1} + a_2 a_{n-2} + \cdots + a_{n-2} a_2 + a_{n-1} a_1.$$

Hence

$$n(a_n - 2a_{n-1}) = n \sum_{k=3}^{n-1} a_{k-1} a_{n-k+1}.$$

It follows that $2(n-3)a_{n-1} = n(a_n - 2a_{n-1})$. Again we have $a_n = \frac{4n-6}{n} a_{n-1}$.

Finally, we establish a one-to-one correspondence between the bracketed expressions and the binary sequences with running totals of 0s not falling behind running totals of 1s. We put an extra set of brackets around the entire expression. Each left bracket is converted to a 0 and each variable except the last one is converted into a 1. Thus $(((x_1 x_2)x_3)x_4)$ becomes 000111 and 010011 becomes $(x_1((x_2 x_3)x_4))$.

Examples

Example 4.3.1.
In a tight two-way election, each candidate ended up getting exactly 50 votes. The results are announced vote by vote, and running totals are kept. It was expected that the lead would change hands frequently. What was the probability that it never did?

Solution:
The total number of possible sequences announced was $\binom{100}{50}$. The number of sequences in which one candidate's running total never fell behind that of the other was $C_{50} = \frac{1}{51}\binom{100}{50}$. Hence the desired probability was $\frac{2}{51}$ because either candidate may be the one who never fell behind.

Example 4.3.2.
A corporation has only presidents. Each is subordinate to exactly one other president, except for the President at the top. Each has either zero or exactly two subordinates, and the two subordinates are called the associate and the assistant. A president with no subordinates is called a lowly president. The diagram below shows one of the 5 different corporate structures which have exactly 4 lowly presidents. How many different corporate structures can have exactly n lowly presidents?

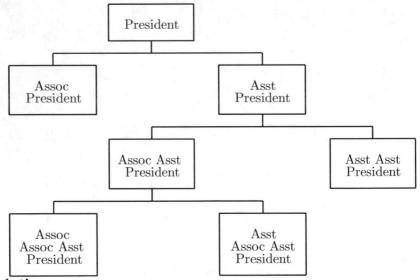

Solution:
Let a_n be the number of such corporate structures. We have $a_1 = a_2 = 1$. For $n \geq 2$, the President must have two subordinates. Each of these two presidents may be considered as the President of a subsidiary corporation. If the first corporation has k lowly presidents, then the second one must have $n - k$ lowly presidents, where $1 \leq k \leq n - 1$. It follows that we have $a_n = a_1 a_{n-1} + a_2 a_{n-2} + \cdots + a_{n-1} a_1$. From the initial conditions, $a_n = C_{n-1}$.

Example 4.3.3.

In the Royal Mint are display cases of coins. Exactly n coins in each display case are on the bottom row, each touching two others except for the two at the ends. The remaining coins in the display case are in the upper rows, each resting on two adjacent coins in the row below. The diagram below shows one possible configuration for $n = 8$. How many different configurations can be in these display cases?

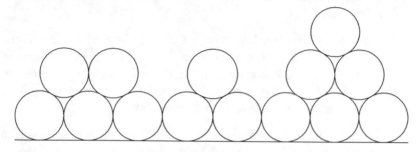

Solution:

Denote by a_n the number of possible configurations. Then $a_0 = a_1 = 1$. There are $n - 1$ spots for coins in the second row from the bottom. Create an extra n-th spot at the end. This forces at least one spot to be unoccupied. Let the k-th spot, $1 \le k \le n$, be the first unoccupied one. Then we have $k - 1$ coins in a row before this gap, and there are a_{k-1} possible configurations based on them. There are a_{n-k} possible configurations based on the last $n - k$ coins on the bottom row. Hence $a_n = a_0 a_{n-1} + a_1 a_{n-2} + \cdots + a_{n-1} a_0$. From the initial conditions, $a_n = C_n$.

Remark:

This problem may also be solved by establishing a one-to-one correspondence of the legal configurations with the legal paths for the little girl, as illustrated in the diagram below.

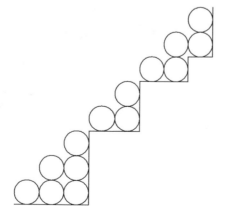

Exercises

1. A class has $2n$ students of different heights, where n is a positive integer. The desks in the classroom are arranged in a $2 \times n$ array. There is a blackboard to the front, and another to the right. Each student must therefore be taller than anyone in front or to the right in order to see both blackboards. How many different seat plans satisfy this condition?

2. The Tower of Bangkok has three pegs. On the first peg are n disks numbered 1 to n from top to bottom. In each move, the top disk of a peg may be moved to become the top disk of a higher-numbered peg. Eventually, all the disks are on the third peg. How many different permutations of the disks can be formed there?

3. How many different possible ways are there for $2n$ people at a round table to shake hands in pairs without crossing arms?

4. At a charity function, $2n$ people were at a round table. The organizer, who was also at the table, put $100 into a hat and passed it round the table. Each person put in an amount which was at least $100, but differing from the preceding donation by exactly $10. The last person put in $110 before returning the hat to the organizer. How many possible sequences of donations could there be?

5. How many sequences $\{y_1, y_2, \ldots, y_n\}$ of integers are there such that $0 = y_1 \leq y_2 \leq \cdots \leq y_n$ and $y_k < k$ for $1 \leq k \leq n$?

6. How many sequences $\{y_1, y_2, \ldots, y_n\}$ of integers are there such that $1 \leq y_1 \leq y_2 \leq \cdots \leq y_n \leq n$ and $y_k = k$ for exactly one value of k, $1 \leq k \leq n$?

Section 4.4. Stirling Numbers

In Section 1.5, we treated Case (2) of the Distribution Problem of Section 0.1. We are distributing n distinct objects into k distinct boxes, with empty boxes not permitted. The answer was given as a sum of multinomial coefficients. We now call it a **Stirling** number of the **third** kind, and denote it by the symbol $S_3(n,k)$.

There are three kinds of Stirling numbers, which we will encounter in reverse order. They are named after the British mathematician James Stirling, who lived primarily in the eighteenth century.

For illustration, let $n=4$ and $k=2$. Direct counting yields the distributions $(1)(234)$, $(12)(34)$, $(13)(24)$, $(14)(23)$, $(123)(4)$, $(124)(3)$, $(134)(2)$, and those obtained by interchanging the two boxes. This is reconfirmed by the calculation in Section 1.5, so that $S_3(4,2) = \binom{4}{3,1} + \binom{4}{2,2} + \binom{4}{1,3} = 4+6+4 = 14$.

Clearly, $S_3(n,1) = 1$, $S_3(n,k) = 0$ for $k > n$ and $S_3(n,n) = n!$. For $k = 2$, we may put i of the objects into the first box and the other $n-i$ objects into the second box. Hence the number of distributions is given by $S_3(n,2) = \binom{n}{1} + \binom{n}{2} + \cdots + \binom{n}{n-1} = 2^n - 2 = \binom{2}{0}2^n - \binom{2}{1}1^n + \binom{2}{2}0^n$. In particular, $S_3(4,2) = 2^4 - 2 = 14$. It can be proved that

$$S_3(n,3) = \binom{3}{0}3^n - \binom{3}{1}2^n + \binom{3}{2}1^n - \binom{3}{3}0^n,$$

$$S_3(n,4) = \binom{4}{0}4^n - \binom{4}{1}3^n + \binom{4}{2}2^n - \binom{4}{3}1^n + \binom{4}{4}0^n,$$

$$S_3(n,5) = \binom{5}{0}5^n - \binom{5}{1}4^n + \binom{5}{2}3^n - \binom{5}{3}2^n + \binom{5}{4}1^n - \binom{5}{5}0^n,$$

and so on. Using these, we can compute the following table of values for $S_3(n,k)$.

1					
1	2				
1	6	6			
1	14	36	24		
1	30	150	240	120	
1	62	540	1560	1800	720

The formulae above for $S_3(n,k)$ for small values of k suggested that the general formula is given by $S_3(n,k) = \sum_{i=0}^{k}(-1)^i\binom{k}{i}(k-i)^n$. This can be proved by mathematical induction. However, we will make use of the Principle of Inclusion-Exclusion instead.

Let S be the set of all k^n distributions of n distinct objects into k distinct boxes, empty cells permitted. For $1 \leq i \leq k$, let A_i be the set of distributions in which the i-th box is empty. Then $S_3(n,k) = |\overline{A_1} \cap \overline{A_2} \cap \cdots \cap \overline{A_k}|$. We have $|A_i| = (k-1)^n$ since each of the n objects may be in any of the other $k-1$ boxes. Similarly, $|A_i \cap A_j| = (k-2)^n$, and so on. By the Principle of Inclusion-Exclusion,

$$
\begin{aligned}
S_3(n,k) &= |S| - (|A_1| + |A_2| + \cdots + |A_k|) \\
&\quad + (|A_1 \cap A_2| + |A_1 \cap A_3| + \cdots + |A_{k-1} \cap A_k|) - \cdots \\
&= \sum_{i=0}^{k} (-1)^i \binom{k}{i} (k-i)^n.
\end{aligned}
$$

Alternatively, for each box, it may contain any non-zero number of the objects. The exponential generating function for this cell is $\sum_{n=1}^{\infty} \dfrac{x^n}{n!} = e^x - 1$. The overall exponential generating function is

$$
\begin{aligned}
(e^x - 1)^k &= \sum_{i=0}^{k} \binom{k}{i} (-1)^i \left(\sum_{n=1}^{\infty} (k-i)^n \frac{x^n}{n!} \right) \\
&= \sum_{n=1}^{\infty} \left(\sum_{i=0}^{k} (-1)^i \binom{k}{i} (k-i)^n \right) \frac{x^n}{n!}.
\end{aligned}
$$

It indeed generates exponentially the sequence $\left\{ \sum_{i=0}^{k} (-1)^i \binom{k}{i} (k-i)^n \right\}$.

The Stirling numbers of the third kind may also be computed via a recurrence relation for $S_3(n,k)$. Let $1 < k < n$. Consider the first object. It can go into any of the k boxes. The distributions may then be classified into two types. In the first type, the first object is all by itself. Then the remaining $n-1$ objects are to be distributed into the remaining $k-1$ boxes, with empty boxes forbidden. In the second type, the first object is not by itself. Hence the remaining $n-1$ objects are to be distributed into all k cells, with empty boxes again forbidden. It follows that

$$
S_3(n,k) = k(S_3(n-1, k-1) + S_3(n-1, k)).
$$

We now turn to Case (6) of the Distribution Problem in Section 0.1. The **Stirling** numbers of the **second** kind, $S_2(n,k)$, count the numbers of ways of distributing n distinct objects into k identical boxes with empty boxes forbidden.

Clearly, $S_2(n,k) = \frac{1}{k!} S_3(n,k)$ since we no longer distinguish among the boxes. From the properties of $S_3(n,k)$, we can derive corresponding properties of $S_2(n,k)$. We have $S_2(n,1) = 1$, $S_2(n,k) = 0$ for $k > n$, $S_2(n,n) = 1$, $S_2(n,k) = S_2(n-1,k-1) + kS_2(n-1,k)$ for $1 < k < n$, and

$$
S_2(n,k) = \frac{1}{k!} \sum_{i=0}^{k} (-1)^i \binom{k}{i} (k-i)^n.
$$

The recurrence relation $S_2(n, k) = S_2(n - 1, k - 1) + kS_2(n - 1, k)$ may be proved by a combinatorial argument similar to the one used to establish the corresponding recurrence relation for $S_3(n, k)$. If a particular object is all by itself in a cell, then the remaining $n - 1$ objects can be distributed among the other $k - 1$ cells in $S_2(n - 1, k - 1)$ ways. If it is not alone, we first distribute the remaining $n - 1$ objects among the k cells in $S_2(n - 1, k)$ ways, and then add the particular object to any of the k cells.

The following table of values for $S_2(n, k)$ may be generated using this recurrence relation, or obtained from the corresponding values for $S_3(n, k)$.

$$
\begin{array}{ccccccc}
 & & & 1 & & & \\
 & & 1 & & 1 & & \\
 & 1 & & 3 & & 1 & \\
1 & & 7 & & 6 & & 1 \\
\end{array}
$$

$$
\begin{array}{ccccccc}
1 & 15 & 25 & 10 & 1 & & \\
1 & 31 & 90 & 65 & 15 & 1 & \\
\end{array}
$$

The ordinary generating function for the Stirling numbers of the second kind is

$$
\begin{aligned}
G_k(x) &= \sum_{n=1}^{\infty} S_2(n, k)x^n \\
&= x\sum_{n=1}^{\infty} S_2(n - 1, k - 1)x^{n-1} + kx\sum_{n=1}^{\infty} S_2(n - 1, k)x^{n-1} \\
&= xG_{k-1}(x) + kxG_k(x).
\end{aligned}
$$

It follows that $G_k(x) = \frac{x}{1-kx}G_{k-1}(x) = \cdots = \frac{x^k}{(1-x)(1-2x)\cdots(1-kx)}$.

On the other hand, we have $\displaystyle\sum_{k=0}^{n} S_2(n, k)x(x - 1)\cdots(x - n + 1) = x^n$.
The k^n ways of distributing n distinct objects into k distinct boxes may be classified according to exactly how many boxes are non-empty, yielding
$$
k^n = \sum_{i=0}^{n}\binom{k}{i}S_3(n, i) = \sum_{i=0}^{n} S_2(n, i)k(k - 1)\cdots(k - n + 1).
$$
Replacing k by x yields the desired result.

We now tackle Case (5) of the Distribution Problem in Section 0.1. Here, we count the numbers of ways of distributing n distinct objects into k identical boxes with empty boxes allowed. The answer is given by the sum $S_2(n, 1) + S_2(n, 2) + \cdots + S_2(n, k)$.

The special case where $k = n$ gives rise to the the **Bell** numbers, which are defined by $B_n = S_2(n, 1) + S_2(n, 2) + \cdots + S_2(n, n)$. They are named after Eric Temple Bell, a mathematician who lived primarily in the twentieth century in the United States.

The Bell numbers count how many ways n distinct objects may be distributed into n identical boxes with empty boxes permitted. We may also think of B_n as the number of ways of partitioning an n-element set into an arbitrary number of subsets of arbitrary sizes.

The first few values are $B_0 = 1$, $B_1 = 1$, $B_2 = 2$ and $B_3 = 5$. We have $B_4 = 15$ since a 4-element set has the following partitions:

$$
\begin{array}{ccccc}
(1234) & (1)(234) & (12)(34) & (13)(24) & (14)(23) \\
(123)(4) & (124)(3) & (134)(2) & (1)(2)(34) & (1)(3)(24) \\
(1)(4)(23) & (12)(3)(4) & (13)(2)(4) & (14)(2)(3) & (1)(2)(3)(4)
\end{array}
$$

The Bell numbers satisfy the recurrence relation $B_{n+1} = \sum\limits_{k=0}^{n} \binom{n}{k} B_k$. The term $\binom{n}{k} B_k$ counts those partitions in which the new element $n + 1$ belongs to a subset of size $n + 1 - k$. There are $\binom{n}{k}$ ways of choosing those elements which will not appear together with $n+1$. They may be partitioned in B_k ways. The remaining $n - k$ elements will form a $(k + 1)$-st subset together with $n + 1$.

We illustrate with $B_4 = \binom{3}{0} B_0 + \binom{3}{1} B_1 + \binom{3}{2} B_2 + \binom{3}{3} B_3$. The partition counted under the first term is (1234). The partitions counted under the second term are $(1)(234)$, $(2)(134)$ and $(3)(124)$. The partitions counted under the third term are $(1)(2)(34)$, $(12)(34)$, $(1)(3)(24)$, $(13)(24)$, $(2)(3)(14)$ and $(23)(14)$. The partitions counted under the fourth term are $(1)(2)(3)(4)$, $(1)(23)(4)$, $(2)(13)(4)$, $(3)(12)(4)$ and $(123)(4)$.

Using this recurrence relation, we compute the next few values of the Bell numbers. They are $B_5 = 52$, $B_6 = 203$, $B_7 = 877$ $B_8 = 4140$, $B_9 = 21147$, $B_{10} = 114975$, $B_{11} = 673570$, $B_{12} = 3940712$ and $B_{13} = 27644437$

How many sequences of n letters are there if a letter may not appear until every letter which precedes it alphabetically has already appeared? For instance, there are 15 such sequences for $n = 4$. They are listed below along with a matching distribution of 4 distinct objects into at most 4 identical boxes.

AAAA	(1234)	AABC	(12)(3)(4)	ABCD	(1)(2)(3)(4)
AAAB	(123)(4)	ABAA	(134)(2)	ABAC	(13)(2)(4)
AABB	(12)(34)	ABBC	(1)(23)(4)	ABCB	(1)(24)(3)
ABBB	(1)(234)	ABBA	(14)(23)	ABAB	(13)(24)
AABA	(124)(3)	ABCC	(1)(2)(34)	ABCA	(14)(2)(3)

Label the boxes A, B, C and so on. The sequence of letters shows the box into which each object is distributed. Since the unlabeled boxes are identical, we use the next available box when an object is to be distributed into a box not occupied by any preceding objects. This one-to-one correspondence shows that the number of sequences is given by the Bell numbers.

Examples

Example 4.4.1.

Use $S_3(n, 2) = \sum\limits_{i=0}^{2}(-1)^i \binom{2}{i}(2-i)^n$ to prove that

$$S_3(n, 3) = \sum_{i=0}^{3}(-1)^i \binom{3}{i}(3-i)^n.$$

Solution:

We may put i of the objects into the first box and the other $n-i$ objects into the remaining two boxes, with neither of them empty. Then the number of distributions is given by

$$
\begin{aligned}
S_3(n, 3) &= \sum_{i=1}^{n}\binom{n}{i}S_3(n-i, 2) \\
&= \sum_{i=1}^{n-2}\binom{n}{i}(2^{n-i} - 2) \\
&= \sum_{i=1}^{n-2}\binom{n}{i}2^{n-i} - 2\sum_{i=1}^{n-2}\binom{n}{i} \\
&= 3^n - \binom{n}{0}2^n - \binom{n}{n-1}2^1 - \binom{n}{n}2^0 \\
&\quad -2\left(2^n - \binom{n}{0}1^n - \binom{n}{n-1}1^1 - \binom{n}{n}1^0\right) \\
&= 3^n - 3\cdot 2^n + 3\cdot 1^n - \binom{n}{n-1}S_3(1, 2) - \binom{n}{n}S_3(0, 2) \\
&= \sum_{i=0}^{3}(-1)^i \binom{3}{i}(3-i)^n.
\end{aligned}
$$

Example 4.4.2.

Prove combinatorially that $S_2(m+n+1, m) = \sum\limits_{k=1}^{m} kS_2(n+k, k)$.

Solution:

The term $kS_2(n+k, k)$ counts the number of the $S_2(m+n+1, m)$ partitions into m subsets of $\{1, 2, \ldots, m+n+1\}$ where $n+k+1$ is the largest element which is not alone in a subset. There are $(m+n+1) - (n+k+1) = m-k$ elements larger than $n+k+1$, and they must be alone. Hence the $n+k$ elements smaller than $n+k+1$ must be partitioned into k subsets. This may be done in $S_2(n+k, k)$ ways, and the element $n+k+1$ itself may be added to any of the k subsets.

Example 4.4.3.

Use the formula $S_3(n, k) = \sum_{i=0}^{k}(-1)^i \binom{k}{i}(k-i)^n$ to prove directly that

(a) $S_3(n, 1) = 1$;

(b) $S_3(n, k) = 0$ for $k > n$.

Solution:

(a) We have $S_3(n, 1) = \sum_{i=0}^{1}(-1)^i \binom{1}{i}(1-i)^n = 1 - 0 = 1$.

(b) For $k > n$, we prove that $S_3(n, k) = 0$ by induction on n. We have

$$S_3(0, k) = \sum_{i=0}^{k}(-1)^i \binom{k}{i} = (1 + (-1))^k = 0.$$

Suppose $S_3(j, k-1) = 0$ for $0 \le j \le n-1 < k-1$. Then

$$
\begin{aligned}
S_3(n, k) &= \sum_{i=0}^{k}(-1)^i \binom{k}{i}(k-i)^n \\
&= k^n + k\sum_{i=1}^{k}(-1)^i \frac{k-i}{k}\binom{k}{i}(k-i)^{n-1} \\
&= k\sum_{i=0}^{k}(-1)^i \binom{k-1}{i}(1 + (k-1-i))^{n-1} \\
&= k\sum_{i=0}^{k}(-1)^i \binom{k-1}{i}\sum_{j=0}^{n-1}\binom{n-1}{j}(k-1-i)^j \\
&= k\sum_{j=0}^{n-1}\binom{n-1}{j}\sum_{i=0}^{k}(-1)^i \binom{k-1}{i}(k-1-i)^j \\
&= k\sum_{j=0}^{n-1}\binom{n-1}{j}S_3(j, k-1) \\
&= 0.
\end{aligned}
$$

Exercises

1. Use $S_3(n, 3) = \sum_{i=0}^{3}(-1)^i \binom{3}{i}(3-i)^n$ to prove that

$$S_3(n, 4) = \sum_{i=0}^{4}(-1)^i \binom{4}{i}(4-i)^n.$$

2. Use $S_3(n, 4) = \sum_{i=0}^{4}(-1)^i \binom{4}{i}(4-i)^n$ to prove that

$$S_3(n, 5) = \sum_{i=0}^{5}(-1)^i \binom{5}{i}(5-i)^n.$$

3. Prove combinatorially that $S_2(n+1, m+1) = \sum_{k=m}^{n} \binom{n}{k} S_2(k, m)$.

4. Prove combinatorially that $S_2(n+1, m+1) = \sum_{k=m}^{n} (m+1)^{n-k} S_2(k, m)$.

5. Use the formula $S_3(n, k) = \sum_{i=0}^{k}(-1)^i \binom{k}{i}(k-i)^n$ to prove directly that $S_3(n, n) = n!$.

6. Use the formula $S_3(n, k) = \sum_{i=0}^{k}(-1)^i \binom{k}{i}(k-i)^n$ to prove directly that, for $1 < k < n$,

$$S_3(n, k) = k(S_3(n-1, k-1) + S_3(n-1, k)).$$

Section 4.5. Other Special Numbers

The **Stirling** numbers of the **first** kind, $S_1(n, k)$, count the numbers of ways of seating n people at k round tables, none unoccupied, such that the cyclic order of seating is also taken into consideration.

For instance, $S_1(4, 2) = 11$. Suppose the four people are 1, 2, 3 and 4. There are three ways of partitioning them into two tables of two, namely, $(12)(34)$, $(13)(24)$ and $(14)(23)$. Since each table has only two people, there is only one cyclic order, so that the number of seatings in pairs is 3. On the other hand, there are four ways of partitioning them into one group of one and one group of three, namely $(1)(234)$, $(2)(134)$, $(3)(124)$ and $(4)(123)$. In each case, there are two cyclic orders in the table of three. This yields four other seating arrangements, namely, $(1)(243)$, $(2)(143)$, $(3)(142)$ and $(4)(132)$, for a total of eleven.

Clearly, $S_1(n, 1) = (n - 1)!$, $S_1(n, k) = 0$ if $n < k$ and $S_1(n, n) = 1$. We now prove that $S_1(n, k) = S_1(n - 1, k - 1) + (n - 1)S_1(n - 1, k)$. If a particular person is all by herself at a table, then the remaining $n - 1$ people can be seated among the other $k - 1$ tables in $S_1(n - 1, k - 1)$ ways. If she is not alone, we first seat the remaining $n - 1$ people at k tables in $S_1(n - 1, k)$ ways, and she can then be seated to the left of any of the other $n - 1$ people.

The following table of values for $S_1(n, k)$ may be generated using this recurrence relation.

			1				
		1		1			
	2		3		1		
6		11		6		1	
24		50		35		10	1
120	274		225		85	15	1

The astute reader may notice that the sum of the numbers in the n-th row is $n!$. In other words,

$$\sum_{k=1}^{n} S_1(n, k) = n!.$$

Clearly, the sum of the left side counts the number of seating of n people in any number of tables which are identical. Suppose the people are 1, 2, ..., n. Now 1 can sit anywhere. There is only 1 choice since the tables are identical. Then 2 can choose to sit to the left of 1 or start a new table, yielding 2 choices. Similarly, 3 can choose to sit to the left of 1, to the left of 2, or start a new table, yielding 3 choices. Continuing this way, we have a total of $n!$ ways.

The alternating sum of the numbers in the n-th row of the above table is 0 when $n \geq 2$. In other words,

$$\sum_{k=1}^{n}(-1)^{k}S_1(n, k) = 0.$$

We shall show that with at least two people, the total number of seatings at any odd number of tables is equal to the total number of seatings at any even number of tables. Consider two people 1 and 2. They are either at the same table or at different tables. There is a one-to-one correspondence between these two kinds of seatings.

In the case where 1 and 2 are together, they divide the table into two parts, the first starting from 1 and going left until it reaches 2, and the second part starting from 2 and going left until it reaches 1. Let the people in the first part stay put and the people in the second part move to a new table, maintaining their internal order. In the case where 1 and 2 are not together, their tables can be merged by reversing the above process.

The generating function for the Stirling numbers of the first kind is

$$x(x + 1)(x + 2) \cdots (x + n - 1) = \sum_{k=1}^{n} S_1(n, k)x^{k}.$$

The two identities we have just proved follow easily by setting $x = 1$ and $x = -1$ respectively. They can also be proved using induction and

$$S_1(n, k) = S_1(n - 1, k - 1) + (n - 1)S_1(n - 1, k).$$

For now, let the sequence generated by $x(x + 1)(x + 2) \cdots (x + n - 1)$ be denoted by $f(n, k)$. The linear term yields $f(n, 1) = (n - 1)!$ while the x^{n} term yields $f(n, n) = 1$. Clearly, $f(n, k) = 0$ when $k > n$. Once we have proved that $f(n, k) = f(n - 1, k - 1) + (n - 1)f(n - 1, k)$, we can conclude that $f(n, k) = S_1(n, k)$, so that $x(x + 1)(x + 2) \cdots (x + n - 1)$ is indeed the generating function for $S_1(n, k)$.

Note that $f(n-1, k-1)$ is the coefficient of the x^{k-1} term and $f(n-1, k)$ is the coefficient of the x^{k} term in $x(x + 1)(x + 2) \cdots (x + n - 2)$. Hence $x(x + 1)(x + 2) \cdots (x + n - 1)$ may be expressed as

$$(x + n - 1)(\cdots + f(n - 1, k - 1)x^{k-1} + f(n - 1, k)x^{k} + \cdots)$$
$$= \cdots + f(n - 1, k - 1)x^{k} + (n - 1)f(n - 1, k)x^{k} + \cdots.$$

It follows that the coefficient of x^{k} in $x(x + 1)(x + 2) \cdots (x + n - 1)$ is $f(n, k) = f(n - 1, k - 1) + (n - 1)f(n - 1, k)$.

The Stirling numbers of the first and second kind are related by two remarkable orthogonality formulae. Let $m \leq n$. Then

$$\sum_{k=m}^{n} (-1)^k S_1(n,k) S_2(k,m) = (-1)^n \delta_{m,n}$$

and

$$\sum_{k=m}^{n} (-1)^k S_1(k,m) S_2(n,k) = (-1)^n \delta_{m,n}.$$

Here, $\delta_{m,n}$ is the Kronecker delta, which is equal to 1 when $m = n$ and equal to 0 whenever $m \neq n$.

We first illustrate the first formula for $m = 2$ and $n = 4$ in the chart below.

$S_1(4,2)S_2(2,2)$ $= 11 \times 1 = 11$	$S_1(4,3)S_2(3,2)$ $= 6 \times 3 = 18$	$S_1(4,4)S_2(4,2)$ $= 1 \times 7 = 7$
(12) + (34)	(1)(2) + (34)	
(13) + (24)	(1)(3) + (24)	
(14) + (23)	(1)(4) + (23)	
(1) + (234)	(1) + (2)(34)	
(1) + (243)	(1) + (24)(3)	
(123) + (4)	(1)(23) + (4)	
(132) + (4)	(13)(2) + (4)	
(124) + (3)	(1)(24) + (3)	
(142) + (3)	(14)(2) + (3)	
(134) + (2)	(1)(34) + (2)	
(143) + (2)	(14)(3) + (2)	
	(1) + (23)(4)	(1) + (2)(3)(4)
	(12)(3) + (4)	(1)(2)(3) + (4)
	(12)(4) + (3)	(1)(2)(4) + (3)
	(13)(4) + (2)	(1)(3)(4) + (2)
	(12) + (3)(4)	(1)(2) + (3)(4)
	(13) + (2)(4)	(1)(3) + (2)(4)
	(14) + (2)(3)	(1)(4) + (2)(3)

In general, we seat n people at k identical tables in $S_1(n,k)$ ways, and then distribute the tables into m identical rooms in $S_2(k,m)$ ways. We claim that the total number of ways in which k is odd is equal to the total number of ways in which k is even, except when $m = n$. In that case, everyone is at a table in a room by herself, so that $S_1(n,n) = S_2(n,n) = 1$. It follows that we indeed have $(-1)^n S_1(n,n) S_2(n,n) = (-1)^n \delta_{n,n}$.

For $n > m$, let the people be 1, 2, ..., n. There are at least two people in the same room. Choose the pair so that they have the lowest numbers. If they are at the same table, we split the table, and if they are at different tables, we merge the tables, using the process before. This yields a one-to-one correspondence between the arrangements with an odd number of tables and those with an even number of tables.

We now illustrate the second formula for $m = 2$ and $n = 4$ in the chart below.

$S_2(4,2)S_1(2,2)$ $= 7 \times 1 = 7$	$S_2(4,3)S_1(3,2)$ $= 6 \times 3 = 18$	$S_2(4,4)S_1(4,2)$ $= 1 \times 11 = 11$
(12) + (34)	(1)(2) + (34)	
(13) + (24)	(1)(3) + (24)	
(14) + (23)	(1)(4) + (23)	
(1) + (234)	(1) + (2)(34)	
(123) + (4)	(1)(23) + (4)	
(124) + (3)	(1)(24) + (3)	
(134) + (2)	(1)(34) + (2)	
	(1) + (24)(3)	(1) + (2)(4)(3)
	(1) + (23)(4)	(1) + (2)(3)(4)
	(12)(3) + (4)	(1)(2)(3) + (4)
	(13)(2) + (4)	(1)(3)(2) + (4)
	(12)(4) + (3)	(1)(2)(4) + (3)
	(14)(2) + (3)	(1)(4)(2) + (3)
	(13)(4) + (2)	(1)(3)(4) + (2)
	(14)(3) + (2)	(1)(4)(3) + (2)
	(12) + (3)(4)	(1)(2) + (3)(4)
	(13) + (2)(4)	(1)(3) + (2)(4)
	(14) + (2)(3)	(1)(4) + (2)(3)

In general, we partition n people into k subsets in $S_2(n,k)$ ways, and then arrange them around m identical tables in $S_1(k,m)$ ways. We claim that the total number of ways in which k is odd is equal to the total number of ways in which k is even, except when $m = n$. In that case, everyone is at a table in a room by herself, so that $S_1(n,n) = S_2(n,n) = 1$. It follows that we indeed have $(-1)^n S_2(n,n)S_1(n,n) = (-1)^n \delta_{n,n}$.

For $n > m$, let the people be 1, 2, ..., n. There are at least two people at the same table. Choose the pair so that they have the lowest numbers. If they are in the same subset, we split the subset, and if they are in different subsets, we merge the subsets, using the process before. This yields a one-to-one correspondence between the arrangements with an odd number of subsets and those with an even number of subsets.

The Euler numbers are named after Leonhard Euler, an eighteenth century Swiss mathematician who spent most of his productive life in Germany under the patronage of Frederick the Great, and in Russia under the patronage of Catherine the Great. He was one of the greatest mathematicians of all time. There is an Euler Formula in practically every branch of mathematics. The Euler numbers are related to the Stirling numbers of the third kind, though we leave the exploration of this to the exercises.

The Euler numbers $E(n, k)$ are defined as follows. In a permutation $x_1, x_2, \ldots, x_n$ of $1, 2, \ldots, n$, an adjacent pair (x_i, x_{i+1}) is called an up-pair if $x_i < x_{i+1}$, and a down-pair if $x_i > x_{i+1}$. Then $E(n, k)$ counts the number of such permutations with exactly $k - 1$ up-pairs. Note that $E(n, 0) = 0$ and $E(n, k) = 0$ if $n < k$.

We have $E(4, 1) = 1$ since only the permutation in which the elements appear in reverse order has no up-pairs. Similarly, $E(4, 4) = 1$ since only the permutation in which the elements appear in order has 3 up-pairs. In general, $E(n, 1) = 1$ and $E(n, n) = 1$.

We have $E(4, 2) = 11$, the permutations with exactly 1 up-pair being 1432, 2143, 2431, 3142, 3214, 3241, 3421, 4132, 4213, 4231 and 4312. Since there are 4!=24 permutations of 1,2,3,4 altogether, we see that $E(4, 3) = 11$. On the other hand, if we reverse each of the 11 permutations counted in $E(4, 2)$, we will have the 11 permutations counted in $E(4, 3)$. More generally, $E(n, k) = E(n, n - k + 1)$.

We now prove that $E(n, k) = (n - k + 1)E(n - 1, k - 1) + kE(n - 1, k)$. Remove the element n from a permutation counted in $E(n, k)$. If n was at the beginning, the number of up-pairs has not changed, and the resulting permutation is one of those counted in $E(n - 1, k)$. If n was at the end, the number of up-pairs has gone down by 1, and the resulting permutation is one of those counted in $E(n - 1, k - 1)$. If n is between an up-pair, then the number of up-pairs is unchanged, but if it is between a down-pair, then the number of up-pairs goes down by 1. Again, the resulting permutation is one counted in $E(n - 1, k)$ or $E(n - 1, k - 1)$.

Add n to a permutation counted in $E(n - 1, k - 1)$. In order to increase the number of up-pairs by 1, we must add n at the end or between a down-pair. Since there are $n - k$ down-pairs, the total number of places where n may be added is $n - k + 1$, yielding a count of $(n - k + 1)E(n - 1, k - 1)$. Now add n to a permutation counted in $E(n - 1, k)$. In order not to increase the number of up-pairs, we must add n at the beginning or between an up-pair. Since there are $k - 1$ up-pairs, the total number of places for adding n is k, yielding a count of $kE(n - 1, k)$.

The following table of values for $E(n, k)$ may be generated using this recurrence relation. Note that the sum of all numbers in the n-th row is $n!$.

$$
\begin{array}{ccccccccc}
& & & & 1 & & & & \\
& & & 1 & & 1 & & & \\
& & 1 & & 4 & & 1 & & \\
& 1 & & 11 & & 11 & & 1 & \\
1 & & 26 & & 66 & & 26 & & 1 \\
1 & 57 & & 302 & & 302 & & 57 & 1
\end{array}
$$

The Euler numbers provide a link between ordinary powers and consecutive binomial coefficients via the formula $m^n = \sum_{k=1}^{n} \binom{m+k-1}{n} E(n, k)$.

We now prove this by induction. For $n = 1$, both sides are equal to m. Suppose the formula holds for some $n \geq 1$. Then

$$
\sum_{k=1}^{n+1} \binom{m+k-1}{n+1} E(n+1, k)
$$

$$
= \binom{m}{n+1} E(n+1, 1) + \binom{m+n}{n+1} E(n+1, n+1)
$$

$$
+ \sum_{k=2}^{n} \binom{m+k-1}{n+1} E(n+1, k)
$$

$$
= \binom{m}{n+1} E(n, 1) + \binom{m+n}{n+1} E(n, n)
$$

$$
+ \sum_{k=2}^{n} \binom{m+k-1}{n+1} ((n-k+2) E(n, k-1) + k E(n, k))
$$

$$
= \sum_{k=1}^{n} E(n, k) \left(k \binom{m+k-1}{n+1} + (n-k+1) \binom{m+k}{n+1} \right)
$$

$$
= m \sum_{k=1}^{n} \binom{m+k-1}{n} E(n, k)
$$

$$
= m^{n+1}.
$$

We now turn to a problem known as the problem of Simon Newcombe. The n cards in a deck are labeled 1 to n. After a thorough shuffle, they are dealt one at a time. The first card starts a new pile. If a subsequent card has a higher label than the card just dealt, it is placed on top in the same pile. If not, it starts a new pile. How many of the $n!$ permutations will result in exactly k piles?

The number of piles is equal to 1 plus the number of down-pairs in the permutation. If the number of piles is k, then the number of down-pairs is $k - 1$ and the number of up-pairs is $n - k$. It follows that the answer we seek is $E(n, n - k + 1) = E(n, k)$.

Examples

Example 4.5.1.

Prove that $S_1(m + n + 1, m) = \sum_{k=1}^{m} (n + k)S_1(n + k, k)$.

Solution:

In each of the $S_1(m+n+1, m)$ seatings of $\{1, 2, \ldots, m+n+1\}$ at m tables, pick out the largest element which is not alone at a table. Let this element be $n + k + 1$. Then $m - k$ larger elements are all alone at individual tables. Seat the $n + k$ smaller elements at k tables. Now $n + k + 1$ can be added to one of these k tables by seating to the left of any of the $n + k$ elements.

Example 4.5.2.

Prove that $\dfrac{n!}{m!}\dbinom{n - 1}{m - 1} = \sum_{k=m}^{n} S_1(n, k)S_2(k, m)$.

Solution:

The people $1, 2, \ldots, n$ may be permuted in $n!$ ways. Among the $n-1$ spaces between adjacent elements, we choose $m - 1$ of them to insert a partition marker. This can be done in $\binom{n-1}{m-1}$ ways. We disregard the ordering of the resulting m lines, and divide the total by $m!$, yielding the expression on the left side. On the other hand, we can start by seating the n people at k tables in $S_1(n, k)$ ways. Then we put these tables arbitrarily into m identical rooms, with at least one table in each room. This yields the sum of the right side. To establish a one-to-one correspondence, we convert the tables in each room into a line. Pick the person with the smallest number from each table. Start with the person with the highest such number, and have the people at that table follow in order. Then the person with the second highest such number joins the line, followed by the people at that table in order, and so on. Conversely, suppose we have a line. Seat the first person at a table, and have the other people join him round the table in order until the last person with a higher number is seated. If there is anyone left, a new table is started and filled according to the same rule.

Example 4.5.3.

Prove that $E(n, m) = \sum_{k=0}^{m-1} (-1)^k \binom{n + 1}{k}(m - k)^n$.

Solution:

We use induction on n for all $m \leq n$. In the basis, $n = m = 1$, and both sides are equal to 1. Suppose the result holds for some $n \geq 1$ and all $m \leq n$.

Then

$$
\begin{aligned}
E(n+1, m) &= (n - m + 2)E(n, m - 1) + mE(n, m) \\
&= (n - m + 2)\sum_{k=0}^{m-2}(-1)^k \binom{n+1}{k}(m - 1 - k)^n \\
&\quad + m\sum_{k=0}^{m-1}(-1)^k \binom{n+1}{k}(m - k)^n \\
&= (n - m + 2)\sum_{k=1}^{m-1}(-1)^{k-1}\binom{n+1}{k-1}(m - k)^n \\
&\quad + m^{n+1} + m\sum_{k=1}^{m-1}(-1)^k \binom{n+1}{k}(m - k)^n \\
&= m^{n+1} + \sum_{k=1}^{m-1}(-1)^k (m - k)^n \frac{(n+1)!}{k!(n - k + 2)!}(m - k)(n + 2) \\
&= \sum_{k=0}^{m-1}(-1)^k \binom{n+2}{k}(m - k)^{n+1}.
\end{aligned}
$$

Exercises

1. Prove that $S_1(n+1, m+1) = \sum_{k=m}^{n} \binom{k}{m} S_1(n, k)$.

2. Prove that $S_1(n+1, m+1) = \sum_{k=m}^{n} \frac{n!}{k!} S_1(k, m)$.

3. Prove that $(-1)^m \binom{n}{m} = \sum_{k=m}^{n} (-1)^k S_1(k, m) S_2(n+1, k+1)$.

4. Prove that $(-1)^m \binom{n}{m}(n - m)! = \sum_{k=m}^{n} (-1)^k S_1(n+1, k+1) S_2(k, m)$.

5. Prove that $S_3(n, m) = \sum_{k=1}^{n} \binom{k-1}{n-m} E(n, k)$.

6. Prove that $(-1)^{n-m} E(n, m) = \sum_{k=1}^{n} (-1)^{k-1} \binom{n-k}{m-1} S_3(n, k)$.

Section 4.6. Ramsey Numbers

Unlike the other kinds of numbers studied in this chapter, the Ramsey numbers do not arise from enumeration problems but from extremal problems.

Prove that in a party of six people, there are three such that either every two of them know each other or no two of them know each other. (Knowing is assumed to be symmetric but not assumed to be transitive.)

This is one of the classic problems in combinatorics, dating back to 1930. It was featured in the 1947 Kürschák Competition in Hungary, and transplanted across the Atlantic into the 1953 Putnam Competition.

The solution is elegant and instructive. Pick a particular person A. The other five people may be put into two rooms, the first containing those who know A, and the second those who do not. By the Mean Value Principle, one of the rooms must contain at least three people. By symmetry, we may assume that it is the first room. If no two of them know each other, there is nothing further to prove. If two of them know each other, then along with A, they form a trio in which every two know each other.

This is a special case of a more general result known as Ramsey's Theorem, due to the twentieth century British mathematician/logician Frank Ramsey. It is a profound generalization of the Mean Value Principle. The much easier counterpart of the above problem is the following, which yields immediately to an application of the Mean Value Principle.

Prove that in a party of five people, there are three such that either every one of them is happy to be there, or no one of them is happy to be there.

Both of these problems, as stated, are existence problems. However, it is natural to frame them as extremal problems. We can ask whether the number of people at the party is as small as possible. In both cases, that turns out to be so.

In the latter problem, let people who are happy to be there wear yellow hats and people who are not happy to be there wear blue hats. If there are only four people in the party instead of five, two of them may be wearing yellow hats and two blue hats, and we do not have three people who satisfy the condition of the problem.

In the former problem, let two people who know each other hold a yellow ribbon between them, and two people who do not know each other hold a blue ribbon between them. If there are only five people in the party instead of six, let them form a pentagon. Let the edges of the pentagon be yellow ribbons and the diagonals be blue ribbons. Then we do not have three people who satisfy the condition of the problem.

Let us restate the Mean Value Principle in the extremal form. Suppose elements are classified into k kinds. Let $n_1, n_2, \ldots, n_k$ be positive integers. Then there exists a smallest positive integer $p = p(n_1, n_2, \ldots, n_k)$ such that every set of size p contains a subset of size n_i where every element is in the i-th class for some i, $1 \leq i \leq k$.

We now give its generalization.

Ramsey's Theorem.
Suppose pairs of elements are classified into k kinds. Let $n_1, n_2, \ldots, n_k$ be positive integers. Then there exists a smallest positive integer

$$R = R(n_1, n_2, \ldots, n_k)$$

such that every set of size R contains a subset of size n_i where every pair of elements is in the i-th class for some i, $1 \leq i \leq k$.

Ramsey's Theorem can in turn be generalized in many different ways, some natural and others esoteric. For instance, we may have every three people holding a triangular flag of some color. However, we will not make such investigations. Perhaps the ultimate generalization comes from the field of behavioral science. *If you have enough of it, something will happen.*

The numbers $R(n_1, n_2, \ldots, n_k)$ are called the classical **Ramsey** numbers, or just Ramsey numbers. It is clear that $R(m, n) = R(n, m)$ so that we can assume that $m \leq n$. The trivial values are $R(1, n) = 1$ and $R(2, n) = n$. From our introductory example, we have $R(3, 3) = 6$. Even for $k = 2$, very very few non-trivial values of the Ramsey numbers have been determined. They are $R(3, 4) = 9$, $R(3, 5) = 14$, $R(3, 6) = 18$, $R(3, 7) = 23$, $R(3, 8) = 28$, $R(3, 9) = 36$, $R(4, 4) = 18$ and $R(4, 5) = 25$.

We only know that $43 \leq R(5, 5) \leq 49$. The great Hungarian combinatorialist Paul Erdős once said that if some aliens threatened to destroy Earth unless we would reveal the value of $R(5, 5)$, the world should get together and figure it out. However, if the aliens demanded the value of R(6,6), the world should get together and figure out a way to destroy the aliens.

We will continue to use the party setting. Call a group of people a **clique** if every two people are holding a ribbon between them. A ribbon may be yellow or blue. For the Ramsey number $R(m, n)$, we always associate the color yellow with the first parameter m, and the color blue with the second parameter n. A group of people such that every two are holding a yellow ribbon between them is called a yellow clique, and a blue clique is similarly defined. A clique is said to be monochromatic if it is either a yellow clique or a blue clique.

The lower bounds such as $43 \leq R(5, 5)$ are obtained by explicit constructions. The upper bounds such as $R(5, 5) \leq 49$ are obtained by general arguments.

The Mean Value Principle yields the following recursive upper bound.

Greenwood-Gleason Inequality. $R(m, n) \leq R(m - 1, n) + R(m, n - 1)$.

We prove this in a way similar to the solution of the original problem. Pick a particular person A. The other $R(m - 1, n) + R(m, n - 1) - 1$ may be holding yellow ribbons or blue ribbons with A. By the Mean Value Principle, we either have $R(m - 1, n)$ people holding yellow ribbons or $R(m, n - 1)$ people holding blue ribbons with A. By symmetry, we need only consider the first case. By the definition of $R(m - 1, n)$, we either have a yellow clique of size $m - 1$ or a blue clique of size n. In the second subcase, there is nothing further to prove. In the first subcase, if we add A, we will have a yellow clique of size m.

This result can be improved slightly under certain circumstances. If both $R(m - 1, n)$ and $R(m, n - 1)$ are even, then the inequality is strict. With a total of only $R(m - 1, n) + R(m, n - 1) - 1$ people, there is now an additional case where we have $R(m - 1, n) - 1$ people holding yellow ribbons with A, and $R(m, n - 1) - 1$ people holding blue ribbons with A. However, this cannot be the case with every choice of A. Otherwise, the total number of people holding yellow ribbons, counting multiplicities, would be

$$(R(m - 1, n) - 1)(R(m - 1, n) + R(m, n - 1) - 1),$$

which is an odd number. However, the number of people holding yellow ribbons, counting multiplicities, must be even. We have a contradiction.

The first application of the strengthened inequality yields

$$R(3, 4) \leq R(2, 4) + R(3, 3) - 1 = 9.$$

It then follows that

$$R(3, 5) \leq R(2, 5) + R(3, 4) = 14$$

and

$$R(4, 4) \leq R(3, 4) + R(4, 3) = 18.$$

These bounds are in fact sharp.

However, $R(3, 6) \leq R(2, 6) + R(3, 5) - 1 = 19$ is not sharp. Suppose to the contrary that an 18-clique lower bound exists for $R(3, 6)$. By Example 4.6.2, each person is holding exactly 5 yellow ribbons. We claim that there cannot be two people Y and Z who hold yellow ribbons with three people U, V and W. Suppose to the contrary that they exist. Each of Y and Z is holding yellow ribbons with 2 other people. Taking these 4 people away along with U, V, W, Y and Z, we are left with $9 = R(3, 4)$ people. Since we do not have a yellow trio, we must have a blue quartet. However, this becomes a blue sextet when Y and Z are included. This contradiction justifies the claim.

Consider a particular person X. By Example 4.6.2, X is holding yellow ribbons with exactly 5 people J, K, L, M and N, and blue ribbons with exactly 12 people A, B, C, D, E, F, G, H, P, Q, R and S. Each member of the 5-clique is holding exactly 4 yellow ribbons with the 12-clique, so that the total number of such ribbons is 20. By Example 4.6.1(a), each member of the 12-clique is holding 1 or 2 yellow ribbons with the 5-clique. Hence 8 of them are holding 2 yellow ribbons while the other 4 are holding 1 yellow ribbon with the 5-clique. We may assume that the latter four are P, Q, R and S.

We cannot have both A and B holding yellow ribbons with both J and K. This is because V is also holding yellow ribbons with both J and K, and we have a contradiction to the claim above. It follows that we may assume that each of A, B, C, D, E, F, G and H is holding yellow ribbons with a different pair of J, K, L, M and N.

Suppose both P and Q are holding yellow ribbons with J. Then they cannot be holding a yellow ribbon between them. Hence P, Q, K, L, M and N will form a blue sextet, contrary to our assumption. It follows that we may assume that P, Q, R and S are holding yellow ribbons with K, L, M and N, respectively.

We may also assume that J is holding yellow ribbons with A, B, C and D. Note that no two of A, B, C and D can hold yellow ribbons with the same one of K, L, M and N, as otherwise we again have a contradiction to our claim above. Hence each of K, L, M and N is holding yellow ribbons with two of E, F, G and H. We may assume that E is holding yellow ribbons with M and N, F with N and K, G with K and L, and H with L and M.

Now none of G, P and Q is holding a yellow ribbon with J, M and N. In order for these six people not to form a blue sextet, P must be holding a yellow ribbon with Q. Similarly, Q is holding a yellow ribbon with R, R is holding a yellow ribbon with S, and S is holding a yellow ribbon with P.

Each of P, Q, R and S is holding a yellow ribbon with one of E, F, G and H. To avoid a yellow trio, P must be holding it with E or H. We may assume that the pairings are P, Q, R and S with E, F, G and H respectively.

It remains to determine with whom A, B, C and D are holding yellow ribbons apart from J. Each is holding one with one of K, L, M and N, one of P, Q, R and S, and two of E, F, G and H. Still without loss of generality, we may assume that one set of pairings consist of A and K, B and L, C and M as well as D and N.

Now P cannot hold a yellow ribbon with A, as otherwise A, K and P will form a yellow trio. P cannot hold a yellow ribbon with D either, as otherwise P and N will be holding yellow ribbons with D, E and S, contrary to our claim above. Hence P is holding a yellow ribbon with either B or C.

Similar considerations of Q, R and S show that there are two possibilities for people holding yellow ribbons with P, Q, R and S.

P	K,Q,S,E,B		P	K,Q,S,E,C
Q	L,R,P,F,C		Q	L,R,P,F,D
R	M,S,Q,G,D		R	M,S,Q,G,A
S	N,P,R,H,A		S	N,P,R,H,B

In either case, E cannot be holding a yellow ribbon with C, as otherwise C, E and M will form a yellow trio. E cannot be holding a yellow ribbon with D either, as otherwise D, E and N will form a yellow trio. Since E is holding a yellow ribbon with two of A, B, C and D, these two must be A and B. However, in the first case, E and S will be holding yellow ribbons with A, N and P, contrary to the above claim. In the second case, E and S will be holding yellow ribbons with B, N and P. This final contradiction establishes the upper bound $R(3,6) \leq 18$.

If the number of colors is greater than two, the only value we know is $R(3,3,3) = 17$. Let the colors be yellow, blue and red. Pick a particular person A. The other 16 may be holding yellow, blue or red ribbons with A. By the Mean Value Principle, we must have six people holding ribbons of the same color with A, say red. If any two of them hold a red ribbon between them, they form with A a red clique of size 3. If not, every two are holding either a yellow or a blue ribbon. Since $R(3,3) = 6$, we must have either a yellow or a blue clique of size 3. This establishes the upper bound.

For the lower bound, let sixteen people be labeled with the sixteen quartets of 0s and 1s. We divide the fifteen non-zero quartets into the following three sets.

Red:	(0,0,0,1)	(0,0,1,0)	(0,1,0,0)	(1,0,0,0)	(1,1,1,1)
Yellow:	(1,1,1,0)	(0,1,1,1)	(0,1,1,0)	(1,1,0,0)	(0,0,1,1)
Blue:	(1,1,0,1)	(1,0,1,1)	(1,0,0,1)	(1,0,1,0)	(0,1,0,1)

Addition of quartets is performed component by component, modulo 2. In each of these three sets, the sum of any two quartets is in a different set. Let two people hold a red ribbon between them if the sum of their quartets is in the red set, and so on. We claim that we have no red, yellow or blue cliques of size 3. Suppose x, y and z are the quartets of three people in say a red clique. This means that all of $x + y$, $x + z$ and $y + z$ are in the red set. However, since $(x + y) + (x + z) = y + z$, this contradicts the fact that the red set is sum-free. Similarly, we cannot have a yellow or a blue clique of size 3. This establishes the lower bound.

We now give an alternative solution to the original problem which yields a somewhat stronger result. We now prove that there are at least two monochromatic cliques of size 3.

Two yellow ribbons being held by the same person are said to form a yellow arrow, and two blue ribbons being held by the same person are said to form a blue arrow. Let us count the number of arrows in two ways. First, we count them person by person. Now a person holds five ribbons in total. They may be split 5:0, 4:1 or 3:2 in colors, yielding 10, 6 or 4 arrows respectively. It follows that the total number of arrows is at least 24.

There are $\binom{6}{3} = 20$ cliques of size 3 among the six people. Let x of them be monochromatic cliques. Each such clique contains 3 arrows, while each of the remaining cliques contains 1 arrow. Hence the total number of arrows is exactly $3x + 20 - x$. From $2x + 20 \geq 24$, we have $x \geq 2$.

Suppose we change one of the fifteen ribbons into a red one. It would appear that we must still have a monochromatic clique of size 3. This is because we have at least two monochromatic cliques, and changing one ribbon may not destroy both of them. However, this is not the case, as shown by the following surprising result.

Golomb-Taylor Theorem.
There exists a clique of size $R(m, n)$ in which one ribbon is red and which contains no yellow cliques of size m and no blue cliques of size n.

Proof:
Start with a clique of size $R(m, n) - 1$ so that it does not contain either a yellow clique of size m or a blue clique of size n. Let the twin of one of the people be added to the clique, with the twins holding a red ribbon between them. The new arrival holds the same color ribbon with other people as the other twin. Clearly, they cannot both be in a yellow or blue clique of any size. However, if they are not, then the new arrival may as well not be there, and as before, we have no yellow cliques of size m or blue cliques of size n.

Returning to the party of six people, if we allow one pair of them not to disclose whether they know each other or not, we need not have three people who acknowledge knowing one another or three people who acknowledge not knowing one another. This is demonstrated in the diagram below, with the two vertices at the top representing the two designated people.

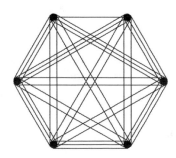

Examples

Example 4.6.1.

(a) Prove that in a 12-clique lower bound for $R(3,5)$, everyone is holding either 3 or 4 yellow ribbons.

(b) Prove that in a 13-clique lower bound for $R(3,5)$, everyone is holding exactly 4 yellow ribbons.

Solution:

(a) Consider a particular person X. Suppose X is holding a yellow ribbons and b blue ribbons. Then $a + b = 11$. Suppose $a \leq 2$. Then we have $b \geq 9 = R(3,4)$. Thus among the 9-clique of people holding blue ribbons with X, there is either a yellow trio or a blue quartet, and the latter becomes a blue quintet when X is included. This is a contradiction. Suppose $a \geq 5$. If there is a yellow ribbon among the 5-clique of people holding yellow ribbons with X, then we have a yellow trio. Otherwise, we have a blue quintet. Again we have a contradiction. It follows that $a = 3$ or 4. Since X is arbitrarily chosen, everyone is holding either 3 or 4 yellow ribbons.

(b) Consider a particular person X. Suppose X is holding a yellow ribbons and b blue ribbons. Then $a + b = 12$. Suppose $a \leq 3$. Then we have $b \geq 9 = R(3,4)$. Thus among the 9-clique of people holding blue ribbons with X, there is either a yellow trio or a blue quartet, and the latter becomes a blue quintet when X is included. This is a contradiction. Suppose $a \geq 5$. If there is a yellow ribbon among the 5-clique of people holding yellow ribbons with X, then we have a yellow trio. Otherwise, we have a blue quintet. Again we have a contradiction. It follows that $a = 4$. Since X is arbitrarily chosen, everyone is holding exactly 4 yellow ribbons.

Example 4.6.2.

Prove that if an 18-clique lower bound for $R(3,6)$ exists, then everyone is holding exactly 5 yellow ribbons.

Solution:

Consider a particular person X. Suppose X is holding a yellow ribbons and b blue ribbons. Then $a + b = 17$. Suppose $a \leq 3$. Then $b \geq 14 = R(3,5)$. Thus among the 13-clique of people holding blue ribbons with X, there is either a yellow trio or a blue quintet, and the latter becomes a blue sextet when X is included. This is a contradiction. Suppose $a \geq 6$. If there is a yellow ribbon among the 6-clique of people holding yellow ribbons with X, then we have a yellow trio. Otherwise, we have a blue sextet. Again we have a contradiction. It follows that $a = 4$ or 5, and this is true of everyone since X is arbitrarily chosen.

Suppose $a = 4$ for X. Let X be holding yellow ribbons with K, L, M and N. Now the 13-clique of people holding blue ribbons with X must form a lower bound for $R(3, 5)$. By Example 4.6.1(b), each of them is holding exactly 4 yellow ribbons within this 13-clique. This means that each can be holding a yellow ribbon with at most one of K, L, M and N. Each of these four is holding 3 or 4 yellow ribbons with people in the 13-clique, and since the number of such ribbons is at most 13, at least one, say K, is holding yellow ribbons with exactly 3 people in the 13-clique, say R, S and T. Clearly, these three hold blue ribbons among themselves to avoid forming a yellow trio with K. Now L, M, N, R, S and T is a blue sextet, which is a contradiction. It follows that $a = 5$. Since X is arbitrarily chosen, everyone is holding exactly 5 yellow ribbons.

Example 4.6.3.
In a clique of size $2n$, what is the maximum number of yellow ribbons such that there are no yellow cliques of size 3?

Solution:
For $n = 1$, we have two people A and B holding a yellow ribbon between them. There are no yellow cliques of size 3, and the number of yellow ribbons is $n^2 = 1$, the largest possible. For $n > 1$, imagine a scenario in which fertility drugs run amok and each of A and B is in a separate set of n-tuples. Two people in the same family hold a blue ribbon between them, and two people in different families hold a yellow ribbon between them. Again there are no yellow cliques of size 3, and the number of yellow ribbons is n^2. We now show that this is indeed maximum. There are finitely many cliques of size $2n$ with no yellow cliques of size 3. By the Extremal Value Principle, there exists one for which the number of yellow ribbons is maximum. We claim that in this clique, there do not exist three people P, Q and R such that P and Q hold a yellow ribbon between them, while P and R hold a blue ribbon between them, and Q and R also hold a blue ribbon between them. Suppose to the contrary that they do exist. Let the number of yellow ribbons held by R be a. We consider two cases.

Case 1. Either P or Q holds more than a yellow ribbons.

By symmetry, we may assume that P holds more than a yellow ribbons. Replace R by a twin P' of P. P' and P hold a blue ribbon between them, and otherwise, P' holds exactly the same color pattern of ribbons as P. In this process, we lose the a yellow ribbons formerly held by R, but gain back more because P holds more than a yellow ribbons. This is a contradiction as the initial clique contains the maximum number of yellow ribbons.

Case 2. Both P and Q hold at most a yellow ribbons.

Replace P and Q by R' and R", who form triplets with R. They hold blue ribbons among the three of them, but otherwise each holds exactly the same color pattern of ribbons as R. In this process, we lose at most $2a - 1$ ribbons held by P and Q, because the yellow ribbon held by P and Q is counted once on each end. On the other hand, we gain back at least $2a$ yellow ribbons held by R' and R". This is again a contradiction.

We have justified the claim, which implies that the relation of not holding a yellow ribbon between two people has the transitive property. Since it obviously has the reflexive and symmetric properties, the people are partitioned into families so that two people within the same family hold a blue ribbon between them, and two people from different families hold a yellow ribbon between them. Since there are no yellow cliques of size 3, the number of families is exactly two. Moreover, if their sizes are not equal, a transfer of one person from the larger one to the smaller one will once again result in an increase in the number of yellow ribbons. It follows that the clique with the maximum number of yellow ribbons is unique, and consists of exactly two families of equal sizes. If the original clique is of size $2n + 1$, then one family is of size n and the other is of size $n + 1$.

Remark:

We could have used red ribbons instead of blue ones. Then it becomes clear that we are actually applying the Golomb-Taylor argument to a one-color Ramsey problem.

Exercises

1. Prove that $R(3,4) = 9$.

2. Prove that $R(3,5) = 14$.

3. Prove that $R(4,4) = 18$.

4. Prove that $R(3,6) = 18$.

5. Show that in a clique of size 10, if 5 of the ribbons are red while the others are yellow and blue, it is possible not to have a monochromatic clique of size 3.

6. Show that in a clique of size 19, if 3 of the ribbons are red while the others are yellow and blue, it is possible not to have a monochromatic clique of size 4.

Chapter Five: Counting under Symmetries

Section 5.1. Finite Groups

A **group** is a set S of elements with a binary operation $\odot$ having the following properties.

(1). The Closure Property.
If a and b are elements in S, then $a \odot b$ is also an element in S.

(2). The Associativity Property.
If a, b and c are elements in S, $(a \odot b) \odot c = a \odot (b \odot c)$.

(3). The Identity Property.
There is an element e in S, called the **identity** of the group, such that $a \odot e = a = e \odot a$ for any element a in S.

(4). The Inverse Property.
For any element a in S, there exists an element a^{-1} in S, called the **inverse** of the element a, such that $a \odot a^{-1} = e = a^{-1} \odot a$.

A **subgroup** of a group is a subset of its elements which also form a group with the same binary operation. Note that the Commutativity Property, which states that $a \odot b = b \odot a$ for any elements a and b in S, is not a requirement in the definition of a group. However, many groups do have this property. Such groups are called **commutative** or **Abelian** groups. In many familiar situations, $\odot$ is often replaced by $+$ in Abelian groups and by $\times$ in all kinds of groups.

Why are we interested in groups? As an attempt to answer this question, we consider a two-player game played with nine cards numbered from 1 to 9. The players alternately take a card that has not been chosen, until either player has three cards which add up to fifteen, or until all the cards are gone. In the former case, the player with the three cards adding up to fifteen wins. In the latter case, the game is a draw.

Faced with an unfamiliar game, we are often at a loss on how to start. We can first find all the winning combinations. As it happens, there are exactly eight of them, namely, 1+5+9=15, 1+6+8=15, 2+4+9=15, 2+5+8=15, 2+6+7=15, 3+4+8=15, 3+5+7=15, and 4+5+6=15. These can be incorporated into a diagram called a Magic Square. The winning combinations are precisely the three rows, the three columns and the two diagonals. So, we have been playing the familiar game Tic-Tac-Toe, where taking a number corresponds to marking a symbol in the square where that number is.

2	7	6
9	5	1
4	3	8

© The Author(s), under exclusive license to Springer Nature Switzerland AG 2021
S. W. Golomb, A. Liu, *Solomon Golomb's Course on Undergraduate Combinatorics*,
https://doi.org/10.1007/978-3-030-72228-9_6

Thus the two games are **isomorphic**, meaning having the same form underneath, despite superficial non-resemblance. Knowing this makes it unnecessary to analyze the Game of Fifteen, but isomorphism is far more important than just that. When we recognize that many different entities are isomorphic to the same structure, we know we have a very important structure on our hands. Groups are such structures, cropping up in many unexpected places.

A group may have infinitely many elements. The set of all integers under addition is such an example. However, we are primarily interested in finite groups. Then we can write down the entire operation table, often called the **Cayley table** of the group.

Some properties of a finite group are immediately obvious from its Cayley table. If the elements are symmetrically situated about the main diagonal, the group is Abelian. For each element a, find the identity on its row. The element a^{-1} is the one on whose column this copy of the identity lies. In particular, copies of the identity on the main diagonal correspond to elements which are their own inverses.

All entries in each row and in each column of any Cayley table are distinct. This is because of the following result.

Cancellation Property.
For group elements a, b and c such that $ab = ac$, we must have $b = c$.

Proof:
We have $b = eb = (a^{-1}a)b = a^{-1}(ab) = a^{-1}(ac) = (a^{-1}a)c = ec = c$.

The Associativity Property is not immediately obvious from the Cayley table. Fortunately, in two large classes of examples, verification is not required. The first class consists of groups whose elements are numbers and whose operations are addition or multiplication. The associativity is inherited from the properties of the number systems.

For any positive integer n, let Z_n be the set $\{0,1,2,\ldots,n-1\}$ under addition modulo n. This is an example of a **cyclic** group, in that every element is generated by a single element, namely 1: $1=1$, $2=1+1$, $3=2+1$, $4=3+1$, and so on. The set Z_n^*, consisting of all positive integers less than or equal to n and relatively prime to n, also forms a group under multiplication modulo n. The Cayley table for the case $n = 8$ is shown below.

$\times_8$	1	3	5	7
1	1	3	5	7
3	3	1	7	5
5	5	7	1	3
7	7	5	3	1

In the second class of groups for which the Associativity Property is easy to verify, the elements are functions or transformations and the operation is composition. Let A, B and C be transformations. Both $(AB)C$ and $A(BC)$ boil down to performing each of A, B and C in turn, and must have the same result.

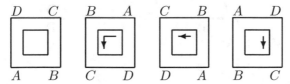

It turns out that every group can be interpreted as a group of transformations on something. For example, the integers with respect to addition can be interpreted as shifts of the number line. Each positive integer k shifts the line k places to the right while each negative integer k shifts the line $|k|$ places to the left.

The $n!$ permutations of $1, 2, \ldots, n$ constitute a transformation group on these n elements. It is called the **symmetric group** of order n, and is denoted by S_n. The subgroup consisting of all the even permutations is called the **alternating group** of order n, and is denoted by A_n.

If we rotate a square about its center, it will coincide with itself after $90°$, $180°$, $270°$ and $360°$. A transformation which maps a geometric object onto itself is called a **symmetry** of the geometric object. In the plane, there are only two kinds of symmetries, rotations and reflections. The $360°$ rotation, after which every point of the square returns to its original position, is a special symmetry called the **identity**. We denote this by I, and the other three rotational symmetries of the square by R ($180°$ rotation or half-turn), A (anticlockwise $90°$ rotation) and C (clockwise $90°$ rotation). Rotations always form a cyclic group. The pictograms of these symmetries and their effects are shown in the diagram below.

D C	B A	C B	A D
□	↱	←	↓
A B	C D	D A	B C

In addition to rotational symmetries, the square also possesses reflectional symmetries. Rotations are direct symmetries in that if the vertices of the square are labeled $ABCD$ in anticlockwise order, they will still be in anticlockwise order under such a symmetry. Reflections, on the other hand, are opposite symmetries which change the labels into clockwise order. For the square, there are four reflectional symmetries, H (reflection about the horizontal axis), V (reflection about the vertical axis), U (reflection about the up-diagonal) and D (reflection about the down-diagonal). The pictograms of these symmetries and their effects are shown in the diagram below.

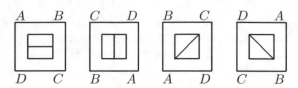

Does the square have any more symmetries? The answer is no. Consider a particular vertex. It can land on any of the four vertices. Then the remaining vertices must follow in order, either anticlockwise or clockwise. Thus there are only $4 \times 2 = 8$ symmetries in all.

When we perform two symmetries in a row, the end result is another symmetry. This operation is called a **composition**. For instance, the diagram below shows that the net result of performing H followed by A is the same as performing U. We write $HA = U$.

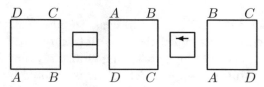

On the other hand, the diagram below shows that $AH = D$, so that this group is non-Abelian.

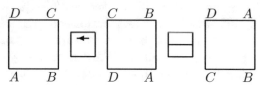

In a similar way, we can compute other compositions and come up with the following operation table.

	I	R	A	C	H	V	U	D
I	I	R	A	C	H	V	U	D
R	R	I	C	A	V	H	D	U
A	A	C	R	I	D	U	H	V
C	C	A	I	R	U	D	V	H
H	H	V	U	D	I	R	A	C
V	V	H	D	U	R	I	C	A
U	U	D	V	H	C	A	I	R
D	D	U	H	V	A	C	R	I

The eight symmetries of the square, along with the operation of composition, constitute the **dihedral group** of the square. The symmetry I naturally serves as the identity. Each element is its own inverse except for C and A, which are inverses of each other. The four symmetries I, R, A and C form the **rotational group** of the square. While the dihedral group of the square is non-Abelian, the rotational group of the square is Abelian.

Two groups G and K are said to be **isomorphic** if their elements can be put in one-to-one correspondence such that, if $g_1 \in G$ corresponds to $k_1 \in K$ and $g_2 \in G$ corresponds to $k_2 \in K$, then $g_1 g_2 \in G$ corresponds to $k_1 k_2 \in K$. For finite groups, this means that their Cayley tables can be arranged to be identical except for the names of the elements.

A necessary condition for two finite groups to be isomorphic is that they must have the same number of elements. However, this condition is not sufficient. For instance, Z_8 is not isomorphic to the dihedral group of the square since the former is Abelian and the latter is not.

On the other hand, $Z_4 = \{0,1,2,3\}$ under addition, $Z_5^* = \{1,2,3,4\}$ under multiplication, $\{1, -1, i, -i\}$ under multiplication, and $\{I, R, A, C\}$ under composition, are all isomorphic to one another. Thus the group operation can be changed in an isomorphism. The Cayley tables of Z_5^* and X are shown below. The Cayley table of the last one is contained in the Cayley table of the dihedral group of the square.

$\times_5$	1	2	3	4
1	1	2	3	4
2	2	4	1	3
3	3	1	4	2
4	4	3	2	1

$\times$	1	-1	i	$-i$
1	1	i	$-i$	-1
i	i	-1	1	$-i$
$-i$	$-i$	1	-1	i
-1	-1	$-i$	i	1

Examples

Example 5.1.1.

The set $Z_2 \times Z_4$ consists of the elements (0,0), (0,1), (0,2), (0,3), (1,0), (1,1), (1,2) and (1,3).

(a) Show that it forms a group with respect to addition modulo 2 in the first component and addition modulo 4 in the second, by constructing the Cayley table.

(b) Determine if it is isomorphic to either Z_8 or the dihedral group of the square.

Solution:

(a) The Cayley table is shown below.

$+_{2/4}$	(0,0)	(0,1)	(0,2)	(0,3)	(1,0)	(1,1)	(1,2)	(1,3)
(0,0)	(0,0)	(0,1)	(0,2)	(0,3)	(1,0)	(1,1)	(1,2)	(1,3)
(0,1)	(0,1)	(0,2)	(0,3)	(0,0)	(1,1)	(1,2)	(1,3)	(1,0)
(0,2)	(0,2)	(0,3)	(0,0)	(0,1)	(1,2)	(1,3)	(1,0)	(1,1)
(0,3)	(0,3)	(0,0)	(0,1)	(0,2)	(1,3)	(1,0)	(1,1)	(1,2)
(1,0)	(1,0)	(1,1)	(1,2)	(1,3)	(0,0)	(0,1)	(0,2)	(0,3)
(1,1)	(1,1)	(1,2)	(1,3)	(1,0)	(0,1)	(0,2)	(0,3)	(0,0)
(1,2)	(1,2)	(1,3)	(1,0)	(1,1)	(0,2)	(0,3)	(0,0)	(0,1)
(1,3)	(1,3)	(1,0)	(1,1)	(1,2)	(0,3)	(0,0)	(0,1)	(0,2)

(b) This Abelian group is not isomorphic to the dihedral group of the square, which is non-Abelian. It is also not isomorphic to Z_8 because it has four self-inverse elements whereas Z_8 has only two.

Example 5.1.2.

The set $Z_{24}^* = \{1, 5, 7, 11, 13, 17, 19, 23\}$ consists of the positive integers less than 24 and relatively prime to 24.

(a) Show that they form a group with respect to multiplication modulo 24 by constructing the Cayley table.

(b) Determine if it is isomorphic to either Z_8, $Z_2 \times Z_4$ or the dihedral group of the square.

Solution:

(a) The Cayley table is shown below.

$\times_{24}$	1	5	7	11	13	17	19	23
1	1	5	7	11	13	17	19	23
5	5	1	11	7	17	13	23	19
7	7	11	1	5	19	23	13	17
11	11	7	5	1	23	19	17	13
13	13	17	19	23	1	5	7	11
17	17	13	23	19	5	1	11	7
19	19	23	13	17	7	11	1	5
23	23	19	17	13	11	7	5	1

(b) This Abelian group is not isomorphic to the dihedral group of the square, which is non-Abelian. It is also not isomorphic to Z_8 or $Z_2 \times Z_4$ because it has eight self-inverse elements whereas the others have only two or four.

Example 5.1.3.
Consider a particular vertex of an equilateral triangle. Under a symmetry, it can land on any of the three vertices. The remaining vertices must follow in order, either anticlockwise or clockwise. Thus there are only $3 \times 2 = 6$ symmetries for the equilateral triangle. They are the identity I, a $120°$ anticlockwise rotation J, a $120°$ clockwise rotation K and three reflections X, Y and Z. Their pictograms are shown in the diagram below.

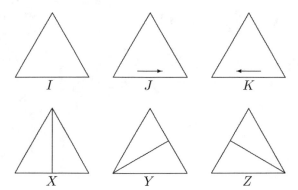

(a) Show that these six symmetries form a group, called the **dihedral group** of the equilateral triangle, with respect to composition, by constructing the Cayley table.

(b) Determine if it is isomorphic to Z_6.

Solution:

(a) The Cayley table is shown below.

$\circ$	I	J	K	X	Y	Z
I	I	J	K	X	Y	Z
J	J	K	I	Y	Z	X
K	K	I	J	Z	X	Y
X	X	Z	Y	I	K	J
Y	Y	X	Z	J	I	K
Z	Z	Y	X	K	J	I

(b) This group is non-Abelian, and is therefore not isomorphic to Z_6.

Remark:

The dihedral group of the equilateral triangle is isomorphic to the symmetric group S_3, consisting of all six permutations of the three vertices.

Exercises

1. The set $Z_2 \times Z_3$ consists of the elements $(0,0)$, $(0,1)$, $(0,2)$, $(1,0)$, $(1,1)$ and $(1,2)$.

   (a) Show that it forms a group with respect to addition modulo 2 in the first component and addition modulo 3 in the second, by constructing the Cayley table.

   (b) Determine if it is isomorphic to either Z_6 or the dihedral group of the equilateral triangle.

2. The set $Z_2 \times Z_2 \times Z_2$ consists of the elements $(0,0,0)$, $(0,0,1)$, $(0,1,0)$, $(0,1,1)$, $(1,0,0)$, $(1,0,1)$, $(1,1,0)$ and $(1,1,1)$.

   (a) Show that it forms a group with respect to addition modulo 2 in each component by constructing the Cayley table.

   (b) Determine if it is isomorphic to either Z_8, $Z_2 \times Z_4$, Z_{24}^* or the dihedral group of the square.

3. The set $Z_{15}^* = \{1, 2, 4, 7, 8, 11, 13, 14\}$ consists of positive integers less than 15 and relatively prime to 15.

   (a) Show that they form a group with respect to multiplication modulo 15 by constructing the Cayley table.

   (b) Determine if it is isomorphic to either Z_8, $Z_2 \times Z_4$, Z_{24}^* or the dihedral group of the square.

4. The **Quaternion group** consists of the elements 1, -1, i, $-i$, j, $-j$, k and $-k$, such that $i^2 = j^2 = k^2 = -1$, $ij = k$, $jk = i$, $ki = j$, $ji = -k$, $kj = -i$ and $ik = -j$.

   (a) Show that it forms a group with respect to multiplication by constructing the Cayley table.

   (b) Determine if it is isomorphic to either Z_8, $Z_2 \times Z_4$, Z_{24}^* or the dihedral group of the square.

5. Consider the set of symmetries of a non-square rectangle. It has only four elements. They are I, R, H and V, analogous to the corresponding symmetries for the square.

   (a) Show that they form a group with respect to composition by constructing the Cayley table.

   (b) Determine if it is isomorphic to either Z_4 or $Z_2 \times Z_2$.

6. Let $a(x) = \frac{1}{1-x}$, $b(x) = \frac{x-1}{x}$, $c(x) = 1 - x$, $d(x) = \frac{x}{x-1}$, $e(x) = x$ and $f(x) = \frac{1}{x}$.

   (a) Show that they form a group with respect to composition by constructing the Cayley table.

   (b) Determine if it is isomorphic to Z_6 or the dihedral group of the equilateral triangle.

Section 5.2. Designs and Patterns

If we join the midpoints of opposite sides of a square, we divide it into 4 cells. We paint each of them black or white. We have $2^4 = 16$ distinct **designs** because each of the 4 cells can be painted in either of the 2 colors.

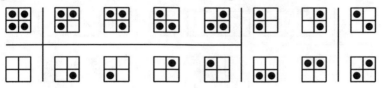

Under rotational symmetries of the squares, various designs become indistinguishable from one another and merge into a **pattern**. The relation "of the same pattern" among designs has the following properties.

(1) **Reflexivity Property.**

Each design is of the same pattern as itself. It is mapped onto itself by I.

(2) **Transitivity Property.**

If design A is of the same pattern as design B, and design B is of the same pattern as design C, then design A is of the same pattern as design C. The symmetry which maps A to C is the product of the symmetries which map A to B and B to C.

(3) **Symmetry Property.**

If design A is of the same pattern as design B, then design B is of the same pattern as design A. The symmetry which maps B to A is the inverse of the symmetry which maps A to B.

Such a relation is called an **equivalence relation**. It partitions the design into equivalence classes each of which is represented by a pattern.

In the problem of the 2×2 square, there are six patterns altogether, consisting of
(1) the design with all four cells black;
(2) the design with all four cells white;
(3) two designs with two diagonally opposite cells black and the other two white;
(4) four designs with two adjacent cells black and the other two white;
(5) four designs with three cells black and the other one white;
(6) four designs with one cell black and the other three white.

However, this is counting the hard way. While we can be reasonably satisfied that we have not missed out on anything here, such confidence would be misplaced in a slightly more complicated problem of counting patterns when a 3×3 square is divided into 9 cells. We seek a better approach in which the task does not become significantly more difficult when the size of the problem increases.

When reflectional symmetries are also under consideration, we assume that colors permeate through to both sides of the square, so that we have four pairs of back-to-back cells in the same color. It is routine to verify that even with reflections, the number of patterns is still 6.

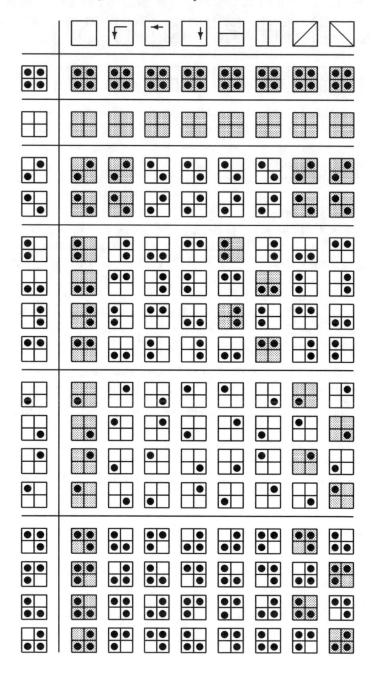

Our study of finite groups in the preceding section provides some mathematical background that would prove helpful. However, we still need a break-through idea which would allow us to bring this powerful knowledge into play.

We now examine carefully the action of each of the eight symmetries on each of the sixteen designs, shown on the preceding page. Some of the results catch our attention immediately, in that they are exactly the same as before. A design is said to be **invariant** under a symmetry if the same design results from performing the symmetry. The invariant entries are shaded.

Clearly, every design is invariant under I, and the designs consisting of four black cells and of four white cells are invariant under every symmetry. Our symmetry group is non-Abelian, and there are pairs of distinct elements which occupy symmetric positions in the Cayley table with respect to the main diagonal. These pairs are A and C, H and V, and D and U. Each element in a pair generates the same cyclic subgroup. The number of invariants under A is equal to the number of invariants under C. The same holds between H and V, and between D and U.

Looking at the situation overall, we make the following Observation: *The number of invariant entries within each pattern is equal to the total number of symmetries under consideration.*

We give an illustration with pattern (6), consisting of the four designs in which one cell is black and the other three white. We choose arbitrarily one design as the Home Base and call the others the First Base, the Second Base and the Third Base.

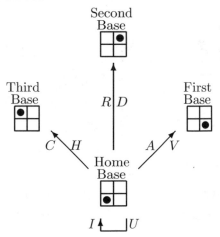

Two designs belong to the same pattern if and only if there is a symmetry which takes one to the other. From any one design, each symmetry takes it either back to itself or to one of the other designs in the same pattern.

The Home Base is invariant under two symmetries, namely, I and U. The symmetries A and V take it to the First Base, the symmetries R and D take it to the Second Base and the symmetries C and H take it to the Third Base.

We claim that the number of symmetries under which another Base is invariant is equal to the number of symmetries going to it from the Home Base. Since each symmetry takes the Home Base either back to itself or to another Base, the total number of invariants within this pattern must be eight.

Consider for example the First Base. One of the symmetries which takes it back to the Home Base is $A^{-1} = C$. Now the symmetry CS takes the First Base to the same destination as the symmetry S takes the Home Base. Moreover, the symmetries CI, CR, CA, CC, CH, CV, CU and CD are distinct by the Cancellation Law. As a matter of fact, they are C, A, I, R, U, D, V and H respectively.

Since the Home Base is invariant under I and U, the symmetries $CI = C$ and $CU = V$ take the First Base to the Home Base. Similarly, the symmetries $CR = A$ and $CD = H$ go from the First Base to the Second Base, the symmetries $CC = R$ and $CH = U$ go from the First Base to the Third Base, and the First Base is invariant under the symmetries $CA = I$ and $CV = D$, and only under these two. This can also be verified directly.

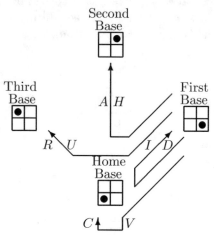

We can analyse the Second Base and the Third Base in the same way. This justifies our claim that there are exactly eight invariant entries within this pattern.

This illustrative example may be reworded into a formal proof of the following result, which had at various times been attributed to Pólya, Burnside, Frobenius and Cauchy, in reverse chronological order.

The Main Theorem.

The number of patterns is given by the total number of invariant entries under all symmetries divided by the number of symmetries.

The fact that the invariants are divided equally among the designs within a pattern is not essential to the proof, though it would have been most surprising were it not the case. This can be established easily. In the example above, the Home Base has two invariants. Then there are exactly two symmetries going from each of the other Bases to the Home Base, because each of those Bases could have been chosen as the Home Base. Now the number of symmetries which takes a Base to the Home Base must be equal to the number of symmetries which takes the Home Base to that Base. Hence there are exactly two symmetries from the Home Base to every other Base. It then follows that every Base has two invariants.

Examples

Example 5.2.1.

If we join the midpoints of the sides of an equilateral triangle to one another, we divide it into 4 cells. We paint each of them black or white.

(a) Draw all the designs.

(b) Find the number of patterns under rotational symmetries.

(c) Find the number of patterns under all symmetries.

Solution:

(a) We have $2^4 = 16$ distinct designs because each of the 4 cells can be painted in either of the 2 colors. These are shown in the diagram below.

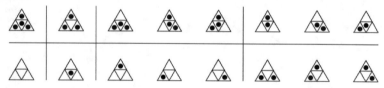

(b) There are eight patterns altogether, consisting of
   (1) the design with all four cells black;
   (2) the design with all four cells white;
   (3) three designs with two corner cells black and the other two white;
   (4) three designs with two corner cells white and the other two black;
   (5) the design with the center cell black and the other three white;
   (6) the design with the center cell white and the other three black;
   (7) three designs with one corner cell black and the other three white;
   (8) three designs with one corner cell white and the other three black.
   These are marked in the diagram above.

(c) We have exactly the same eight patterns as before.

Example 5.2.2.

If we join the midpoints of the sides of an equilateral triangle to one another, we divide it into 4 cells. We paint each of them black or white. Verify directly that the number of invariant entries within each pattern is equal to six, the number of symmetries.

Solution:

This is shown in the diagram below, where the invariant entries are shaded.

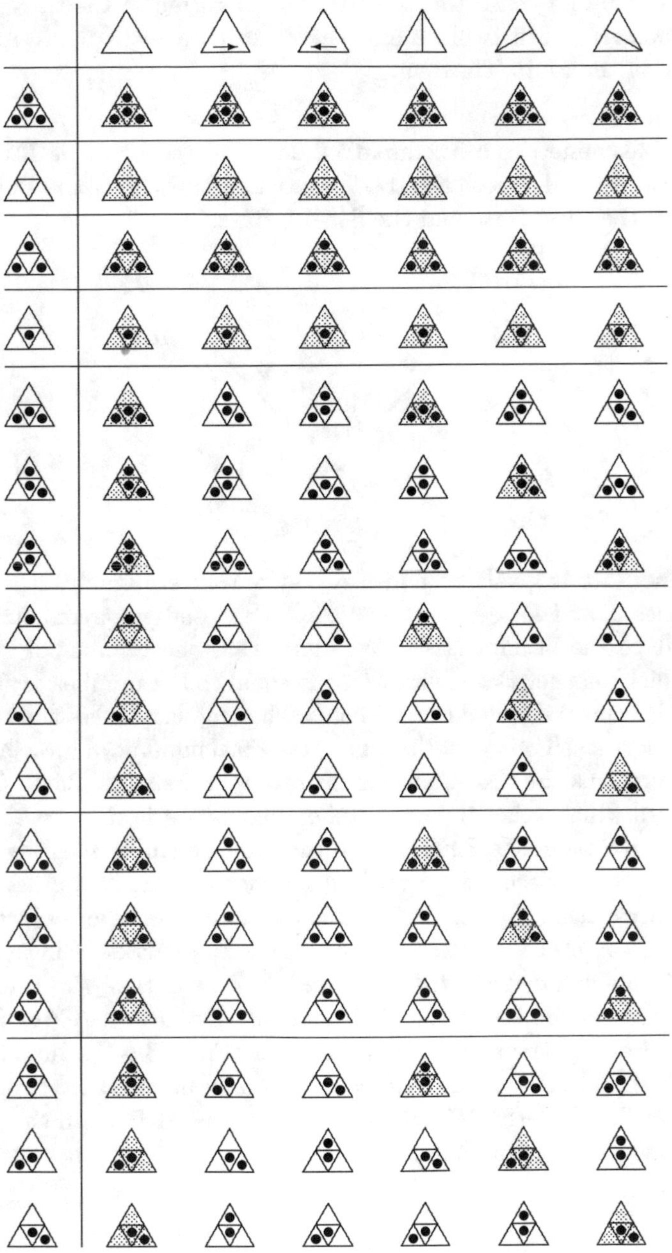

Example 5.2.3.

If we join the midpoints of the sides of an equilateral triangle to one another, we divide it into 4 cells. Use the pattern consisting of the three designs in which one corner cell is black and the other three cells white to illustrate the proof of the Main Theorem.

Solution:

Let the six symmetries in Example 5.2.2 be denoted by I, J, K, X, Y and Z respectively. We choose arbitrarily one design as the Home Base and call the others the First Base and the Second Base.

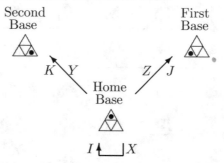

The Home Base is invariant under two symmetries, namely, I and X. The symmetries J and Z take it to the First Base and the symmetries K and Y take it to the Second Base. We claim that the number of symmetries under which another Base is invariant is equal to the number of symmetries going to it from the Home Base. Since each symmetry takes the Home Base either back to itself or to another Base, the total number of invariants within this pattern must be six. Consider for example the First Base. One of the symmetries which takes it back to the Home Base is $J^{-1} = K$. Now the symmetry KS takes the First Base to the same destination as the symmetry S takes the Home Base. Moreover, the symmetries KI, KJ, KK, KX, KY and KZ are distinct by the Cancellation Law. As a matter of fact, they are K, I, J, Z, X and Y respectively. Since the Home Base is invariant under I and X, the symmetries $KI = K$ and $KX = Z$ take the First Base to the Home Base. Similarly, the symmetries $KK = J$ and $KY = X$ go from the First Base to the Second Base, and the First Base is invariant under the symmetries $KJ = I$ and $KZ = Y$, and only under these two. This can also be verified directly. We can analyse the Second Base in the same way. This justifies our claim that there are exactly six invariant entries within this pattern.

Exercises

1. If we join the vertices of an equilateral triangle to its center, we divide it into 3 cells. We paint each of them black or white.

   (a) Draw all the designs.

   (b) Find the number of patterns under rotational symmetries.

   (c) Find the number of patterns under all symmetries.

2. A 2×4 rectangle is divided into 4 cells, each a 1×2 rectangle, as shown in the diagram below. We paint each cell black or white.

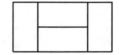

   (a) Draw all the designs.

   (b) Find the number of patterns under rotational symmetries.

   (c) Find the number of patterns under all symmetries.

3. If we join the vertices of an equilateral triangle to its center, we divide it into 3 cells. We paint each of them black or white. Verify directly that the number of invariant entries within each pattern is equal to six, the number of symmetries.

4. A 2×4 rectangle is divided into 4 cells, each a 1×2 rectangle, as shown in the diagram below. We paint each cell black or white. Verify directly that the number of invariant entries within each pattern is equal to four, the number of symmetries.

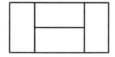

5. If we join the vertices of an equilateral triangle to its center, we divide it into 3 cells. Use the pattern consisting of the three designs in which one cell is black and the other two cells white to illustrate the proof of the Main Theorem.

6. A 2×4 rectangle is divided into 4 cells, each a 1×2 rectangle, as shown in the diagram below. Use the pattern consisting of the four designs in which one end cell and one middle cell are black and the other two cells white to illustrate the proof of the Main Theorem.

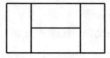

Section 5.3. Cycle Indices

By the Main Theorem in the preceding section, the problem of counting patterns has been converted to the problem counting invariant entries. What is a good way to count invariant entries? It is just as difficult as counting patterns if we do it design by design.

However, it turns out to be a much simpler task if we do it symmetry by symmetry. Let us illustrate with the example where a square is divided into 4 cells by joining the midpoints of opposite sides, and each cell is painted black or white.

Two cells are said to be in the same **cycle** under a given symmetry S if one can be mapped into the other by S, perhaps applied more than once. In order for a design to be invariant under S, all cells in the same cycle must have the same color.

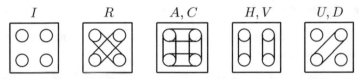

Under I, each cell forms a cycle of length 1. We record this as x_1^4. Under R, opposite cells form two cycles of length 2. We record this as x_2^2. Under C or A, all four cells form a cycle of length 4. We record this as x_4. Under H or V, adjacent cells form two cycles of length 2. We record this as x_2^2. Finally, under U or D, two opposite cells forms a cycle of length 2, while each of the other two cells forms a cycle of length 1. We record this as $x_1^2 x_2$.

Summing these terms and dividing by the number of symmetries, we obtain the **cycle index** for the problem, namely,

$$\frac{x_1^4 + 3x_2^2 + 2x_4 + 2x_1^2 x_2}{8}.$$

Note that $1 \times 4 = 2 \times 2 = 4 \times 1 = 2 \times 1 + 2 = 4$, the number of cells.

In order for a design to be invariant under S, all cells in the same cycle must have the same color. Colors for different cycles may be chosen independently. Hence the number of invariant designs under S is given by n^m, where n is the number of colors and m is the number of cycles under S.

It follows that the total number of invariant entries may be obtained by setting $x_1 = x_2 = x_4 = 2$, which yields $2^4 + 3 \times 2^2 + 2 \times 2 + 2 \times 2^3 = 48$. Dividing by 8, we obtain 6, confirming our direct count in the preceding section.

Without the reflections, the revised cycle index is $\frac{x_1^4 + x_2^2 + 2x_4}{4}$. The number of patterns is $\frac{2^4 + 2^2 + 2 \times 2^1}{4} = 6$ also.

If the cells are painted in red, white or blue instead, the answers may be obtained by setting $x_1 = x_2 = x_4 = 3$. They are respectively 21 and 24.

We now venture to the third dimension, and consider the rotational symmetries of a cube. There are three types of axes of rotations, those joining the centers of opposite faces, those joining the midpoints of opposite edges and those joining opposite vertices. The three types of rotational axes are illustrated in the diagram below.

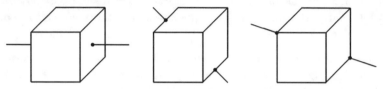

There are altogether 24 rotational symmetries of the cube.

(C1) The identity symmetry.

(C2) Three 180° rotational symmetries about axes joining the centers of opposite faces.

(C3) Six 90° rotational symmetries about axes joining the centers of opposite faces.

(C4) Six 180° rotational symmetries about axes joining the midpoints of opposite edges.

(C5) Eight 120° rotational symmetries about axes joining opposite vertices.

This total may be obtained without a detailed breakdown. Each face of the cube can be mapped onto any of the six faces, each in four different rotational positions. Hence the total number of symmetries of the cube is indeed $6 \times 4 = 24$.

Suppose each face of a $2 \times 2 \times 2$ cube is divided into four unit squares by joining the midpoints of opposite sides, and the 24 unit squares are to be painted in black or white. We wish to know how many such cubes are distinguishable.

There are 24 cells each of which is a cycle of length 1 under (C1). Under (C2), they form 12 cycles of length 2. Under (C3), they form 6 cycles of length 4. Under (C4), they form 12 cycles of length 2. Under (C5), they form 8 cycles of length 3. Hence the cycle index is

$$\frac{x_1^{24} + 9x_2^{12} + 6x_4^6 + 8x_3^8}{24}.$$

Setting $x_1 = x_2 = x_3 = x_4 = 2$, $\frac{2^{24} + 9 \times 2^{12} + 6 \times 2^6 + 8 \times 2^8}{24} = 700688$ such cubes are distinguishable.

We now consider the rotational symmetries of the regular tetrahedron. There are two types of axes of rotation, those joining vertices to the centers of opposite faces and those joining the midpoints of opposite edges. The two types of rotational axes are illustrated in the diagram below.

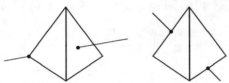

There are altogether 12 rotational symmetries of the regular tetrahedron.

(T1) The identity symmetry.

(T2) Eight 120° rotational symmetries about axes joining vertices to the centers of opposite faces.

(T3) Three 180° rotational symmetries about axes joining the midpoints of opposite edges.

This total may be obtained without a detailed breakdown. Each face of the regular tetrahedron can be mapped onto any of the four faces, each in three different rotational positions. Hence the total number of symmetries of the regular tetrahedron is indeed $4 \times 3 = 12$.

Suppose each face of a regular tetrahedron is divided into four equilateral triangles by joining the midpoints of the sides, and the 16 equilateral triangles are to be painted in red, white or blue. We wish to know how many such regular tetrahedra are distinguishable.

There are 16 cells each of which is a cycle of length 1 under (T1). Under (T2), they form 1 cycle of length 1 and 5 cycles of length 3. Under (T3), they form 8 cycles of length 2. Hence the cycle index is

$$\frac{x_1^{16} + 8x_1 x_3^5 + 3x_2^8}{12}.$$

Setting $x_1 = x_2 = x_3 = 3$, $\frac{3^{16} + 8 \times 3^6 + 3 \times 3^8}{12} = 3589353$ such regular tetrahedra are distinguishable.

All our problems so far involve painting regions. We can also paint segments and points. For instance, we may wish to paint in black or white the vertices of a cube, the midpoints of its edges and the centers of its faces.

There are 26 cells each of which is a cycle of length 1 under (C1). Under (C2), they form 2 cycles of length 1 and 12 cycles of length 2. Under (C3), they form 2 cycles of length 1 and 6 cycles of length 4. Under (C4), they form 2 cycles of length 1 and 12 cycles of length 2. Under (C5), they form 2 cycles of length 1 and 8 cycles of length 3. Hence the cycle index is

$$\frac{x_1^{26} + 9x_1^2x_2^{12} + 6x_1^2x_4^6 + 8x_1^2x_3^8}{24}.$$

Setting $x_1 = x_2 = x_3 = x_4 = 2$, the number of coloring patterns is $\frac{2^{26}+9\times2^{14}+6\times2^8+8\times2^{10}}{24} = 2802752$.

Alternatively, we may wish to paint in red, white or blue the edges of a cube. There are 12 cells each of which is a cycle of length 1 under (C1). Under (C2), they form 6 cycles of length 2. Under (C3), they form 3 cycles of length 4. Under (C4), they form 2 cycles of length 1 and 5 cycles of length 2. Under (C5), they form 4 cycles of length 3. Hence the cycle index is

$$\frac{x_1^{12} + 3x_2^6 + 6x_4^3 + 6x_1^2x_2^5 + 8x_3^4}{24}.$$

Setting $x_1 = x_2 = x_3 = x_4 = 3$, the number of coloring patterns is $\frac{3^{12}+3\times3^6+6\times3^3+6\times3^7+8\times3^4}{24} = 182520$.

Examples

Example 5.3.1.
A square is divided into nine squares by two pairs of lines parallel to the sides. Each cell is colored black or white. Determine the number of patterns

(a) under all symmetries;

(b) under only the rotational symmetries.

Solution:

(a) Under I, each of the 9 cells forms a cycle of length 1. We record this as the term x_1^9. Under R, the central cell forms a cycle of length 1 while the outside cells form 4 cycles of length 2. We record this as the term $x_1 x_2^4$. Under A or C, the central cell forms a cycle of length 1 while the outside cells form 2 cycles of length 4. We record this as $x_1 x_4^2$. Under H, V, U or D, the 3 cells on the axis of reflection form 3 cycles of length 1 while the other cells form 3 cycles of length 2. We record this as $x_1^3 x_2^3$. The cycle index is

$$\frac{x_1^9 + x_1 x_2^4 + 2x_1 x_4^2 + 4x_1^3 x_2^3}{8}.$$

The number of coloring patterns is $\frac{2^9 + 2^5 + 2 \times 2^3 + 4 \times 2^6}{8} = 102$ when we set $x_1 = x_2 = x_4 = 2$.

(b) Without the reflectional symmetries, $\frac{x_1^9 + x_1 x_2^4 + 2x_1 x_4^2}{4}$ is the revised cycle index. Setting $x_1 = x_2 = x_4 = 2$, the number of coloring patterns is $\frac{2^9 + 2^5 + 2 \times 2^3}{4} = 140$.

Example 5.3.2.
Each face of a regular octahedron is divided into four equilateral triangles by joining midpoints of the sides, and the 32 equilateral triangles are to be painted in black or white. Determine how many such regular octahedra are distinguishable.

Solution:
We first consider the rotational symmetries of a regular octahedron. There are three types of axes of rotations, those joining the centers of opposite faces, those joining the midpoints of opposite edges and those joining opposite vertices. The three types of rotational axes are illustrated in the diagram below.

There are 24 rotational symmetries of the regular octahedron.

(O1) The identity symmetry.

(O2) Three 180° rotational symmetries about axes joining opposite vertices.

(O3) Six 90° rotational symmetries about axes joining opposite vertices.

(O4) Six 180° rotational symmetries about axes joining the midpoints of opposite edges.

(O5) Eight 120° rotational symmetries about axes joining the centers of opposite faces.

This total may be obtained without a detailed breakdown. Each face of the regular octahedron can be mapped onto any of the eight faces, each in three different rotational positions. Hence the total number of symmetries of the regular octahedron is indeed $8 \times 3 = 24$.

There are 32 cells each of which is a cycle of length 1 under (O1). Under (O2), they form 16 cycles of length 2. Under (O3), they form 8 cycles of length 4. Under (O4), they form 16 cycles of length 2. Under (O5), they form 2 cycles of length 1 and 10 cycles of length 3. Hence the cycle index is

$$\frac{x_1^{32} + 9x_2^{16} + 6x_4^8 + 8x_1^2 x_3^{10}}{24}.$$

Setting $x_1 = x_2 = x_3 = x_4 = 2$, $\frac{2^{32}+9\times 2^{16}+6\times 2^8+8\times 2^{12}}{24} = 178982976$ such regular octahedra are distinguishable.

Remark:
The astute reader may notice the one-to-one correspondence between the five classes of symmetries of the cube and the five classes of symmetries of the regular octahedron. There is a good reason for this. The two solids are **duals** of each other, in that if we join with edges the centers of all pairs of adjacent faces of one solid, we obtain the other one. A similar relationship exists between the dodecahedron and the icosahedron while the tetrahedron is the dual of itself.

Example 5.3.3.
How many coloring patterns are there if

(a) the vertices of a regular tetrahedron and the midpoints of its edges are to be painted in red, white or blue;

(b) each half of the edges of a regular tetrahedron is to be painted in red, white or blue?

Solution:

(a) There are 10 cells each of which is a cycle of length 1 under (T1). Under (T2), they form 1 cycle of length 1 and 3 cycles of length 3. Under (T3), they form 2 cycles of length 1 and 4 cycles of length 2. Hence the cycle index is

$$\frac{x_1^{10} + 8x_1x_3^3 + 3x_1^2x_2^4}{12}.$$

Setting $x_1 = x_2 = x_3 = 3$, there are $\frac{3^{10}+8\times3^4+3\times3^6}{12} = 5157$ coloring patterns.

(b) There are 12 cells each of which is a cycle of length 1 under (T1). Under (T2), they form 4 cycles of length 3. Under (T3), they form 6 cycles of length 2. Hence the cycle index is

$$\frac{x_1^{12} + 8x_3^4 + 3x_2^6}{12}.$$

Setting $x_1 = x_2 = x_3 = 3$, there are $\frac{3^{12}+8\times3^4+3\times3^6}{12} = 44523$ coloring patterns.

Exercises

1. A square is divided into nine squares by two pairs of lines parallel to the sides. Each cell is colored red, white or blue. Determine the number of patterns

   (a) under all symmetries;

   (b) under only the rotational symmetries.

2. A square is divided into sixteen squares by three pairs of lines parallel to the sides. Each cell is colored black or white. Determine the number of patterns

   (a) under all symmetries;

   (b) under only the rotational symmetries.

3. (a) Determine the rotational symmetries of the regular dodecahedron.

   (b) Determine the cycle index if the faces of the regular dodecahedron are to be painted.

4. (a) Determine the rotational symmetries of the regular icosahedron.

   (b) Determine the cycle index if the faces of the regular icosahedron are to be painted.

5. How many coloring patterns are there if the vertices of a regular octahedron and the midpoints of its edges are to be painted in red, white or blue?

6. How many coloring patterns are there if each half of the 12 edges of a regular octahedron is to be painted in black or white?

Section 5.4. Inventory Functions

At one point, Americans were fond of making Polish joke. They named the $3 \times 3 \times 3$ version of Rubik's cube with all six faces red the Polish Cube. In reprisal, the Poles named the six-colored $1 \times 1 \times 1$ cube the American Cube.

Under the rotational symmetries of a cube, how many distinguishable American Cubes are there? Suppose the six colors are red, orange, yellow, green, blue and violet. We may fix the red face. Then there are 5 choices for the color of the opposite face. By symmetry, we need only consider the case where the red face is opposite to the blue face. We may now fix the yellow face. Then there are 3 choices for the color of the opposite face, and the remaining two faces may be painted in 2 different ways. Hence the total count is $5 \times 3 \times 2 = 30$.

Let us solve this problem in a different way. There are 6 cells each of which is a cycle of length 1 under (C1). Under (C2), they form 2 cycles of length 1 and 2 cycles of length 2. Under (C3), they form 2 cycles of length 1 and 1 cycle of length 4. Under (C4), they form 3 cycles of length 2. Under (C5), they form 2 cycles of length 3. Hence the cycle index is

$$\frac{x_1^6 + 3x_1^2 x_2^2 + 6x_1^2 x_4 + 6x_2^3 + 8x_3^2}{24}.$$

However, it would not work if we simply set $x_1 = x_2 = x_3 = x_4 = 6$, because we would be counting many other cubes besides the American Cubes, including the Polish-American Cube, the $1 \times 1 \times 1$ cube with all six faces red!

We turn to the technique of generating functions. Using the initial letters to represent the colors, we set $x_1 = r+o+y+g+b+v$ since a single face is red, orange, yellow, green, blue or violet. We set $x_2 = r^2 + o^2 + y^2 + g^2 + b^2 + v^2$ because both cells in the cycle must be of the same color. Similarly, we set $x_3 = r^3 + o^3 + y^3 + g^3 + b^3 + v^3$ and $x_4 = r^4 + o^4 + y^4 + g^4 + b^4 + v^4$.

Substituting these into the cycle index, we obtain the **inventory function**, which is

$$\frac{1}{24}((r+o+y+g+b+v)^6 + 3(r+o+y+g+b+v)^2(r^2+o^2+y^2+g^2+b^2+v^2)^2$$

$$+6(r+o+y+g+b+v)^2(r^4+o^4+y^4+g^4+b^4+v^4)+6(r^2+o^2+y^2+g^2+b^2+v^2)^3$$

$$+8(r^3 + o^3 + y^3 + g^3 + b^3 + v^3)^2).$$

We seek the coefficient of the term $roygbv$, and it comes only from the first term in the numerator. The final answer is

$$\frac{1}{24}\binom{6}{1}\binom{5}{1}\binom{4}{1}\binom{3}{1}\binom{2}{1}\binom{1}{1} = 30.$$

We now return to the problem where each face of a $2 \times 2 \times 2$ cube is divided into four unit squares by joining the midpoints of opposite sides, and the 24 unit squares are to be painted in red, white or blue. The cycle index is

$$\frac{x_1^{24} + 3x_2^{12} + 6x_4^6 + 6x_2^{12} + 8x_3^8}{24}.$$

Suppose we are only interested in those cubes which have exactly six white unit squares. Instead, we let $x_1 = r + w + b$, $x_2 = r^2 + y^2 + b^2$, $x_3 = r^3 + y^3 + b^3$ and $x_4 = r^4 + y^4 + b^4$. The inventory function is

$$\frac{1}{24}((r + w + b)^{24} + 3(r^2 + w^2 + b^2)^{12} + 6(r^4 + w^4 + b^4)^6$$

$$+6(r^2 + w^2 + b^2)^{12} + 8(r^3 + w^3 + b^3)^8).$$

Since we are not interested in the red or blue unit squares, we modify the inventory function by defining $r = b = 1$, so that $x_1 = w + 2$, $x_2 = w^2 + 2$, $x_3 = w^3 + 2$ and $x_4 = w^4 + 2$. The modified inventory function is therefore

$$\frac{1}{24}((w + 2)^{24} + 3(w^2 + 2)^{12} + 6(w^4 + 2)^6 + 6(w^2 + 2)^{12} + 8(w^3 + 2)^8).$$

We seek the coefficient of the term w^6. From the first term of the numerator, we have $\binom{24}{6}w^6 2^{18}$. From the second and the fourth terms, we have $9\binom{12}{3}w^6 2^9$. We get nothing from the third term. From the fifth term, we have $8\binom{8}{2}w^6 2^6$. The final answer is

$$\frac{1}{24}\left(\binom{24}{6}2^{18} + 9\binom{12}{3}2^9 + 8\binom{8}{2}2^6\right) = 1470190080.$$

Suppose we wish to know instead how many distinguishable cubes there are with exactly four red unit squares and two blue unit squares. We obtain

$$\frac{1}{24}((r+b+1)^{24}+3(r^2+b^2+1)^{12}+6(r^4+b^4+1)^6+6(r^2+b^2+1)^{12}+8(r^3+b^3+1)^8)$$

from the inventory function and seek the coefficient of the term $r^4 b^2$. The final answer is

$$\frac{1}{24}\left(\binom{24}{4}\binom{20}{2} + 9\binom{12}{2}\binom{10}{1}\right) = 84370.$$

Finally, suppose we use only red and blue and wish to know how many distinguishable cubes there are with an equal number of red and blue cells. The inventory function is

$$\frac{1}{24}((r+b)^{24} + 3(r^2+b^2)^{12} + 6(r^4+b^4)^6 + 6(r^2+b^2)^{12} + 8(r^3+b^3)^8).$$

We seek the coefficient of the term $r^{12}b^{12}$. The final answer is

$$\frac{1}{24}\left(\binom{24}{12} + 9\binom{12}{6} + 6\binom{6}{3} + 8\binom{8}{4}\right) = 113048.$$

Suppose that the edges of a regular octahedron are to be painted in red, white or blue. We wish to know the number of coloring patterns with exactly four blue edges.

The cycle index is

$$\frac{1}{24}(x_1^{12} + 8x_3^4 + 2x_1^2x_2^5 + 3x_2^6 + 6x_4^3).$$

The inventory function is

$$\frac{1}{24}((b+2)^{12} + 8(b^3+2)^4 + 6(b+2)^2(b^2+2)^5 + 3(b^2+2)^6 + 6(b^4+2)^3).$$

We seek the coefficient of the b^4 term. The final answer is

$$\frac{1}{24}\left(\binom{12}{4}2^8 + 6\left(\binom{2}{2}\binom{5}{1}2^4 + \binom{2}{0}2^2\binom{5}{2}2^3\right) + 3\binom{6}{2}2^4 + 6\binom{3}{1}2^2\right) = 5413.$$

Examples

Example 5.4.1.
Each face of a regular octahedron is divided into four equilateral triangles by joining midpoints of the sides, and the 32 equilateral triangles are to be painted in red, white or blue. Determine the number of coloring patterns with exactly four red cells.

Solution:
From Example 5.3.2, the cycle index is

$$\frac{1}{24}(x_1^{32} + 9x_2^{16} + 6x_4^8 + 8x_1^2x_3^{10}).$$

Using the variable r, the inventory function is

$$\frac{1}{24}((r+2)^{32} + 9(r^2+2)^{16} + 6(r^4+2)^8 + 8(r+2)^2(r^3+2)^{10}).$$

We seek the coefficient of the r^4 term. We have $\binom{32}{4}2^{28}$ from $(r+2)^{32}$, $9\binom{16}{2}2^{14}$ from $9(r^2+2)^{16}$, $8\binom{2}{1}2\binom{10}{1}2^9$ from $8(r+2)^2(r^3+2)^{10}$, and $6\binom{8}{1}2^7$ from $6(r^4+2)^8$. The final answer is

$$\frac{1}{24}\left(\binom{32}{4}2^{28} + 9\binom{16}{2}2^{14} + 8\binom{2}{1}\binom{10}{1}2^{10} + 6\binom{8}{1}2^7\right) = 98932311.$$

Example 5.4.2.
The vertices of a cube are to be painted red, yellow, green and blue. Determine the number of distinguishable cubes with exactly two cells of each color.

Solution:
There are 8 cells each of which is a cycle of length 1 under (C1). Under (C2), they form 4 cycles of length 2. Under (C3), they form 3 cycles of length 4. Under (C4), they form 4 cycles of length 2. Under (C5), they form 2 cycles of length 1 and 2 cycles of length 3. Hence the cycle index is

$$\frac{1}{24}(x_1^8 + 9x_2^4 + 6x_4^2 + 8x_1^2x_3^2).$$

Using the variables r, y, g and b, the inventory function is

$$\frac{1}{24}((r+y+g+b)^8 + 9(r^2+y^2+g^2+b^2)^4 + 6(r^4+y^4+g^4+b^4)^2$$

$$+8(r+y+g+b)^2(r^3+y^3+g^3+b^3)^2).$$

We seek the coefficient of the term $r^2y^2g^2b^2$. From the first two terms of the numerator, we have

$$\frac{1}{24}\left(\binom{8}{2}\binom{6}{2}\binom{4}{2}\binom{2}{2} + 9\binom{4}{1}\binom{3}{1}\binom{2}{1}\binom{1}{1}\right) = 114.$$

Example 5.4.3.

Determine the number of non-equivalent ways of placing two identical markers on a 4×4 board by

(a) direct counting of invariant entries;

(b) using the inventory function.

Solution:

(a) There are $\binom{16}{2} = 120$ ways of placing the markers, all of which are invariant under I. Under R, H or V, one of the markers can be on any of the 8 squares in a half board. To be invariant, the other marker has to be in the corresponding position in the other half board. Under C and A, there are no invariants. Under D or U, first we may have both markers in the reflecting diagonal, yielding $\binom{4}{2} = 6$ ways. Second, we may have one marker on each side of the reflecting diagonal in corresponding positions, yielding 6 ways. It follows that the desired number of ways is $\frac{1}{8}(120 + 8 + 0 + 0 + 8 + 8 + 12 + 12) = 21$.

(b) The problem is equivalent to painting the 16 squares in black and white with exactly two squares black. The cycle index is

$$\frac{1}{8}(x_1^{16} + x_2^8 + 2x_4^4 + 2x_2^8 + 2x_1^4 x_2^6).$$

The inventory function is

$$\frac{1}{8}((b+1)^{16} + (b^2+1)^8 + 2(b^4+1)^4 + 2(b^2+1)^8 + 2(b+1)^4(b^2+1)^6).$$

The final answer is

$$\frac{1}{8}\left(\binom{16}{2} + \binom{8}{1} + 2\binom{8}{1} + 2\left(\binom{4}{2}\binom{6}{0} + \binom{4}{0}\binom{6}{1}\right)\right) = 21.$$

Exercises

1. Each face of a regular tetrahedron is divided into four equilateral triangles by joining the midpoints of the sides, and the 16 equilateral triangles are to be painted in black or white. Determine the number of coloring patterns with exactly three black cells.

2. Each face of a regular tetrahedron is divided into three isosceles triangles by joining the midpoints of the sides to the center, and the 12 isosceles triangles are to be painted in red, white or blue. Determine the number of coloring patterns with exactly four red cells.

3. The edges of a cube are to be painted in red, white or blue. Determine the number of coloring patterns with exactly three white edges.

4. The vertices of a regular octahedron, along with the midpoints of its edges, are to be painted in black or white. Determine the number of coloring patterns with exactly 5 black cells.

5. Determine the number of non-equivalent ways of placing three identical markers on a 4×4 board by

   (a) direct counting of invariant entries;

   (b) using the inventory function.

6. Determine the number of non-equivalent ways of placing four identical markers on a 4×4 board by

   (a) direct counting of invariant entries;

   (b) using the inventory function.

Section 5.5. Necklaces

A necklace consists of m beads evenly spaced around a circular loop. Each bead may be of any of n colors. Clearly there are n^m distinct designs. We wish to know the number of distinct patterns when rotations are allowed. We will not consider reflections for now.

The problem of painting in two colors a square divided into a 2×2 configuration, is the special case $m = 4$ and $n = 2$. As another special case, let $m = 6$ and $n = 2$. As pointed out above, there are $2^6 = 64$ different designs. If we may rotate the necklace, a number $(1, 2, 3$ or $6)$ of designs will collapse into a single pattern. The diagram below shows that the number of patterns is 14.

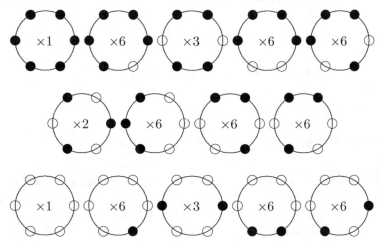

This answer may be obtained more formally by applying the Main Theorem. Under the identity transformation, each bead is a cycle, and all $2^6 = 64$ designs are invariant. Under the 180° rotation, opposite beads form a cycle, and there are $2^3 = 8$ invariants. Under the two 120° rotations (clockwise and anticlockwise), alternate beads form a cycle, and there are $2 \times 2^2 = 8$ invariants. Under the two 60° rotations, all six beads form a single cycle, and there are $2 \times 2^1 = 4$ invariants. Hence the number of patterns is $\frac{64+8+8+4}{6} = 14$.

Of these patterns, 9 of them have no symmetry other than the trivial identity. They contain a full complement of 6 designs. We can calculate this number by means of the Principle of Inclusion-Exclusion.

The non-trivial rotations are those of 180°, 120° and 60°. Under the 180° rotation, diametrically opposite beads form 3 cycles of length 2, and they can be painted in $2^3 = 8$ ways. Under the 120° rotations, alternate beads form 2 cycles of length 3, and they can be painted in $2^2 = 4$ ways. Under the 60° rotations, all the beads form 1 cycle of length 6, and they can be painted in $2^1 = 2$ ways.

In our attempt to eliminate invariants with symmetries, note that those counted under the $60°$ rotation have already been counted once under the $180°$ and once under the $120°$ rotations. Hence they must be added back according to the Principle of Inclusion-Exclusion. Thus the number of patterns we seek is $\frac{64-8-4+2}{6} = 9$.

Let us now give a general formulation. Suppose m is a composite number. Let d be any divisor of m. A rotation through d beads divides the m beads into d cycles. The number of rotations which divide the beads in this way is $\phi(\frac{m}{d})$. For each of these rotations, there are n^d invariant designs. Hence the number of patterns is given by $\frac{1}{m} \sum_{d|m} \phi\left(\frac{m}{d}\right) n^d$.

For $m = 6$, this formula yields

$$\frac{1}{6}(\phi(6)2^1 + \phi(3)2^2 + \phi(2)2^3 + \phi(1)2^6) = 14.$$

For $m = 12$, this formula yields

$$\frac{1}{12}(\phi(12)2^1 + \phi(6)2^2 + \phi(4)2^3 + \phi(3)2^4 + \phi(2)2^6 + \phi(1)2^{12}) = 352.$$

The special case where the number of beads is a prime number p is particularly easy. There are n designs in which all beads are of the same color. Each is the unique design of a pattern. Of the remaining $n^p - n$ designs, p of them merge into a single design. Hence the total number of patterns is $n + \frac{n^p - n}{p}$.

Alternatively, from the Main Theorem, all n^p designs are invariant under the identity, while only the n monochromatic designs are invariant under each of the $p - 1$ non-trivial rotations. Hence the number of patterns is $\frac{n^p + (p-1)n}{p} = n + \frac{n^p - n}{p}$.

We now have a pleasant surprise. Since the number of patterns must be an integer, we have just proved the following result in number theory.

Fermat's Little Theorem.
Let p be a prime number and n be a positive integer. Then $n^p - n$ is divisible by p. Alternatively, $n^p \equiv n \pmod{p}$.

The number of patterns with no non-trivial symmetries is given by $\frac{1}{m} \sum_{d|m} \mu\left(\frac{m}{d}\right) n^d$. Compare this with the upper bound for the size of comma-free dictionaries in Section 3.6. When $m = 6$, this formula yields

$$\frac{1}{6}(\mu(6)2^1 + \mu(3)2^2 + \mu(2)2^3 + \mu(1)2^6) = 9,$$

as observed before. When $m = 12$, this formula yields

$$\frac{1}{12}(\mu(12)2^1 + \mu(6)2^2 + \mu(4)2^3 + \mu(3)2^4 + \mu(2)2^6 + \mu(1)2^{12}) = 335.$$

When m is a square-free number, that is, a number not divisible by the square of a prime number, the formula follows directly from the Principle of Inclusion-Exclusion as we have seen earlier when $m = 6$. When m is not square-free, such as $m = 12$, the formula suggests that four-fold and twelve-fold symmetries can be ignored. This is because patterns with four-fold symmetries also have two-fold symmetries, and those with twelve-fold symmetries also have six-fold symmetries.

We can verify directly that when $m = 12$, there are indeed $352 - 335 = 17$ patterns with non-trivial symmetries. These are sketched in the diagram below except for the two monochromatic necklaces.

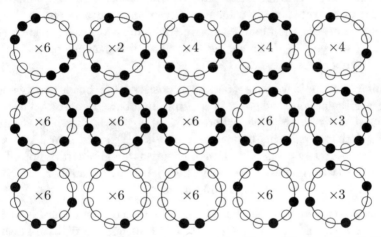

Finally, we take reflections into consideration. We have to distinguish between whether the number of beads m is odd or even. There are m axes of symmetry in either case.

When m is odd, each axis passes through a bead at one end and between two beads at the other end. Thus the beads form 1 cycle of length one and $\frac{m-1}{2}$ cycles of length two. Hence the number of patterns is given by

$$\frac{1}{2m}\left(\sum_{d|m}\phi\left(\frac{m}{d}\right)n^d + mn^{\frac{m+1}{2}}\right).$$

When m is even, half of the axes pass through two beads and the other half of the axes pass between two pairs of beads. In the first case, the beads form 2 cycles of length one and $\frac{m-2}{2}$ cycles of length two. In the second case, they form $\frac{m}{2}$ cycles of length two. Hence the number of patterns is given by $\dfrac{1}{2m}\left(\sum_{d|m}\phi\left(\dfrac{m}{d}\right)n^d+\dfrac{m}{2}\left(n^{\frac{m}{2}}+n^{\frac{m+2}{2}}\right)\right)$.

When $m=6$, this formula yields

$$\frac{1}{12}(\phi(6)2^1+\phi(3)2^2+\phi(2)2^3+\phi(1)2^6+3(2^3+2^4))=13.$$

When $m=12$, this formula yields

$$\frac{1}{24}(\phi(12)2^1+\phi(6)2^2+\phi(4)2^3+\phi(3)2^4+\phi(2)2^6+\phi(1)2^{12}+6(2^6+2^7))=200.$$

A **Catalan** necklace consists of $2n+1$ beads evenly spaced around a circular loop. Exactly n beads are yellow and the rest are blue. Clearly there are $\binom{2n+1}{n}$ distinct designs. We wish to know the number of distinct patterns when rotations are allowed.

All designs are invariant under the identity transformation. Under each of the $2n$ other rotations, we have no invariants since n and $n+1$ are relatively prime. By the Main Theorem, the number of patterns is given by $\frac{1}{2n+1}\binom{2n+1}{n}=\frac{1}{n+1}\binom{2n}{n}$. Thus we have given an independent proof that the Catalan numbers are integers.

Similarly, suppose we have a necklace consisting of p beads evenly spaced around a circular loop, where p is a prime number. Exactly k beads are yellow and the rest are blue, where $1\le k\le p-1$. Then the number of distinguishable patterns under rotations is $\frac{1}{p}\binom{p}{k}$ since k and $p-k$ are necessarily relatively prime. Thus we have given a combinatorial proof that on the p-th row of the Pascal triangle where p is a prime number, the entries except the 1s at the ends are divisible by p.

Alternatively, suppose the beads come in k colors and there is at least one bead of each color, where $1\le k\le p$. Then the number of distinguishable patterns under rotations is $\frac{1}{p}S_3(p,k)$, proving combinatorially that on the p-th row of the triangle for the Stirling numbers of the third kind, where p is prime, the entries except the 1 at the beginning are divisible by p.

In a similar manner, we can show that the triangles for the Stirling numbers of the second kind and of the first kind also have the same property. We illustrate with the case $p=5$ and $k=3$. The $S_2(5,3)=25$ partitions form 4 patterns each consisting of 5 designs, as shown in the diagram below.

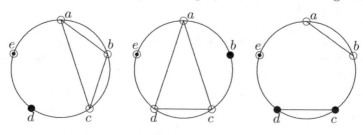

$$\{a,b,c\},\{d\},\{e\} \qquad \{a,c,d\},\{b\},\{e\} \qquad \{a,b\},\{c,d\},\{e\}$$
$$\{b,c,d\},\{e\},\{a\} \qquad \{b,d,e\},\{c\},\{a\} \qquad \{b,c\},\{d,e\},\{a\}$$
$$\{c,d,e\},\{a\},\{b\} \qquad \{c,e,a\},\{d\},\{b\} \qquad \{c,d\},\{e,a\},\{b\}$$
$$\{d,e,a\},\{b\},\{c\} \qquad \{d,a,b\},\{e\},\{c\} \qquad \{d,e\},\{a,b\},\{c\}$$
$$\{e,a,b\},\{c\},\{d\} \qquad \{e,b,c\},\{a\},\{d\} \qquad \{e,a\},\{b,c\},\{d\}$$

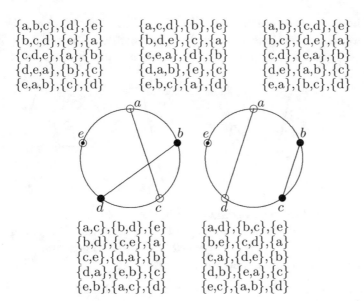

$$\{a,c\},\{b,d\},\{e\} \qquad \{a,d\},\{b,c\},\{e\}$$
$$\{b,d\},\{c,e\},\{a\} \qquad \{b,e\},\{c,d\},\{a\}$$
$$\{c,e\},\{d,a\},\{b\} \qquad \{c,a\},\{d,e\},\{b\}$$
$$\{d,a\},\{e,b\},\{c\} \qquad \{d,b\},\{e,a\},\{c\}$$
$$\{e,b\},\{a,c\},\{d\} \qquad \{e,c\},\{a,b\},\{d\}$$

The $\binom{5}{3} = 10$ subsets form 2 patterns each consisting of 5 designs. These may be seen as the first two patterns with the singleton subsets omitted. The $S_3(5,3) = 150$ ordered partitions form 30 patterns each consisting of 5 designs. Each of the five patterns featured generates a set of six by permuting the subsets among themselves. Finally, the $S_1(5,3) = 35$ partitions of cycles form 7 patterns each consisting of 5 designs. Each of the first two patterns generates another one by reversing the order of the elements in the 3-cycle.

For the Stirling numbers of the second kind, the entries at both ends of row p are 1s. For the Stirling numbers of the first kind, the entry at the end of row p is 1, but the entry at the beginning is $(p-1)!$. This is clearly not divisible by p. In fact, we have the following well-known result in number theory.

Wilson's Theorem.
Let p be a prime number. Then $(p-1)!+1$ is divisible by p. Alternatively, $(p-1)! \equiv -1 \pmod{p}$.

This is clearly true for $p = 2$. Let p be an odd prime number. Consider p points evenly spaced around a circle. The points are joined to one another by segments to form a closed and possibly self-intersecting polygon with p sides. Under rotations, how many distinct patterns are there?

Since we have a closed polygon, we may choose any of the p points as the first point. Then there are $p-1$ choices for the second point, $p-2$ choices for the third point, and so on. Thus there are $(p-1)!$ designs.

Like the necklace problem, groups of p designs merge into single patterns.

Before, the exceptions are the necklaces in which all beads have the same color. Here, the exceptions are closed paths which are the regular polygon and star polygons, as shown in the diagram below for the case $p = 7$.

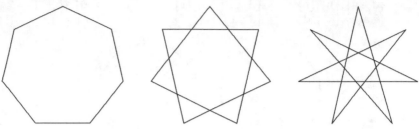

Since each of these may be traversed in the clockwise and the anticlockwise directions, their total number is $p-1$. It follows that the total number of patterns is $p - 1 + \frac{(p-1)! - (p-1)}{p} = p - 2 + \frac{(p-1)! + 1}{p}$. The same answer may be obtained from the Main Theorem.

We now have a second pleasant surprise, a combinatorial proof of Wilson's Theorem.

Examples

Example 5.5.1.
A necklace is formed of 14 beads in 2 colors.

  (a) Determine the total number of patterns.

  (b) Determine the number of patterns with no non-trivial symmetries.

  (c) Sketch those patterns counted in (a) but not in (b).

Solution:

  (a) The total number of patterns is given by

$$\frac{1}{14}\sum_{d|14}\phi\left(\frac{14}{d}\right)2^d = \frac{\phi(14)2^1 + \phi(7)2^2 + \phi(2)2^7 + \phi(1)2^{14}}{14}$$

$$= \frac{12 + 24 + 128 + 16384}{14}$$

$$= 1182.$$

  (b) The number of patterns with no non-trivial symmetries is given by

$$\frac{1}{14}\sum_{d|14}\mu\left(\frac{14}{d}\right)2^d = \frac{\mu(14)2^1 + \mu(7)2^2 + \mu(2)2^7 + \mu(1)2^{14}}{14}$$

$$= \frac{2 - 4 - 128 + 16384}{14}$$

$$= 1161.$$

  (c) The 21 patterns counted in (a) but not in (b) are sketched below, except for the two monochromatic necklaces.

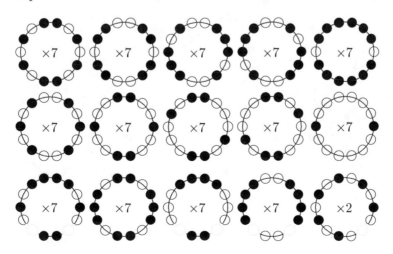

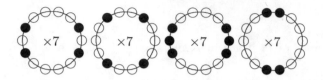

Example 5.5.2.

A necklace is formed of 15 beads in 3 colors. Determine the total number of patterns if

(a) only rotations are taken into consideration;

(b) reflections are also taken into consideration.

Solution:

(a) The total number of patterns if only rotations are taken into consideration is given by

$$\frac{1}{15} \sum_{d|15} \phi\left(\frac{15}{d}\right) 3^d = \frac{\phi(15)3^1 + \phi(5)3^3 + \phi(3)3^5 + \phi(1)3^{15}}{15}$$

$$= \frac{24 + 108 + 486 + 14348907}{15}$$

$$= 956635.$$

(b) The total number of patterns if reflections are also taken into consideration is given by

$$\frac{1}{30}\left(\sum_{d|15} \phi\left(\frac{15}{d}\right) 3^d + 15(3^8)\right)$$

$$= \frac{\phi(15)3^1 + \phi(5)3^3 + \phi(3)3^5 + \phi(1)3^{15} + 15(3^8)}{30}$$

$$= \frac{24 + 108 + 486 + 14348907 + 98415}{30}$$

$$= 481598.$$

Example 5.5.3.

Consider the congruence $x^2 \equiv -1 \pmod{p}$ where p is prime.

(a) Prove that it has no solutions if p is of the form $p = 4t + 3$.

(b) Prove that it has solutions if p is of the form $p = 4t + 1$.

Solution:

(a) Suppose to the contrary $x^2 \equiv -1$ (mod p) has a solution $x = n$. Then $n \not\equiv 0$ (mod p). We have $n^{p-1} = n^{4t+2} = (n^2)^{2t+1} \equiv (-1)^{2t+1} = -1$ (mod p) or $n^p \equiv -n$ (mod p). This contradicts Fermat's Little Theorem since $-n \not\equiv n$ (mod p) when $p > 2$.

(b) By Wilson's Theorem,

$$
\begin{aligned}
-1 &\equiv (p-1)! \\
&= (4t)(4t-1)\cdots(2t+1)(2t)! \\
&\equiv (-1)(-2)\cdots(-2t)(2t)! \\
&= (-1)^{2t}(2t)!(2t)! \\
&= ((2t)!)^2 \quad (\text{mod } p).
\end{aligned}
$$

Thus $x = (2t)!$ is a solution to $x^2 \equiv -1$ (mod p), where $t = \frac{p-1}{4}$.

Exercises

1. A necklace is formed of 9 beads in 3 colors.

    (a) Determine the total number of patterns.

    (b) Determine the number of patterns with no non-trivial symmetries.

    (c) Sketch those patterns counted in (a) but not in (b).

2. A necklace is formed of 10 beads in 2 colors.

    (a) Determine the total number of patterns.

    (b) Determine the number of patterns with no non-trivial symmetries.

    (c) Sketch those patterns counted in (a) but not in (b).

3. A necklace is formed of 8 beads in 4 colors. Determine the total number of patterns if

    (a) only rotations are taken into consideration;

    (b) reflections are also taken into consideration.

4. A necklace is formed of 21 beads in 2 colors. Determine the total number of patterns if

    (a) only rotations are taken into consideration;

    (b) reflections are also taken into consideration.

5. Prove the converse of Wilson's Theorem, that if n is composite, then $(n-1)! + 1$ is not divisible by n.

6. A game involves $p-1$ boys wearing jerseys numbered from 1 to $p-1$, and $p-1$ cards numbered also from 1 to $p-1$, where p is an odd prime. Each boy is dealt a card, multiplies the number on the card with the number on his jersey, and finds the remainder when this product is divided by p. Prove that at least two boys have the same final answer.

Section 5.6. Combinatorial Symmetries

Up to now, we have been dealing with geometric symmetries, such as those of the square. If we join the midpoints of opposite sides of a square, we divide it into 4 cells. We paint each of them black or white.

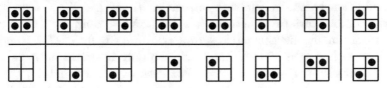

Recall from Section 5.2 that there are six patterns altogether, consisting of

(1) the design with all four cells black;
(2) the design with all four cells white;
(3) two designs with two diagonally opposite cells black and the other two white;
(4) four designs with two adjacent cells black and the other two white;
(5) four designs with three cells black and the other one white;
(6) four designs with one cell black and the other three white.

Suppose a color-blind person cannot distinguish between black and white though he can tell that white is different from black. This person will confuse pattern (1) with pattern (2), and pattern (5) with pattern (6), so that to him, there are only four different patterns.

How can we handle this variation on our problem? We have to introduce a new symmetry called color-reversal, which changes every white cell into a black cell and vice versa. For reasons which will soon become clear, we denote this symmetry by I'. This is a new type of symmetry, which we will call **combinatorial symmetry**.

The rotation group of the square consists of the four symmetries I, R, A and C. When we expand it to the dihedral group, we add four reflectional symmetries. This is forced. We cannot have added only H. In order to preserve the Closure property, we must also add, in addition to $IH = H$, the symmetries $RH = V$, $AH = D$ and $CH = U$. Similarly, we must now add $II' = I'$, $RI' = R'$, $AI' = A'$, $CI' = C'$, $HI' = H'$, $VI' = V'$, $UI' = U'$ and $VI' = V'$.

With color reversal, cells in the same cycle must have alternate colors. Hence the invariants in R, A, C, H and V are retained while those in I, U and D are lost. The number of different patterns is therefore $\frac{48+4+2+2+4+4}{16} = 4$.

Consider an equilateral triangle divided into four equilateral triangles by joining the midpoints of the sides. Each triangle is colored red, yellow or blue. To a person who cannot distinguish between yellow and blue but can tell they are different, how many different patterns are there?

In addition to the geometric symmetries I, J, K, X, Y and Z, we have the combinatorial symmetries I', J', K', X', Y' and Z'. The number of invariants in the six geometric symmetries are 81, 9, 9, 27, 27 and 27, respectively. For each of I', J' and K', the only invariant entry is when all four equilateral triangles are red. For each of X', Y' and Z', the two equilateral triangles on the axis of symmetry must be red. The other two are either both red, or one yellow and one blue. Thus there are 3 invariants. Hence the number of different patterns is $\frac{180+3+9}{12} = 16$.

In addition to the eight cases of the Distribution Problem in Section 0.1, there are cases where the objects are neither all distinct nor all identical, and similarly for the boxes. Empty boxes are allowed.

Suppose we wish to find the number of distributions of two identical red objects and a blue object into three distinct boxes. Let us for now assume that the two red objects are distinct. We have two symmetries, the identity and the switching of the two red objects. In the first case, each object is an independent cycle, and the contribution to the cycle index is x_1^3. In the second case, the two red objects being switched form a single cycle, and the contribution to the cycle index is $x_1 x_2$. Hence the cycle index is $\frac{x_1^3 + x_1 x_2}{2}$. Setting $x_1 = x_2 = 3$, we have $\frac{27+9}{2} = 18$. This is the number of distributions we seek.

To help us understand this solution more clearly, the 27 designs are written out in the chart below, in 9 rows and 3 columns. The boxes are denoted by B1, B2 and B3, the red objects by r_1 and r_2 and the blue object by b.

Column	A			B			C		
Row	B1	B2	B3	B1	B2	B3	B1	B2	B3
1	all			r_2, b	r_1		r_2, b		r_1
2	r_1, r_2	b		r_2	r_1, b		r_2	b	r_1
3	r_1, r_2		b	r_2	r_1	b	r_2		r_1, b
4	r_1, b	r_2		b	r_1, r_2		b	r_2	r_1
5	r_1	r_2, b			all			r_2, b	r_1
6	r_1	r_2	b		r_1, r_2	b		r_2	r_1, b
7	r_1, b		r_2	b	r_1	r_2	b		r_1, r_2
8	r_1	b	r_2		r_1, b	r_2		b	r_1, r_2
9	r_1		r_2, b		r_1	r_2, b			all

All 27 designs are invariants under the identity. The invariants under the switching are A1, A2, A3, B4, B5, B6, C7, C8 and C9. The merged designs are A4=B1, A5=B2, A6=B3, A7=C1, A8=C2, A9=C3, B7=C4, B8=C5 and B9=C6.

Suppose we wish to find the number of distributions of four distinct objects into two identical black boxes and one white box. Let us for now assume that the two black boxes are distinct. We have two symmetries, the identity and the switching of the two black boxes. In the first case, each object is an independent cycle, and the number of invariants is $3^4 = 81$. In the second case, the two black boxes must be empty because the objects are distinct. Hence the number of invariants is $1^4 = 1$. The number of patterns is $\frac{81+1}{2} = 41$. This is the number of distributions we seek.

Examples

Example 5.6.1.

A two-sided square is divided into 9 cells in a 3×3 configuration. Each cell is painted in green, yellow or blue. A color-blind person cannot distinguish between green and blue, but can tell whether two cells have the same color or different colors. How many different parterns can this color-blind person distinguish?

Solution:

There are two relevant color permutations, $(g)(b)$ and (gb). The first means all colors stay put, while the second means green and blue are interchanged. Combining with the usual eight symmetries of the square I, R, A, C, H, V, D and U, we have a total of sixteen symmetries. We calculate the total number of invariants in the following chart.

Invariants	$(g)(b)$	(gb)
I	3^9	3^0
R	3^5	3^4
A	3^3	3^2
C	3^3	3^2
H	3^6	3^3
V	3^6	3^3
D	3^6	3^3
U	3^6	3^3

In all invariants under (gb), the central cell must be yellow. In combination with I, all cells are yellow. In combination with H, V, D and U, the two cells on the axis of symmetry apart from the central cell must also be yellow. The total number of distinct patterns is $\frac{19683+243+54+2916+1+81+18+108}{16} = 1444$.

Example 5.6.2.

Determine the number of distributions of three identical red objects, two identical yellow objects and one blue object into three distinct boxes.

Solution:

Denote the red objects by r_1, r_2 and r_3 and the yellow objects by y_1 and y_2. There are 6 symmetries among the red objects and 2 symmetries among the yellow objects, intertwining to yield 12 symmetries in all. The cycle index is $\frac{x_1^6+4x_1^4x_2+3x_1^2x_2^2+2x_1x_2x_3+2x_1^3x_3}{12}$. Setting $x_1 = x_2 = x_3 = 3$, the number of desired distributions is $\frac{729+972+81+54+162}{12} = 180$.

Red Symmetries	Yellow Symmetries	
	$(y_1)(y_2)$	$(y_1 y_2)$
$(r_1)(r_2)(r_3)$	x_1^6	$x_1^4 x_2$
$(r_1)(r_2 r_3)$	$x_1^4 x_2$	$x_1^2 x_2^2$
$(r_2)(r_1 r_3)$	$x_1^4 x_2$	$x_1^2 x_2^2$
$(r_3)(r_1 r_2)$	$x_1^4 x_2$	$x_1^2 x_2^2$
$(r_1 r_2 r_3)$	$x_1^3 x_3$	$x_1 x_2 x_3$
$(r_1 r_3 r_2)$	$x_1^3 x_3$	$x_1 x_2 x_3$

Example 5.6.3.
Determine the number of distributions of three distinct objects into three identical black boxes and one white box.

Solution:
Let us for now assume that the three black boxes are distinct. We have six symmetries, the identity, three pairwise switchings of the three black boxes and two cyclic switching of the three black boxes. In the first case, each object is an independent cycle, and the number of invariants is $4^3 = 64$. In the second case, the two black boxes being switched must be empty because the objects are distinct. Hence the number of invariants is $2^3 = 8$. In the last case, all three black boxes must be empty, and the number of invariants is $1^3 = 1$. The number of patterns is $\frac{64+3\times8+2\times1}{6} = 15$. This is the number of distributions we seek.

Exercises:

1. Determine the number of distinct patterns of a square divided into nine squares by two pairs of lines parallel to the sides, with each square colored green, yellow or blue, to a person who cannot distinguish between the colors but can tell they are different.

2. Determine the number of distinct patterns of an equilateral triangle divided into nine triangles by three pairs of lines parallel to the sides, with each triangle colored green, yellow or blue, to a person who cannot distinguish between green and blue but can tell they are different.

3. Determine the number of distributions of three identical red objects and one blue object into three distinct boxes.

4. Determine the number of distributions of three distinct objects into two identical black boxes and two identical white boxes.

5. Determine the number of distributions of two identical red objects and one yellow object into two identical black boxes and one white box.

6. Determine the number of distributions of two identical red objects and one yellow object into two identical black boxes and two identical white boxes.

Chapter Six: Combinatorial Structures

Section 6.1. Finite Fields and Geometries

A **field** is a set S of elements with two binary operations $+$ and $\times$ having the following properties. By convention, the $\times$ sign is usually suppressed.

(1) The Closure Property.
If a and b are elements in S, then $a + b$ and ab are also elements in S.

(2) Commutativity Property.
If a and b are elements in S, $a + b = b + a$ and $ab = ba$.

(3) The Associativity Property.
If a, b and c are elements in S, $(a + b) + c = a + (b + c)$ and $(ab)c = a(bc)$.

(4) The Distributivity Property.
If a, b and c are elements in S, $a(b + c) = ab + ac$.

(5) The Identity Property.
There is an element 0 in S such that $a + 0 = a = 0 + a$ for any element a in S, and an element 1 in S such that $a1 = a = 1a$ for any element a in S. The same element may not serve as both 0 and 1.

(6) The Inverse Property.
For any element a in S, there exists an element $-a$ in S such that $a + (-a) = 0$, and for any element $a \neq 0$ in S, there exists an element a^{-1} such that $aa^{-1} = 1$.

Naturally, the element 0 is called the additive identity, and 1 the multiplicative identity of the field. Moreover, the element $-a$ is called the additive inverse, and the element a^{-1} the multiplicative inverse of a. Note that a field is an Abelian group under addition, and the non-zero elements of the field also form an Abelian group under multiplication.

A field may have infinitely many elements. The set of all rational numbers or the set of all real numbers, under addition and multiplication, is such an example. However, we are primarily interested in finite fields. A field with n elements is called a **Galois** field of order n, and denoted by $\mathbf{GF}(n)$.

When n is a prime, $\mathbf{GF}(n)$ is simply $\mathbf{Z}_n$, the integers modulo n. When n is not a prime, $\mathbf{Z}_n$ is not even a field. For $n > 1$, divisors of n between 1 and n have no multiplicative inverses. For instance, in $\mathbf{Z}_4$, the element 2 has no multiplicative inverse. However, a finite field of order 4 does exist. In fact, when n is a power of a prime, we can construct $\mathbf{GF}(n)$.

We now construct $\mathbf{GF}(4)$. Since $4 = 2^2$, we begin with $\mathbf{Z}_2$. All linear polynomials have roots within $\mathbf{GF}(2)$. Of the quadratic polynomials, x^2 has roots 0 and 0, $x^2 + 1$ has roots 1 and 1, and $x^2 + x$ has roots 0 and 1. The remaining quadratic polynomial $x^2 + x + 1$ is irreducible.

Let t be a root of this polynomial, which is necessarily not in $\mathbf{Z}_2$. Since $t^2 + t + 1 = 0$, we have $t^2 = t + 1$. Thus the non-zero elements of $\mathbf{GF}(4)$ are t, $t^2 = t + 1$ and $t^3 = 1$. Below are the addition and multiplication tables of $\mathbf{GF}(4)$. It is routine to verify that this is a field.

$+$	0	1	t	$t+1$
0	0	1	t	$t+1$
1	1	0	$t+1$	t
t	t	$t+1$	0	1
$t+1$	$t+1$	t	1	0

$\times$	0	1	t	$t+1$
0	0	0	0	0
1	0	1	t	$t+1$
t	0	t	$t+1$	1
$t+1$	0	$t+1$	1	t

We now construct $\mathbf{GF}(8)$. Since $8 = 2^3$, we begin with $\mathbf{GF}(2)$. From our consideration of $\mathbf{GF}(4)$, the only irreducible polynomial of degree at most 2 is $x^2 + x + 1$. Of the cubic polynomials, x^3 has roots 0, 0 and 0, $x^3 + x$ has roots 0, 1 and 1, $x^3 + x^2$ has roots 0, 0 and 1, and $x^3 + x^2 + x + 1$ has roots 1, 1 and 1. Now $x^3 + 1$ has a root 1, and may be factored as $(x+1)(x^2 + x + 1)$. Similarly, $x^3 + x^2 + x = x(x^2 + x + 1)$ has a root 0. The only irreducible cubic polynomials are $x^3 + x + 1$ and $x^3 + x^2 + 1$.

We can use either to construct $\mathbf{GF}(8)$, and we choose the former. Let t be a root of $x^3 + x + 1$, which is necessarily not in $\mathbf{GF}(2)$. The non-zero elements of $\mathbf{GF}(8)$ are t, t^2, $t^3 = t+1$, $t^4 = t(t+1) = t^2 + t$, $t^5 = t(t^2 + t) = t^2 + t + 1$, $t^6 = t(t^2 + t + 1) = t^2 + 1$ and $t^7 = 1$. Below is the addition table for $\mathbf{GF}(8)$.

$+$	0	1	t	$t+1$	t^2	t^2+1	t^2+t	t^2+t+1
0	0	1	t	$t+1$	t^2	t^2+1	t^2+t	t^2+t+1
1	1	0	$t+1$	t	t^2+1	t^2	t^2+t+1	t^2+t
t	t	$t+1$	0	1	t^2+t	t^2+t+1	t^2	t^2+1
$t+1$	$t+1$	t	1	0	t^2+t+1	t^2+t	t^2+1	t^2
t^2	t^2	t^2+1	t^2+t	t^2+t+1	0	1	t	$t+1$
t^2+1	t^2+1	t^2	t^2+t+1	t^2+t	1	0	$t+1$	t
t^2+t	t^2+t	t^2+t+1	t^2	t^2+1	t	$t+1$	0	1
t^2+t+1	t^2+t+1	t^2+t	t^2+1	t^2	$t+1$	t	1	0

Using $t^3 = t+1$ and the addition table, we construct the multiplication below. Again, the index row and column are omitted. It is routine to verify that this is a field.

$\times$	0	1	t	$t+1$	t^2	t^2+1	t^2+t	t^2+t+1
0	0	0	0	0	0	0	0	0
1	0	1	t	$t+1$	t^2	t^2+1	t^2+t	t^2+t+1
t	0	t	t^2	t^2+t	$t+1$	1	t^2+t+1	t^2+1
$t+1$	0	$t+1$	t^2+t	t^2+1	t^2+t+1	t^2	1	t
t^2	0	t^2	$t+1$	t^2+t+1	t^2+t	t	t^2+1	1
t^2+1	0	t^2+1	1	t^2	t	t^2+t+1	$t+1$	t^2+t
t^2+t	0	t^2+t	t^2+t+1	1	t^2+1	$t+1$	t	t^2
t^2+t+1	0	t^2+t+1	t^2+1	t	1	t^2+t	t^2	$t+1$

The **minimal** polynomial for an element is the monic polynomial of the lowest degree satisfied by that element. There is a remarkable one-to-one correspondence between the set of minimal polynomials of the elements of $\mathbf{GF}(n^m)$ and the set of patterns under rotation of necklaces consisting of m beads in n colors. This was discovered by **Solomon Golomb**. We illustrate with the case $m = 4$ and $n = 2$.

In $\mathbf{GF}(16)$, there are three irreducible polynomials of degree 4, one of which is $x^4 + x + 1$. Let t be a root of this polynomial. Then the non-zero elements of $\mathbf{GF}(16)$ are t, t^2, t^3, $t^4 = t + 1$, $t^5 = t^2 + t$, $t^6 = t^3 + t^2$, $t^7 = t^3 + t + 1$, $t^8 = t^2 + 1$, $t^9 = t^3 + t$, $t^{10} = t^2 + t + 1$, $t^{11} = t^3 + t^2 + t$, $t^{12} = t^3 + t^2 + t + 1$, $t^{13} = t^3 + t^2 + 1$, $t^{14} = t^3 + 1$ and $t^{15} = 1$. The minimal polynomials are x for 0, $x + 1$ for t^{15}, $x^2 + x + 1$ for t^5 and t^{10}, $x^4 + x + 1$ for t, t^2, t^4 and t^8, $x^4 + x^3 + 1$ for t^3, t^6, t^{12} and $t^{24} = t^9$, and $x^4 + x^3 + x^2 + x + 1$ for t^7, t^{14}, $t^{28} = t^{13}$ and $t^{56} = t^{11}$.

Consider now a necklace with 4 beads in black and white. Let the one in the North position be worth 1, the one in the East position be worth 2, the one in the South position be worth 4 and the one in the West position be worth 8, provided that the bead is black. A white bead is worth 0.

A necklace of total worth k corresponds to the element t^k. Each pattern of necklaces corresponds to a minimal polynomial whose roots are the elements corresponding to the necklaces in the pattern.

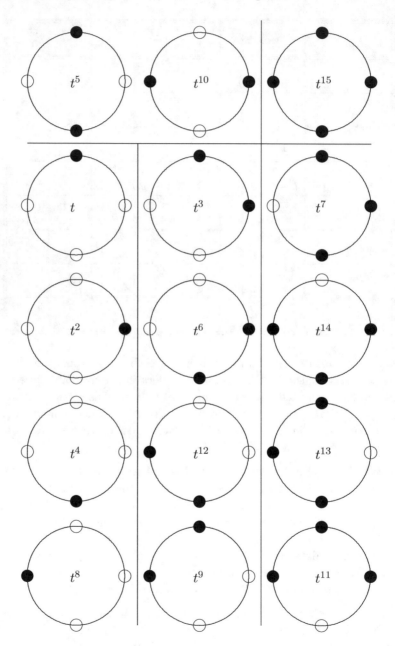

The pattern with 3 black beads corresponds to the minimal polynomial $x^4 + x^3 + 1$. The pattern with 2 adjacent black beads corresponds to the minimal polynomial $x^4 + x^3 + x^2 + x + 1$. The pattern with 1 black bead corresponds to the minimal polynomial $x^4 + x + 1$. These three patterns have no non-trivial symmetries and correspond to minimal polynomials of degree 4.

The pattern with 4 black beads corresponds to the minimal polynomial $x + 1$. The pattern with 2 non-adjacent black beads corresponds to the minimal polynomial $x^2 + x + 1$. The pattern with 0 black beads, which is not shown, is assigned to the minimal polynomial x. These three patterns have nontrivial symmetries and correspond to minimal polynomials of degree less than 4.

From results in the preceding chapter on necklaces, we can prove the following general result on minimal polynomials in finite fields. The formula $\frac{1}{m} \sum_{d|m} \phi\left(\frac{m}{d}\right) n^d$ gives the total number of minimal polynomials for all the elements in $\mathbf{GF}(n^m)$. The formula $\frac{1}{m} \sum_{d|m} \mu\left(\frac{m}{d}\right) n^d$ gives the number of minimal polynomials of degree m in $\mathbf{GF}(n^m)$.

Indeed, there are $\frac{1}{4}(\phi(4)2 + \phi(2)2^2 + \phi(1)2^4) = 6$ minimal polynomials for the elements in $\mathbf{GF}(16)$, of which $\frac{1}{4}(\mu(4)2 + \mu(2)2^2 + \mu(1)2^4) = 3$ are of degree 4.

We now turn our attention to finite geometries. In the standard Cartesian or **affine plane**, points have coordinates (x, y) where x and y are real numbers. Lines have equations $y = mx + b$ or $x = k$, where m, b and k are also real numbers.

Every two points determine a line, but not every two lines determine a point. This is because the two lines may be parallel. In an attempt to create a symmetry between points and lines, we add an **ideal** point to each line such that parallel lines have the same ideal point and non-parallel lines have different ideal points. Thus an ideal point symbolizes the common slope of a family of parallel lines.

Every two lines now determine a point, but we still have not achieved the desired symmetry. Every two ordinary points determine a line as before. An ordinary point and an ideal point also determine a line. However, two ideal points do not yet determine a line. So we add an **ideal** line which passes through all the ideal points. Finally, we have symmetry between points and lines. The resulting structure is called the **projective plane**.

In a finite geometry, we use elements in a finite field as coordinates for the points and coefficients for the lines. It is said to be of **order** n if the finite field has n elements. Such a plane has n^2 points and $n^2 + n$ lines. Each line passes through n points and each point lies on $n + 1$ lines.

The affine plane of order 2 is based on the field $\mathbf{Z}_2$. The four points have coordinates (0,0), (0,1), (1,0) and (1,1). The six lines may be divided into three pairs of parallel lines, listed below with the points on them.

$y = 0$	(0,0) (1,0)
$y = 1$	(0,1) (1,1)
$y = x$	(0,0) (1,1)
$y = x + 1$	(0,1) (1,0)
$x = 0$	(0,0) (0,1)
$x = 1$	(1,0) (1,1)

To obtain a projective plane of a finite order from an affine plane of the same order, we carry out exactly the same transformation as in the case of the infinite plane. Thus a projective plane of order n has $n^2 + n + 1$ points and $n^2 + n + 1$ lines. Each line passes through $n + 1$ points and each point lies on $n + 1$ lines.

For $n = 2$, we add three ideal points (0), (1) and (∞) which symbolize slopes of ordinary lines, and an ideal line passing through these three points. The ideal point (0) is added to $y = 0$ and $y = 1$. The ideal point (1) is added to $y = x$ and $y = x + 1$. The ideal point (∞) is added to $x = 0$ and $x = 1$. In the diagram below, this projective plane is redrawn in a symmetric way, with one line bent into a circle. Any of the 7 lines may be taken as the ideal line.

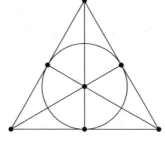

This is often referred to as the **Fano Plane**, named after a distinguished Italian geometer Gino Fano. It is such an important structure that we give it another representation, via **homogeneous** coordinates.

We start with the four points of the affine plane of order 2, but add a third w-coordinate. This coordinate is 1 for any ordinary point and 0 for any ideal point. Thus the four ordinary points are (0,0,1), (0,1,1), (1,0,1) and (1,1,1).

The slope of an ordinary line is $\frac{y}{x}$. When the slope is 0, we have $y = 0$ and $x = 1$. The ideal point representing this slope is therefore $(1,0,0)$. Similarly, the ideal point representing the slope 1 is $(1,1,0)$, and the ideal point representing the slope ∞ is $(0,1,0)$. Note that no point has coordinates $(0,0,0)$. The equation of the ideal line is $w = 0$. The lines $x = 0$ and $y = 0$ retain their equations while the line $y = x$ is rewritten as $x + y = 0$. The line $x = 1$ becomes $x + w = 0$. The line $y = 1$ becomes $y + w = 0$. The line $y = x + 1$ becomes $x + y + w = 0$.

The affine plane of order 3, based on the field $\mathbf{Z}_3$, is essentially the Tic-Tac-Toe board. The nine points have coordinates $(0,0)$, $(0,1)$, $(0,2)$, $(1,0)$, $(1,1)$, $(1,2)$, $(2,0)$, $(2,1)$ and $(2,2)$. The twelve lines may be divided into four classes of parallel lines, listed below with the points on them.

$y = 0$	$(0,0)\ (1,0)\ (2,0)$	$y = x$	$(0,0)\ (1,1)\ (2,2)$
$y = 1$	$(0,1)\ (1,1)\ (2,1)$	$y = x + 1$	$(0,1)\ (1,2)\ (2,0)$
$y = 2$	$(0,2)\ (1,2)\ (2,2)$	$y = x + 2$	$(0,2),\ (1,0)\ (2,1)$
$y = 2x$	$(0,0)\ (1,2)\ (2,1)$	$x = 0$	$(0,0)\ (0,1)\ (0,2)$
$y = 2x + 1$	$(0,1)\ (1,0)\ (2,2)$	$x = 1$	$(1,0)\ (1,1)\ (1,2)$
$y = 2x + 2$	$(0,2)\ (1,1)\ (2,0)$	$x = 2$	$(2,0)\ (2,1)\ (2,2)$

To obtain a projective plane of order 3, we add four ideal points (0), (1), (2) and (∞), and an ideal line passing through these four points. The ideal points are added to each line in the respective classes in the chart above.

An affine plane of order 4 cannot be based on $\mathbf{Z}_4$, which is not a field. Otherwise, the line $y = 2x$ will pass through the four points $(0,0)$, $(1,2)$, $(2,0)$ and $(3,2)$, and will intersect the line $y = 0$ in two points, namely, $(0,0)$ and $(2,0)$. Nevertheless, there is an affine plane of order 4. It is based on the Galois field $\mathbf{GF}(4)$.

Examples

Example 6.1.1.

(a) Compute the number of irreducible monic quadratic polynomials over $\mathbf{Z}_3$.

(b) Determine all of them.

(c) One of them is $x^2 + 1$. Let t be a root of this polynomial. List the 9 elements in a $\mathbf{GF}(9)$ obtained by adding t to $\mathbf{Z}_3$.

(d) Compute the number of minimal polynomials for all these elements.

(e) Determine all of them.

Solution:

(a) The number of irreducible monic quadratic polynomials over $\mathbf{Z}_3$ is given by $\frac{1}{2}(\mu(2)3 + \mu(1)3^2) = 3$.

(b) The chart below shows all reducible monic quadratic polynomials over $\mathbf{Z}_3$.

$\times$	x	$x + 1$	$x + 2$
x	x^2	$x^2 + x$	$x^2 + 2x$
$x + 1$	$x^2 + x$	$x^2 + 2x + 1$	$x^2 + 2$
$x + 2$	$x^2 + 2x$	$x^2 + 2$	$x^2 + x + 1$

Hence the irreducible monic quadratic polynomials over $\mathbf{Z}_3$ are $x^2 + 1$, $x^2 + x + 2$ and $x^2 + 2x + 2$.

(c) The elements are 0, 1, 2, t, $t + 1$, $t + 2$, $2t$, $2t + 1$ and $2t + 2$, where $t^2 + 1 = 0$ or $t^2 = 2$.

Remark:

The extension of $\mathbf{Z}_3$ into $\mathbf{GF}(9)$ by adding t is very much like the extension of the real numbers into the complex numbers by adding $i = \sqrt{-1}$. We can display the nine elements of $\mathbf{GF}(9)$ in the "complex" plane as shown in the diagram below, with 2 replaced by -1 since $-1 \equiv 2 \pmod{3}$.

$-1 + t$	t	$1 + t$
-1	0	1
$-1 - t$	$-t$	$1 - t$

(d) The total number of minimal polynomials is $\frac{1}{2}(\phi(2)3 + \phi(1)3^2) = 6$.

(e) The minimal polynomials are x for 0, $x + 2$ for 1, $x + 1$ for 2, $x^2 + 1$ for t and $2t$, $x^2 + x + 2$ for $t + 1$ and $2t + 1$, and $x^2 + 2x + 2$ for $t + 2$ and $2t + 2$.

Example 6.1.2.
Partition the 20 lines of the affine geometry of order 4 into 5 classes of mutually parallel lines, and list the points on each line.

Solution:
In the chart below, the 20 lines are partitioned into 5 classes of mutually parallel lines. The points on each line are listed.

$y = 0$	$(0,0)\ (1,0)\ (t, 0)\ (t + 1, 0)$
$y = 1$	$(0,1)\ (1,1)\ (t, 1)\ (t + 1, 1)$
$y = t$	$(0, t)\ (1, t)\ (t, t)\ (t + 1, t)$
$y = t + 1$	$(0, t + 1)\ (1, t + 1)\ (t, t + 1)\ (t + 1, t + 1)$
$y = x$	$(0,0)\ (1,1)\ (t, t)\ (t + 1, t + 1)$
$y = x + 1$	$(0,1)\ (1,0)\ (t, t + 1)\ (t + 1, t)$
$y = x + t$	$(0, t)\ (1, t + 1)\ (t, 0)\ (t + 1, 1)$
$y = x + t + 1$	$(0, t + 1)\ (1, t)\ (t, 1)\ (t + 1, 0)$
$y = tx$	$(0,0)\ (1, t)\ (t, t + 1)\ (t + 1, 1)$
$y = tx + 1$	$(0,1)\ (1, t + 1)\ (t, t)\ (t + 1, 0)$
$y = tx + t$	$(0, t)\ (1,0)\ (t, 1)\ (t + 1, t + 1)$
$y = tx + t + 1$	$(0, t + 1)\ (1,1)\ (t, 0)\ (t + 1, t)$
$y = (t + 1)x$	$(0,0)\ (1, t + 1)\ (t, 1)\ (t + 1, t)$
$y = (t + 1)x + 1$	$(0,1)\ (1, t)\ (t, 0)\ (t + 1, t + 1)$
$y = (t + 1)x + t$	$(0, t)\ (1,1)\ (t, t + 1)\ (t + 1, 0)$
$y = (t + 1)x + t + 1$	$(0, t + 1)\ (1,0)\ (t, t)\ (t + 1, 1)$
$x = 0$	$(0,0)\ (0,1)\ (0, t)\ (0, t + 1)$
$x = 1$	$(1,0)\ (1,1)\ (1, t)\ (1, t + 1)$
$x = t$	$(t, 0)\ (t, 1)\ (t, t)\ (t, t + 1)$
$x = t + 1$	$(t + 1, 0)\ (t + 1, 1)\ (t + 1, t)\ (t + 1, t + 1)$

Example 6.1.3.
A certain community of lions and ponies is defined by the following postulates.
(1) There are at least two lions.
(2) Each lion has bitten at least three ponies.
(3) For any pair of lions, there is exactly one pony that both have bitten.
(4) For any pair of ponies, there is at least one lion that has bitten both of them.

Deduce the following results.

(a) For any lion, there is at least one pony that it has not bitten.

(b) For any pony, there is at least one lion that has not bitten it.

Solution:

(a) Suppose there is a lion that has bitten every pony. By Postulate 1, there is another lion. It follows from Postulate 3 that this lion can have bitten only one pony. This contradicts Postulate 2. Hence for any lion, there is at least one pony that it has not bitten.

(b) Suppose there is a pony P that every lion has bitten. By Postulate 1, there is a pair of lions A and B. By Postulate 2, A has bitten a pony Q other than P and B has bitten a pony R other than P. By Postulate 3, A has not bitten R and B has not bitten Q. Hence Q and R are not the same pony. By Postulate 4, there is a lion C that has bitten both Q and R. Hence C and A are two different lions but they have both bitten P and Q. This contradicts Postulate 3. Hence for any pony, there is at least one lion that has not bitten it.

Exercises

1. (a) Compute the number of irreducible monic quadratic polynomials over $\mathbf{Z}_5$.

   (b) Determine all of them.

   (c) One of them is $x^2 + 2$. Let t be a root of this polynomial. List the 25 elements in a $\mathbf{GF}(25)$ obtained by adjoining t to $\mathbf{Z}_5$.

   (d) Compute the number of minimal polynomials for all these elements.

   (e) Determine all of them.

2. (a) Compute the number of irreducible monic cubic polynomials over $\mathbf{Z}_3$.

   (b) Determine all of them.

   (c) One of them is $x^3 + 2x + 1$. Let t be a root of this polynomial. List the 27 elements in a $\mathbf{GF}(27)$ obtained by adjoining t to $\mathbf{Z}_3$.

   (d) Compute the number of minimal polynomials for all these elements.

   (e) Determine all of them.

3. Partition the 56 lines of the affine geometry of order 7 into 8 classes of mutually parallel lines, and list the points on each line.

4. Partition the 30 lines of the affine geometry of order 5 into 6 classes of mutually parallel lines, and list the points on each line.

5. A certain community of lions and ponies is defined by the following postulates.
   (1) There are at least two lions.
   (2) Each lion has bitten at least three ponies.
   (3) For any pair of lions, there is exactly one pony that both have bitten.
   (4) For any pair of ponies, there is at least one lion that has bitten both of them.
   Deduce the following results.

   (a) For any pair of lions, there is at least one pony that neither has bitten.

   (b) For any pair of ponies, there is at least one lion that has not bitten either.

6. A certain community of lions and ponies is defined by the following postulates.

   (1) There are at least two lions.

   (2) Each lion has bitten at least three ponies.

   (3) For any pair of lions, there is exactly one pony that both have bitten.

   (4) For any pair of ponies, there is at least one lion that has bitten both of them.

   Deduce the following results.

   (a) There are at least two ponies.

   (b) Each pony has been bitten by at least three lions.

   (c) For any pair of ponies, there is exactly one lion that has bitten both.

   (d) For any pair of lions, there is at least one pony that both have bitten.

Section 6.2. Latin Squares

A **Latin square** of order n is an $n \times n$ array in which each row and each column is a permutation of a set of n symbols. They derived their name from the fact that they were already known in Roman times. Recently, they have emerged into prominence again because they are the underlying structure on which the popular puzzle Sudoku is based.

The standard form of a Latin square of order n uses the symbols 0, 1, ..., $n-1$, appearing in their natural order in the first row and the first column. The diagram below shows all Latin squares of order up to 4.

$$
\begin{array}{cccccc}
0 & & 0 \ 1 & & 0 \ 1 \ 2 & \\
& & 1 \ 0 & & 1 \ 2 \ 0 & \\
& & & & 2 \ 0 \ 1 &
\end{array}
$$

$$
\begin{array}{cccc}
0\ 1\ 2\ 3 & 0\ 1\ 2\ 3 & 0\ 1\ 2\ 3 & 0\ 1\ 2\ 3 \\
1\ 2\ 3\ 0 & 1\ 0\ 3\ 2 & 1\ 3\ 0\ 2 & 1\ 0\ 3\ 2 \\
2\ 3\ 0\ 1 & 2\ 3\ 1\ 0 & 2\ 0\ 3\ 1 & 2\ 3\ 0\ 1 \\
3\ 0\ 1\ 2 & 3\ 2\ 0\ 1 & 3\ 2\ 1\ 0 & 3\ 2\ 1\ 0
\end{array}
$$

A good source for Latin squares are the Cayley tables of finite groups. In fact, the first four Latin squares above are the Cayley tables of $\mathbf{Z}_1$, $\mathbf{Z}_2$, $\mathbf{Z}_3$ and $\mathbf{Z}_4$, and the last is that of $\mathbf{Z}_2 \times \mathbf{Z}_2$. The fifth and sixth are different renditions of the Cayley table of $\mathbf{Z}_4$, as seen in the diagram below, interchanging 1 and 2 in the former and 2 and 3 in the latter.

$+_4$	0	2	1	3
0	0	2	1	3
2	2	0	3	1
1	1	3	2	0
3	3	1	0	2

$+_4$	0	1	3	2
0	0	1	3	2
1	1	2	0	3
3	3	0	2	1
2	2	3	1	0

However, for $n \geq 5$, the majority of Latin squares of order n are not Cayley tables. It is known that if n and $\phi(n)$ are relatively prime, then there is a unique group of order n. On the other hand, unlike the number C_n of Cayley tables, the number L_n of Latin squares of order n in standard forms grows incredibly rapidly.

Order n	1	2	3	4	5	6	7	8
C_n	1	1	1	2	1	2	1	5
L_n	1	1	1	4	56	9404	16942080	535281401856

A **transversal** of a Latin square of order n is a set of n cells in the Latin square with no two on the same row, no two in the same column, and no two containing the same symbol. Two transversals of the same Latin square are disjoint if no cell of the Latin square is a member of both transversals.

We use a second square to record the positions of the cells of a transversal. In the diagram below, a Latin square is given on the left, with a transversal in boldface. These five cells are marked with 0s in the square in the middle. Similarly, the five cells marked with 1s denote another transversal disjoint from the first one. In fact, five pairwise disjoint transversals can be found, yielding another Latin square on the right.

0	1	**2**	3	4		1		0				1	3	0	2	4
1	2	3	4	**0**			1		0			2	4	1	3	0
2	**3**	4	0	1		0			1			3	0	2	4	1
3	4	0	**1**	2		1	0					4	1	3	0	2
4	0	1	2	3		0		1				0	2	4	1	3

It is possible for a Latin square of order n to have exactly k pairwise disjoint transversals for $0 \leq k \leq n$, except for $k = n - 1$. This is because after $n-1$ pairwise disjoint transversals have been picked out, the remaining n cells automatically form an n-th transversal disjoint from the others.

For the special case that the Latin square of order n is actually a Cayley table, if it has one transversal, then it will have a complete set of n pairwise disjoint transversals. Let G be a finite group consisting of the elements g_1, g_2, $\ldots$, g_n, in the order in which they appear in the Cayley table. This may be considered as a Latin square. If it has a transversal, it may be represented as a set of n equations $h_i g_i = \ell_i$, $1 \leq i \leq n$, where h_1, h_2, $\ldots$, h_n and ℓ_1, ℓ_2, $\ldots$, ℓ_n are permutations of g_1, g_2, $\ldots$, g_n. The former is regarded as the row index while the latter is regarded as the entry in the transversal.

We now consider the effect of multiplying each of these equations, on the left, by a fixed group element p. Recall that if all elements of a group are multiplied by a fixed element of the group, the result is a permutation of the group elements; and unless the fixed element is the identity (which leaves all elements unchanged) no group element is left unchanged. Thus

$$
\begin{array}{lll}
p(h_1 g_1) = p\ell_1 & (ph_1)g_1 = p\ell_1 & f_1 g_1 = m_1 \\
p(h_2 g_2) = p\ell_2 & (ph_2)g_2 = p\ell_2 & f_2 g_2 = m_2 \\
\cdots = \cdots & \cdots = \cdots & \cdots = \cdots \\
p(h_n g_n) = p\ell_n & (ph_n)g_n = p\ell_n & f_n g_n = m_n
\end{array}
$$

where f_1, f_2, $\ldots$, f_n and m_1, m_2, $\ldots$, m_n are permutations of g_1, g_2, $\ldots$, g_n.

It follows that a new disjoint transversal can be obtained from any non-identity element p in the group. It follows that the existence of one transversal implies the existence of a complete set of disjoint transversals.

As an illustration, consider the Latin square in the diagram below. It is the Cayley table of the multiplicative group modulo 16. A transversal is given by the boldfaced entries.

1	3	5	**7**	9	11	13	15
3	9	15	5	11	1	7	**13**
5	15	9	3	13	7	1	11
7	5	3	1	**15**	13	11	9
9	**11**	13	15	1	3	5	7
11	1	7	13	3	**9**	15	5
13	7	**1**	11	5	15	9	3
15	13	11	9	7	5	**3**	1

Here, $g_1 = 1, g_2 = 3, g_3 = 5, g_4 = 7, g_5 = 9, g_6 = 11, g_7 = 13$ and $g_8 = 15$. From the given transversal, $h_1 = 5, h_2 = 9, h_3 = 13, h_4 = 1, h_5 = 7, h_6 = 11, h_7 = 15$ and $h_8 = 3$, while $\ell_1 = 5, \ell_2 = 11, \ell_3 = 1, \ell_4 = 7, \ell_5 = 15, \ell_6 = 9, \ell_7 = 3$ and $\ell_8 = 13$. This transversal is marked by 1s in the diagram below.

13	9	5	1	7	3	15	11
7	11	15	3	5	9	13	1
1	13	9	5	3	15	11	7
11	15	3	7	1	5	9	13
5	1	13	9	15	11	7	3
15	3	7	11	13	1	5	9
9	5	1	13	11	7	3	15
3	7	11	15	9	13	1	5

We now obtain a disjoint transversal using the element $p = 3$. We have $f_1 = 15, f_2 = 11, f_3 = 7, f_4 = 3, f_5 = 5, f_6 = 1, f_7 = 13$ and $f_8 = 9$. Moreover, $m_1 = 15, m_2 = 1, m_3 = 3, m_4 = 5, m_5 = 13, m_6 = 11, m_7 = 9$ and $m_8 = 7$. These are permutations of the group elements, so that we have indeed a transversal disjoint from the given one. We mark it with 3s in the diagram above.

By taking p through other elements of the group, we can find a complete set of disjoint transversals. Note that it is not necessary to calculate the values of ℓ_i or m_i.

When a second Latin square of order n is obtained from a first by recording a complete set of transversals, the n^2 ordered pairs of corresponding elements are all distinct. Two such Latin squares are said to be **orthogonal** to each other. In the example above, the 64 ordered pairs of corresponding elements are shown in the diagram below.

(1,13)	(3,9)	(5,5)	(7,1)	(9,7)	(11,3)	(13,15)	(15,11)
(3,7)	(9,11)	(15,15)	(11,3)	(13,5)	(1,9)	(5,13)	(9,1)
(5,1)	(15,13)	(9,9)	(3,5)	(13,3)	(7,15)	(1,11)	(11,7)
(7,11)	(5,15)	(3,3)	(1,7)	(15,1)	(13,5)	(11,9)	(9,13)
(9,5)	(11,1)	(13,13)	(15,9)	(1,15)	(3,11)	(5,7)	(7,3)
(11,15)	(1,3)	(7,7)	(13,11)	(3,13)	(9,1)	(15,5)	(5,9)
(13,9)	(7,5)	(1,1)	(11,13)	(5,11)	(15,7)	(9,3)	(3,15)
(15,3)	(13,7)	(11,11)	(9,15)	(7,9)	(5,13)	(3,1)	(1,5)

If we interpret the ordered pairs of two orthogonal Latin squares in the diagram below on the left as two-digit numbers in base 5, we get the square in the diagram below on the right, where all row sums and column sums are constant, namely, 60.

(0,1)	(1,3)	(2,0)	(3,2)	(4,4)	1	8	10	17	24
(1,2)	(2,4)	(3,1)	(4,3)	(0,0)	7	14	16	23	0
(2,3)	(3,0)	(4,2)	(0,4)	(1,1)	13	15	22	4	6
(3,4)	(4,1)	(0,3)	(1,0)	(2,2)	19	21	3	5	12
(4,0)	(0,2)	(1,4)	(2,1)	(3,3)	20	2	9	11	18

The study of orthogonal Latin squares began when L. Euler (1707-1783) was in St. Petersburg, Russia, at the court of Catherine the Great. Someone had proposed a military honor guard consisting of 36 officers in a 6×6 array, where the officers represented 6 different ranks, from 6 different regiments. Each row and each column of the array was to have one officer of each rank, and one officer from each regiment; but no one could figure out how to design such an array.

Finally, the "Problem of the 36 Officers" was communicated to Euler, who recognized that the problem was, in our terminology, constructing a pair of orthogonal Latin squares of order 6. He became convinced that this was not possible, but the impossibility was not proved until early in the 20th century. Euler found a way to obtain a pair of orthogonal Latin squares for every n except when n is twice an odd number (i.e. except when $n = 2, 6, 10, 14, 18, 22, 26, \ldots$), and he conjectured that orthogonal Latin squares do not exist for these values of n.

We now know he was right only about $n = 2$ and $n = 6$. He would have been correct had his conjecture dealt only with Latin squares that are also Cayley tables. In 1958, counter-examples (i.e. examples of orthogonal Latin squares) were found for all the other values of n by R. C. Bose, S. S. Shrikhande, and E. T. Parker, whom Martin Gardner dubbed the "Euler spoilers". Parker's example of orthogonal Latin squares of order 10 made the cover of Scientific American for November, 1959.

Once we know of the existence of two orthogonal Latin squares, the natural question is whether we can have three mutually orthogonal Latin squares, and what is the maximum number of mutually orthogonal Latin squares we can have for a given order.

The second question is easier to answer. For Latin squares of order n, there can be at most $n - 1$ mutually orthogonal Latin squares. We may assume that the first row of each Latin square consists of 0, 1, 2, ..., $n-1$. Consider the first entry in the second row. It cannot be 0 as we already have a 0 in the first column. This entry may not be duplicated from Latin square to Latin square as all the doubles are already in the first row. Hence it can take on at most $n - 1$ values, so that there can be at most $n - 1$ mutually orthogonal Latin squares of order n,

When $n = p^k$, where p is a prime number and k is any positive integer, there are $n - 1$ mutually orthogonal Latin squares of order n. We do not know of the existence of $n - 1$ mutually orthogonal Latin squares for any other order n. There are results which rule out infinitely many values of n for having $n - 1$ mutually orthogonal Latin squares. However, there are also infinitely many values of n for which no construction of $n - 1$ mutually orthogonal Latin squares is known, but for which the possibility has not been ruled out. The smallest undecided case is $n = 12$, where a set of 5 mutually orthogonal Latin squares is known.

When n is a power of a prime number, we turn to affine geometry of order n. We illustrate with the case $n = 4$. Recall from Example 6.1.2 that this geometry has 16 points and 20 lines divided into 5 parallel classes. The two classes with slope 0 and slope ∞ are set aside for coordinate references. Each of the other three classes will yield one of the three mutually orthogonal Latin squares we seek.

Note that each of the four lines in a class is a transversal of the Latin square, which we will mark by the constant term of its equation. Consider the class with slope 1. The line $y = x$ passes through $(0,0)$, $(1,1)$, (t,t) and $(t + 1, t + 1)$. These four points will be labeled 0. The line $y = x + 1$ passes through $(0,1)$, $(1,0)$, $(t, t + 1)$ and $(t + 1, t)$. These four points will be labeled 1. The line $y = x + t$ passes through $(0, t)$, $(1, t + 1)$, $(t, 0)$ and $(t + 1, 1)$. These four points will be labeled t. Finally, the line $y = x + t + 1$ passes through $(0, t + 1)$, $(1, t)$, $(t, 1)$ and $(t + 1, 0)$. These four points will be labeled $t + 1$, yielding the Latin square in the diagram below.

$$
\begin{array}{cccc}
t + 1 & t & 1 & 0 \\
t & t + 1 & 0 & 1 \\
1 & 0 & t + 1 & t \\
0 & 1 & t & t + 1
\end{array}
$$

In an analogous manner, the classes with slope t and slope $t + 1$ yield two other Latin squares, shown in the diagram below. Mutual orthogonality among all three Latin squares is assured because two points determine a unique line.

$$
\begin{array}{cccc}
t+1 & 1 & 0 & t \\
t & 0 & 1 & t+1 \\
1 & t+1 & t & 0 \\
0 & t & t+1 & 1
\end{array}
\qquad
\begin{array}{cccc}
t+1 & 0 & t & 1 \\
t & 1 & t+1 & 0 \\
1 & t & 0 & t+1 \\
0 & t+1 & 1 & t
\end{array}
$$

We should mention that this process is reversible, that is, given a set of $n-1$ mutually orthogonal Latin squares, we can construct an affine geometry of order n. Below is a family of four mutually orthogonal Latin squares of order 5.

$$
\begin{bmatrix}
0 & 1 & 2 & 3 & 4 \\
1 & 2 & 3 & 4 & 0 \\
2 & 3 & 4 & 0 & 1 \\
3 & 4 & 0 & 1 & 2 \\
4 & 0 & 1 & 2 & 3
\end{bmatrix}
\qquad
\begin{bmatrix}
0 & 1 & 2 & 3 & 4 \\
3 & 4 & 0 & 1 & 2 \\
1 & 2 & 3 & 4 & 0 \\
4 & 0 & 1 & 2 & 3 \\
2 & 3 & 4 & 0 & 1
\end{bmatrix}
$$

$$
\begin{bmatrix}
0 & 1 & 2 & 3 & 4 \\
2 & 3 & 4 & 0 & 1 \\
4 & 0 & 1 & 2 & 3 \\
1 & 2 & 3 & 4 & 0 \\
3 & 4 & 0 & 1 & 2
\end{bmatrix}
\qquad
\begin{bmatrix}
0 & 1 & 2 & 3 & 4 \\
4 & 0 & 1 & 2 & 3 \\
3 & 4 & 0 & 1 & 2 \\
2 & 3 & 4 & 0 & 1 \\
1 & 2 & 3 & 4 & 0
\end{bmatrix}
$$

From this family, we can construct an affine plane of order 5. Represent the 25 points of this geometry by the following array.

$$
\begin{array}{ccccc}
A & B & C & D & E \\
F & G & H & I & J \\
K & M & N & O & P \\
Q & R & S & T & U \\
V & W & X & Y & Z
\end{array}
$$

There are six groups of 5 lines each. The first group is just the rows and the last group is just the columns. The other four are based on the mutually orthogonal Latin squares. For instance, in the second group, the points A, J, O, S and W are where the 0s are in the first Latin square. The points B, F, P, T and X are where the 1s are, and so on. The six groups are parallel lines of slopes 0, 1, 2, 3, 4 and ∞, respectively.

(A,B,C,D,E)   (A,J,O,S,W)   (A,H,P,R,Y)
(F,G,H,I,J)   (B,F,P,T,X)   (B,I,K,S,Z)
(K,M,N,O,P)   (C,G,K,U,Y)   (C,J,M,T,V)
(Q,R,S,T,U)   (D,H,M,Q,Z)   (D,F,N,U,W)
(V,W,X,Y,Z)   (E,I,N,R,V)   (E,G,O,Q,X)

(A,I,M,U,X)   (A,G,N,T,Z)   (A,F,K,Q,V)
(B,J,N,Q,Y)   (B,H,O,U,V)   (B,G,M,R,W)
(C,F,O,R,Z)   (C,I,P,Q,W)   (C,H,N,S,X)
(D,G,P,S,V)   (D,J,K,R,Y)   (D,I,O,T,Y)
(E,H,K,T,W)   (E,F,M,S,Y)   (E,J,P,U,Z)

Examples.

Example 6.2.1.

(a) Complete the following partially filled square into a Latin square.

(b) Determine the maximum number of disjoint transversals it can have.

0	1			4	5
1	4		2	0	
2			1		
					4
	5				
5		3			1

Solution:

(a) The completed Latin square is shown in the diagram below.

(0)	1	[2]	3	**4**	{5}
1	[4]	(5)	2	{0}	**3**
2	(3)	4	{1}	[5]	0
[3]	{2}	0	**5**	(1)	4
{4}	5	**1**	[0]	3	(2)
5	**0**	{3}	(4)	2	[1]

(b) There are four transversals, and their elements are marked in different styles in the diagram above. We know that there are no two orthogonal Latin squares of order 6, so that we cannot have a complete set of six transversals. We cannot have five transversals either as the remaining elements must form a sixth transversal. Hence four is indeed the maximum number of transversals in any 6×6 Latin square.

Example 6.2.2.
The following is the Cayley table of the multiplicative group modulo 24 considered as a Latin square. A transversal is given by the boldfaced entries. Construct a Latin square orthogonal to this square.

1	5	7	**11**	13	17	19	23
5	1	11	7	17	13	23	**19**
7	11	1	5	19	23	13	17
11	7	5	1	**23**	19	17	13
13	**17**	19	23	1	5	7	11
17	13	23	19	5	**1**	11	7
19	23	**13**	17	7	11	1	5
23	19	17	13	11	7	**5**	1

Solution:

Here, $g_1 = 1, g_2 = 5, g_3 = 7, g_4 = 11, g_5 = 13, g_6 = 17, g_7 = 19$ and $g_8 = 23$. From the given transversal, the values of h_i are recorded in the diagram below on the left as the first column. The subsequent columns are obtained by multiplying the first column with the other elements, yielding the f_i values for each. Note that the rows are just permutations of the rows of the Cayley table. The resulting Latin square orthogonal to the Cayley table is shown in the diagram below on the right.

7	11	5	5	19	23	13	17	7	13	19	1	11	17	23	5
13	17	19	23	1	5	7	11	11	17	23	5	7	13	19	1
19	23	13	17	7	11	1	5	1	19	13	7	5	23	17	11
1	5	7	11	13	17	19	23	5	23	17	11	1	19	13	7
11	7	5	1	23	19	17	13	19	1	7	13	23	5	11	17
17	13	23	19	5	1	11	7	23	5	11	17	19	1	7	13
23	19	17	13	11	7	5	1	13	7	1	19	17	11	5	23
5	1	11	7	17	13	23	19	17	11	5	23	13	7	1	19

Example 6.2.3.

Consider the Cayley table of the multiplicative group modulo p as a Latin square, where $p > 2$ is prime. Prove that this Latin square cannot have any transversals, and therefore no orthogonal mates.

Solution:

If a transversal exists, it may be represented as a set of $p - 1$ equations $h_i i = \ell_i$, $1 \leq i \leq p - 1$, where h_1, h_2, ..., h_n and ℓ_1, ℓ_2, ..., ℓ_n are permutations of 1, 2, ..., $p - 1$. The former is regarded as the row index while the latter is regarded as the entries in the transversal. Then the product of all the terms on the left sides of the equations is $(p - 1)!(p - 1)!$ while the product of the right sides is $(p - 1)!$. Hence $(p - 1)! \equiv 1 \pmod{p}$. However, by Wilson's Theorem, $(p - 1)! \equiv -1 \pmod{p}$. We have a contradiction since $1 \equiv -1 \pmod{p}$ is not possible for $p > 2$.

Exercises

1. For each of the following partially filled squares, try to complete it into a Latin square, and if successful, find a transversal of it.

(a)

0	1	2		
1			0	
	3			
			4	
4		1		

(b)

0	1	2	3	
1		0		
2	3	1		
3	0			

(c)

0	1		3	4
1				
2		0		1
		3		0

2. For each of the following partially filled squares, try to complete it into a Latin square, and if successful, try to find as many disjoint transversals of it as you can.

(a)

0	1	2		4	
1	0		4	3	
2		3		0	4
3			5		
		0			
	2				

(b)

0	1	2	3	4	5
1	2				
2		5			
3			1		
4				3	
5					4

3. The following is the Cayley table of the multiplicative group modulo 15 considered as a Latin square. A transversal is given by the boldfaced entries. Construct a Latin square orthogonal to this square.

1	2	4	**7**	8	11	13	14
2	4	8	14	1	7	11	**13**
4	8	1	13	2	14	7	11
7	14	13	4	**11**	2	1	8
8	**1**	2	11	4	13	14	7
11	7	**14**	2	13	1	8	4
13	11	7	1	14	**8**	4	2
14	13	11	8	7	4	**2**	1

4. The following is the Cayley table of the multiplicative group modulo 20 considered as a Latin square. A transversal is given by the boldfaced entries. Construct a Latin square orthogonal to this square.

$$
\begin{array}{cccccccc}
1 & 3 & 7 & \mathbf{9} & 11 & 13 & 17 & 19 \\
3 & 9 & 1 & 7 & 13 & 19 & 11 & \mathbf{17} \\
\mathbf{7} & 1 & 9 & 3 & 17 & 11 & 19 & 13 \\
9 & 7 & 3 & 1 & \mathbf{19} & 17 & 13 & 11 \\
11 & \mathbf{13} & 17 & 19 & 1 & 3 & 7 & 9 \\
13 & 19 & \mathbf{11} & 17 & 3 & 9 & 1 & 7 \\
17 & 11 & 19 & 13 & 7 & \mathbf{1} & 9 & 3 \\
19 & 17 & 13 & 11 & 9 & 7 & \mathbf{3} & 1 \\
\end{array}
$$

5. Construct six mutually orthogonal 7×7 Latin squares.

6. Construct seven mutually orthogonal 8×8 Latin squares.

Section 6.3. Block Designs

A **balanced incomplete block design**, or simply block design, is a family of b subsets of a v-element set such that
(1) each subset contains exactly k elements, $1 < k < v$;
(2) each element appears in exactly r subsets;
(3) every two elements appear together in exactly λ subsets.
Each subset is called a **block** and the family is called a (b, v, r, k, λ)-design.

As a very simple example of a block design, we take an n-element set and all pairs of elements as blocks. This yields a $(\binom{n}{2}, n, n-1, 2, 1)$-design. Because of (3), all block designs for which $k = 2$ must have the form $(\lambda\binom{n}{2}, n, \lambda(n-1), 2, \lambda)$. For $n = 5$ and $\lambda = 1$, we have a (10,5,4,2,1)-design with the elements 0, 1, 2, 3 and 4, and the blocks (0,1), (0,2), (0,3), (0,4), (1,2), (1,3), (1,4), (2,3) (2,4) and (3,4).

The word "incomplete" refers to the inequality $k < v$. If $k = v$, then every block is the entire v-element set, and the structure is very uninteresting. The word "balanced" refers to the requirement that every element appears the same number of times and that every pair of elements also appears the same number of times.

The five parameters in a (b, v, r, k, λ)-design are not independent. They satisfy the following two equations:
(4) $bk = vr$;
(5) $r(k - 1) = \lambda(v - 1)$.
From these, we can see that (2) is redundant since it can be derived from (1) and (3). However, we retain it partly by tradition, but mostly because it introduces a useful parameter.

Each of the two equations can be proved by the technique of counting something in two different ways. In (4), we count the total number of appearances of the elements in the blocks. First, there are b blocks, and each block has k elements. Hence the total is bk. Second, there are v elements, and each appears r times. Hence the total is vr. It follows that $bk = vr$.

In (5), we count the total number of joint appearances of a fixed element with every other element. First, the fixed element appears in r blocks, and each block has $k-1$ other elements. Hence the total is $r(k-1)$. Second, there are $v - 1$ other elements, and each appears jointly with the fixed element exactly λ times. Hence the total is $\lambda(v-1)$. It follows that $r(k-1) = \lambda(v-1)$.

In order to avoid degenerate block designs, we impose the inequality $1 < k < v$. From this, we can now deduce the inequality $b > r > \lambda > 0$. There is one other, known as **Fisher's Inequality**, which states that $b \geq v$. The proof is surprisingly involved.

We shall prove the equivalent result $r \geq k$. Let B be a fixed block. For $0 \leq i \leq k$, define S_i to be the set of blocks distinct from B which have exactly i elements in common with B, and n_i to be the number of blocks in S_i. Let

$$\alpha = \sum_{i=0}^{k} n_i, \quad \beta = \sum_{i=0}^{k} i n_i \quad \text{and} \quad \gamma = \sum_{i=0}^{k} i^2 n_i.$$

Since each block other than B belongs to exactly one S_i, $\alpha = b - 1$. Each block in S_i contains i elements which belong to B, and each of these k elements belongs to $r - 1$ blocks other than B. Hence $\beta = k(r - 1)$.

Finally, each block in S_i contains $\frac{i(i-1)}{2}$ pairs of elements which belong to B, and each of these $\frac{k(k-1)}{2}$ pairs belongs to $\lambda - 1$ blocks other than B. Hence

$$\sum_{i=0}^{k} \frac{i(i-1)}{2} n_i = \frac{k(k-1)}{2}(\lambda - 1).$$

It follows that $\gamma = k(k-1)(\lambda - 1) + \beta = k((k-1)(\lambda-1) + r - 1)$.

Define $F(x) = \sum_{i=0}^{k} (x - i)^2 n_i = \alpha x^2 - 2\beta x + \gamma$. Then $F(x) \geq 0$ for all real values of x. If $F(x)$ has two distinct real roots $x_1 < x_2$, then $F(x) < 0$ for all x such that $x_1 < x < x_2$. It follows that the discriminant $4\beta^2 - 4\alpha\gamma$ of $F(x)$ must be non-positive. Substituting in the values of α, β and γ, we have $0 \geq k^2(r-1)^2 - (b-1)k((k-1)(\lambda-1)+r-1)$. Using $b = \frac{vr}{k}$ and $\lambda = \frac{r(k-1)}{v-1}$, the right side of the above inequality simplifies to $r(v - k)^2(k - r)$. Since $r > 0$ and $v > k$, we have $r \geq k$, and it follows that $b \geq v$.

For more sophisticated examples of block designs, we turn to the finite geometries.

An affine geometry of order n has n^2 points and $n(n + 1)$ lines. Each point lies on $n + 1$ lines and each line passes through n points. Every two lines intersect in exactly 1 point. Hence the geometry serves as an example of an $(n(n+1), n^2, n+1, n, 1)$-design.

For $n = 3$, we have a $(12,9,4,3,1)$-design with the elements 0, 1, 2, 3, 4, 5, 6, 7 and 8, and the blocks (0,1,2), (3,4,5), (6,7,8), (0,3,6), (1,4,7), (2,5,8), (0,4,8), (1,5,6), (2,3,7), (0,5,7), (1,3,8) and (2,4,6).

Recall that the $n(n + 1)$ lines may be divided into $n + 1$ classes of n mutually parallel lines. Each points lies on exactly one line in each class. In other words, the $n(n + 1)$ blocks may be divided into classes such that each element appears exactly once in each class. Such a block design is said to be **resolvable** and the classes are called **resolution classes**. Affine geometries yield resolvable designs.

A projective geometry of order n has $n^2 + n + 1$ points and $n^2 + n + 1$ lines. Each point lies on $n+1$ lines and each line passes through $n+1$ points. Every two lines intersect in exactly 1 point. Hence the geometry serves as an example of an $(n^2 + n + 1, n^2 + n + 1, n + 1, n + 1, 1)$-design.

For $n = 2$, we have a (7,7,3,3,1)-design with the elements 0, 1, 2, 3, 4, 5 and 6, and the blocks (0,1,3), (1,2,4), (2,3,5), (3,4,6), (4,5,0), (5,6,1) and (6,0,2).

A block design in which $b = v$ is said to be **symmetric**. Projective geometries yield symmetric designs. A symmetric (v, v, k, k, λ)-design has a very nice property, in that every two blocks have exactly λ elements in common. We now prove this result.

We may assume that the elements are 1, 2, $\ldots$, v and that one of the blocks is $\{v - k + 1, v - k + 2, \ldots, v\}$. Delete this block and all appearances of its elements from the other blocks. We are left with $v - 1$ subsets of $\{1, 2, \ldots, v - k\}$ of size s_i for $1 \le i \le v - 1$. The desired result is equivalent to $s_i = k - \lambda$ for $1 \le i \le v - 1$.

Note that we have $\lambda(v-1) = k(k-1)$. Each of the $v-k$ elements appear k times, so that

$$
\begin{aligned}
\sum_{i=1}^{v-1} s_i &= k(v - k) \\
&= k(v - 1) - k(k - 1) \\
&= k(v - 1) - \lambda(v - 1) \\
&= (v - 1)(k - \lambda).
\end{aligned}
$$

Hence the average size of s_i for $1 \le i \le v - 1$ is $k - \lambda$.

Every two of the $v - k$ elements appear together λ times, so that

$$
\begin{aligned}
\sum_{i=1}^{v-1} \binom{s_i}{2} &= \lambda \binom{v - k}{2} \\
&= \frac{1}{2k} k(v - k)\lambda(v - 1 - k) \\
&= \frac{1}{2k}(v - 1)(k - \lambda)(k(k - 1) - \lambda k) \\
&= (v - 1)\binom{k - \lambda}{2}.
\end{aligned}
$$

This certainly holds if $s_i = k - \lambda$ for $1 \le i \le v - 1$.

We now show that this cannot hold otherwise. If one s_i moves away from the mean value $k - \lambda$, then another one must move away in the opposite direction. We claim that this will always result in an increase in $\sum_{i=1}^{v-1} \binom{s_i}{2}$, so that it cannot stay equal to $(v - 1)\binom{k-\lambda}{2}$.

Let x and y be positive integers with $x \leq y$. Then

$$\binom{x-1}{2} + \binom{y+1}{2} - \binom{x}{2} - \binom{y}{2}$$

$$= \frac{(x-1)(x-2)}{2} + \frac{y(y+1)}{2} - \frac{x(x-1)}{2} - \frac{y(y-1)}{2}$$

$$= \frac{y((y+1) - (y-1))}{2} - \frac{(x-1)(x - (x-2))}{2}$$

$$= y - (x - 1)$$

$$> 0.$$

This justifies the claim and the desired conclusion follows.

Examples

Example 6.3.1.
Determine the parameters of a block design based on

  (a) an affine geometry of order 4;

  (b) a projective geometry of order 4.

Solution:

  (a) An affine geometry of order 4 has 16 points and 20 lines in 5 parallel classes. Each line passes through 4 points, and each point lies on exactly 1 line in each parallel classes. Every 2 points determine 1 line. Thus we have a (20,16,5,4,1)-design.

  (b) To obtain a projective geometry of order 4 from an affine geometry of order 4, we add an ideal point to each line in the same parallel class, and add an ideal line joining these 5 points. Now each line passes through 5 points and each point lies on 5 lines. Every 2 points still determine 1 line. Thus we have a (21,21,5,5,1)-design.

Example 6.3.2.
Does there exist a symmetric block design that is resolvable?

Solution:
Suppose we have a resolvable (v, v, k, k, λ)-design. Then $v = nk$ for some integer $n > 1$ so that $\lambda nk - \lambda = k(k-1)$. This implies that λ is divisible by k, and is therefore at least as large as k. We have a contradiction. Hence there are no block designs which are both resolvable and symmetric.

Example 6.3.3.
The Chief of a village on Pagan Island was seriously ill. The Oracle revealed that he could only be cured by a potion containing exactly five herbs, at least four of which must be of quintessential nature. Unfortunately, the Oracle did not reveal what a quintessential herb was, and nobody on Pagan Island knew. The Grand Alpharmacist gathered a number of herbs and concocted sixty-eight potions, each containing exactly five herbs. In an effort to include as many combinations as possible, each trio of herbs was used in exactly one potion. The Oracle was consulted again, but it revealed only that each of the potions contained at least one quintessential herb. The Chief's condition had deteriorated so much that further delay would prove fatal. The Grand Alpharmacist therefore administered one dose of each potion, hoping that one of them would contain the necessary four quintessential herbs. What was the fate of the Chief?

Solution:

Let H denote the set of herbs, T the set of triples of herbs, P the set of potions and Q the set of quintessential herbs. A typical element of any of these sets will be represented by the appropriate lower case letter. Let $f(i)$ denote the number of potions containing exactly i quintessential herbs. We claim that $f(4) + 4f(5)$ is not congruent to 0 (mod 5), so that at least one potion contains 4 or 5 quintessential herbs. In human language, the Chief was cured and lived happily ever after! First, since each potion contains $\binom{5}{3} = 10$ triples of herbs and each triple is in exactly one potion, we have $\binom{|H|}{3} = |T| = 680$. It follows that $|H| = 17$. Next, let h be a fixed herb which appears in k potions. Count the number of ordered pairs of the form (t, p) where $h \in t \subset p$. This is equal to $\binom{16}{2} = k\binom{4}{2}$, so that $k = 20$.

A count of the number of ordered pairs (q, p) with $q \in p$ yields

$$20|Q| = f(1) + 2f(2) + 3f(3) + 4f(4) + 5f(5).$$

Let h and h' be two fixed herbs. Consider ordered pairs of the form (t, p) with $\{h, h'\} \in t \subset p$. It follows that each pair of herbs is in five potions. By examining ordered pairs of the form $(\{q, q'\}, p)$ where $\{q, q'\} \subset p$, we obtain

$$5\binom{|Q|}{2} = f(2) + 3f(3) + 6f(4) + 10f(5).$$

Since every triple of herbs is in a unique potion, we obtain

$$\binom{|Q|}{3} = f(3) + 4f(4) + 10f(5).$$

Finally, there are sixty-eight potions altogether, so that

$$|P| = 68 = f(1) + f(2) + f(3) + f(4) + f(5).$$

Combining these equations yields

$$\binom{|Q|}{3} - 5\binom{|Q|}{2} + 20|Q| - 68 = f(4) + 4f(5).$$

As $\binom{|Q|}{3}$ is congruent to either 0, 1 or 4 (mod 5), the left side of the above equation is congruent to 2, 3 or 1 (mod 5), and the claim is justified.

Remark:

Families of 68 five-element subsets of a 17-element set, with each triple of elements occurring in exactly one subset, do in fact exist. Below is an example.

(1,2,3,4,17)	(5,6,7,8,17)	(9,10,11,12,17)	(13,14,15,16,17)
(1,5,9,13,17)	(2,6,10,14,17)	(3,7,11,15,17)	(4,8,12,16,17)
(1,6,11,16,17)	(1,8,10,15,17)	(1,7,12,14,17)	(2,5,12,15,17)
(2,7.9,16,17)	(2,8,11,13,17)	(3,8,9,14,17)	(3,6,12,13,17)
(3,5,10,16,17)	(4,7,10,13,17)	(1,5,11,14,17)	(4,6,9,15,17)
(1,2,5,7,10)	(1,2,6,8,9)	(1,2,11,14,15)	(1,2,12,13,16)
(1,3,5,6,14)	(1,3,7,8,16)	(1,3,9,12,15)	(1,3,10,11,13)
(1,4,5,15,16)	(1,4,6,10,12)	(1,4,7,9,11)	(1,4,8,13,14)
(2,3,5,9,11)	(2,3,6,15,16)	(2,3,7,13,14)	(2,3,8,10,12)
(2,4,5,6,13)	(2,4,9,12,14)	(2,4,7,8,15)	(2,4,10,11,16)
(3,4,9,13,16)	(3,4,10,14,15)	(3,4,6,8,11)	(3,4,5,7,12)
(1,5,8,11,12)	(1,6,7,13,15)	(1,9,10,14,16)	(2,5,8,14,16)
(2,6,7,11,12)	(2,9,10,13,15)	(3,5,8,13,15)	(3,6,7,9,10)
(3,11,12,14,16)	(4,5,8,9,10)	(4,6,7,14,16)	(4,11,12,13,15)
(5,6,9,12,16)	(5,6,10,11,15)	(5,7,9,14,15)	(5,7,11,13,16)
(5,10,12,13,14)	(6,8,10,13,16)	(6,8,12,14,15)	(7,8,9,12,13)
(7,8,10,11,14)	(6,9,11,13,14)	(7,10,12,15,16)	(8,9,11,15,16)

By choosing the quintessential herbs to be 1, 2, 4, 5, 7 ,8, 11 and 13, each of the 68 potions contains at least one but not more than four quintessential herbs.

Exercises

1. Determine the parameters of a block design based on

   (a) an affine geometry of order 5;

   (b) a projective geometry of order 5.

2. Determine the parameters of a block design based on

   (a) an affine geometry of order 7;

   (b) a projective geometry of order 7.

3. (a) Let $B_1, B_2, \ldots, B_v$ be the blocks of a (v, v, k, k, λ)-design. Prove that the $v - 1$ blocks $B_2 - B_1, B_3 - B_1, \ldots, B_v - B_1$ constitute a $(v - 1, v - k, k, k - \lambda, \lambda)$-design.

   (b) Illustrate the construction with the following (15,15,7,7,3)-design.

$$\begin{array}{ll}
B_1=\{1,2,3,4,5,6,7\} & B_2=\{1,2,3,8,9,10,11\} \\
B_3=\{1,2,3,12,13,14,15\} & B_4=\{1,4,5,8,9,12,13\} \\
B_5=\{1,4,5,10,11,14,15\} & B_6=\{1,6,7,8,9,14,15\} \\
B_7=\{1,6,7,10,11,12,13\} & B_8=\{2,4,6,8,10,12,14\} \\
B_9=\{2,4,7,8,11,13,15\} & B_{10}=\{2,5,6,9,11,12,15\} \\
B_{11}=\{2,5,7,9,10,13,14\} & B_{12}=\{3,4,6,9,11,13,14\} \\
B_{13}=\{3,4,7,9,10,12,15\} & B_{14}=\{3,5,6,8,10,13,15\} \\
B_{15}=\{3,5,7,8,11,12,14\} &
\end{array}$$

4. (a) Let $B_1, B_2, \ldots, B_v$ be the blocks of a (v, v, k, k, λ)-design with $\lambda \geq 2$. Prove that the $v - 1$ blocks $B_2 \cap B_1, B_3 \cap B_1, \ldots, B_v \cap B_1$ constitute a $(v - 1, k, k - 1, \lambda, \lambda - 1)$-design.

   (b) Illustrate the construction with the following (15,15,7,7,3)-design.

$$\begin{array}{ll}
B_1=\{1,2,3,4,5,6,7\} & B_2=\{1,2,3,8,9,10,11\} \\
B_3=\{1,2,3,12,13,14,15\} & B_4=\{1,4,5,8,9,12,13\} \\
B_5=\{1,4,5,10,11,14,15\} & B_6=\{1,6,7,8,9,14,15\} \\
B_7=\{1,6,7,10,11,12,13\} & B_8=\{2,4,6,8,10,12,14\} \\
B_9=\{2,4,7,8,11,13,15\} & B_{10}=\{2,5,6,9,11,12,15\} \\
B_{11}=\{2,5,7,9,10,13,14\} & B_{12}=\{3,4,6,9,11,13,14\} \\
B_{13}=\{3,4,7,9,10,12,15\} & B_{14}=\{3,5,6,8,10,13,15\} \\
B_{15}=\{3,5,7,8,11,12,14\} &
\end{array}$$

5. A difficult mathematical competition consisted of a morning session and an afternoon session, with a combined total of 31 problems. Each contestant solved exactly 6 problems altogether. For each pair of problems, exactly one contestant solved both of them. Prove that if every contestant solved at least one problem in the morning, then some contestant solved at most 2 problems in the afternoon.

6. A difficult mathematical competition consisted of a morning session and an afternoon session, with a combined total of 28 problems. Each contestant solved exactly 7 problems altogether. For each pair of problems, exactly two contestants solved both of them. Prove that if no contestant solved all 7 problems in the morning, then some contestant solved at least 4 problems in the afternoon.

Section 6.4. Difference Sets

In the preceding section, we constructed a $(7,7,3,3,1)$-design from the projective plane of order 2. Here is another way. Consider the finite field $\mathbf{Z}_7$ consisting of the integers modulo 7, namely, $\{0, 1, 2, 3, 4, 5, 6\}$. We use $(1,2,4)$ as the starter block. To generate the others, we add each element of $\mathbf{Z}_7$ to the block and reduce the numbers modulo 7. Thus the design is $\{(1,2,4),(2,3,5),(3,4,6),(4,5,0),(5,6,1),(6,0,2),(0,1,3)\}$. It is easy to verify that this does work, but not as easy to see why.

The pairwise differences, modulo 7, of 1, 2 and 4 are $2-1 = 1$, $4-2 = 2$, $4-1 = 3$, $1-4 = 4$, $2-4 = 5$ and $1-2 = 6$. Note that each non-zero element appears exactly once as a difference. From this, we can deduce that every pair of elements will appear together in exactly one block. For instance, since $6-4 = 2$, 4 and 6 can only appear together in the positions corresponding to 2 and 4 in the starter block. Similarly, since $5-1 = 4$, 5 and 1 can only appear together in the positions corresponding to 1 and 4 in the starter block.

In $\mathbf{Z}_v$, the integers modulo v, if each non-zero element appears the same number times as a difference of two elements of a set, this set is called a cyclic difference set, but we will use the shortened term **difference set**. It is called a (v, k, λ)-set (shortened from (v, k, λ)-difference set) if it consists of k of the v elements and each element appears as a difference λ times. Note that we must have $\lambda(v-1) = 2\binom{k}{2} = k(k-1)$. Using a (v, k, λ)-set as a starter block, we can generate a (v, v, k, k, λ)-design.

Thus we seek difference sets in $\mathbf{Z}_v$. In the example above with $v = 7$, note that $1 = 1^2 = 6^2$, $2 = 3^2 = 4^2$ and $4 = 2^2 = 5^2$. Thus $\{1, 2, 4\}$ is the set of non-zero squares. In general, if v is a prime number of the form $4t + 3$, then the non-zero squares in the finite field $\mathbf{Z}_v$ form a difference set.

Note that exactly half of the non-zero elements in $\mathbf{Z}_v$ for any odd prime number v are squares. Clearly, the product of two non-zero squares is another non-zero square. It is easy to see that the product of a non-zero square and a non-square must be a non-square. It then follows that the product of two non-squares must be a non-zero square.

Recall from Example 5.5.3 that -1 is not a square when v is of the form $4t + 3$. Suppose a non-zero square c is expressible as a difference of two other non-zero squares, say $c = a - b$. For any other non-zero square d, we have $dc = da - db$. Hence the square dc appears as a difference the same number of times as the square c. That -1 is not a square means that every non-square is of the form $-c$ where c is a square. Since $-c = b - a$, it also appears this many times. Hence the non-zero squares form a difference set.

As an illustration, the non-zero squares in $\mathbf{Z}_{19}$ are 1, 4, 5, 6, 7, 9, 11, 16 and 17. We have $1 = 5 - 4 = 6 - 5 = 7 - 6 = 17 - 16$ so that the square 1 appears four times as a difference. We claim that the square 4 also appears four times. Multiplying the squares by 4, we obtain 4, 16, 1, 5, 9, 17, 6, 7 and 11. Then we have $4 = 1 - 16 = 5 - 1 = 9 - 5 = 11 - 7$. The non-square 15's four appearances as a difference are $15 = 16 - 1 = 1 - 5 = 5 - 9 = 7 - 11$.

We can now construct the following (19,19,9,9,4)-design. We use as the starter block (1,4,5,6,7,9,11,16,17), and add to it every element of $\mathbf{Z}_{19}$.

(1,4,5,6,7,9,11,16,17)	(11,14,15,16,17,0,2,7,8)
(2,5,6,7,8,10,12,17,18)	(12,15,16,17,18,1,3,8,9)
(3,6,7,8,9,11,13,18,0)	(13,16,17,18,0,2,4,9,10)
(4,7,8,9,10,12,14,0,1)	(14,17,18,0,1,3,5,10,11)
(5,8,9,10,11,13,15,1,2)	(15,18,0,1,2,4,6,11,12)
(6,9,10,11,12,14,16,2,3)	(16,0,1,2,3,5,7,12,13)
(7,10,11,12,13,15,17,3,4)	(17,1,2,3,4,6,8,13,14)
(8,11,12,13,14,16,18,4,5)	(18,2,3,4,5,7,9,14,15)
(9,12,13,14,15,17,0,5,6)	(0,3,4,5,6,8,10,15,16)
(10,13,14,15,16,18,1,6,7)	

When v is a prime number of the form $4t + 1$, the set of non-zero squares do not form a difference set. However, if we throw in the set of non-squares, every non-zero element appears the same number of times as a difference of two squares or two non-squares. These two sets are called **supplementary difference sets**. Using these as starter blocks, we can generate a block design.

As in the earlier case, the non-zero squares appear the same number of times as a difference of two non-zero squares. Suppose a non-square c is expressible as a difference of two non-zero squares, say $c = a - b$. For any non-zero square d, we have $dc = da - db$. Hence the non-square dc appears as a difference of two non-zero squares the same number of times as the non-square c.

For any non-square e, we have $ec = ea - eb$. Hence the non-zero square ec appears as a difference of two non-squares the same number of times as the non-square c as a difference of two non-zero squares. Similarly, a non-square appears as a difference of two non-squares the same number of times as a non-zero square appears as a difference of two non-zero squares. Hence the squares and the non-squares form supplementary difference sets.

As an illustration, the non-zero squares in $\mathbf{Z}_{17}$ are 1, 2, 4, 8, 9, 13, 15 and 16. We can show that each non-square appears 4 times as a difference of two non-zero squares, while each square appears 4 times as the difference of two non-squares.

We can now construct the following (34,17,16,8,7)-design.

$$(1,2,4,8,9,13,15,16)$$

(2,3,5,9,10,14,16,0)	(10,11,13,0,1,5,7,8)
(3,4,6,10,11,15,0,1)	(11,12,14,1,2,6,8,9)
(4,5,7,11,12,16,1,2)	(12,13,15,2,3,7,9,10)
(5,6,8,12,13,0,2,3)	(13,14,16,3,4,8,10,11)
(6,7,9,13,14,1,3,4)	(14,15,0,4,5,9,11,12)
(7,8,10,14,15,2,4,5)	(15,16,1,5,6,10,12,13)
(8,9,11,15,16,3,5,6)	(16,0,2,6,7,11,13,14)
(9,10,12,16,0,4,6,7)	(0,1,3,7,8,12,14,15)

$$(3,5,6,7,10,11,12,14)$$

(4,6,7,8,11,12,13,15)	(12,14,15,16,2,3,4,6)
(5,7,8,9,12,13,14,16)	(13,15,16,0,3,4,5,7)
(6,8,9,10,13,14,15,0)	(14,16,0,1,4,5,6,8)
(7,9,10,11,14,15,16,1)	(15,0,1,2,5,6,7,9)
(8,10,11,12,15,16,0,2)	(16,1,2,3,6,7,8,10)
(9,11,12,13,16,0,1,3)	(0,2,3,4,7,8,9,11)
(10,12,13,14,0,1,2,4)	(1,3,4,5,8,9,10,12)
(11,13,14,15,1,2,3,5)	(2,4,5,6,9,10,11,13)

We make use of both the set of non-zero squares and the set of non-squares here, whereas in the earlier case where the prime v is of the form $4t + 3$, we only use the set of non-zero squares. It turns out that the set of non-squares is also a difference set. In $\mathbf{Z}_7$, $\{3,5,6\}$ is also a (7,3,1)-set because $6 - 5 = 1$, $5 - 3 = 2$, $6 - 3 = 3$, $3 - 6 = 4$, $3 - 5 = 5$ and $5 - 6 = 6$.

To see why this is so in general, observe that if the set of non-zero squares form a (v, k, λ)-set, then the set of all squares, including 0, is a $(v, k+1, \lambda+1)$-set. Clearly, the set has one more element than before. Each square a appears an extra time as a difference, namely, $a - 0$, and each non-square $-a$ also appears an extra time as a difference, namely, $0 - a$.

We now remove the restriction that v is a prime of the form $4t + 3$. For any $\mathbf{Z}_v$, the entire set is a (v, v, v)-set. In the diagram below, the difference table for $\mathbf{Z}_7$ is shown. Since each non-zero difference appears exactly once in each row, we have a difference set.

−	0	1	2	3	4	5	6
0	0	6	5	4	3	2	1
1	1	0	6	5	4	3	2
2	2	1	0	6	5	4	3
3	3	2	1	0	6	5	4
4	4	3	2	1	0	6	5
5	5	4	3	2	1	0	6
6	6	5	4	3	2	1	0

If we eliminate any rows or any columns, each non-zero element still appears the same number of times in the residual table. What happens if we eliminate some rows and some columns? For instance, if we eliminate the 0-th row and the 0-th column, we obtain a trivial $(7,6,5)$-set. This is because each non-zero element appears the same number of times at the intersection of these rows and columns, namely 0 times in the present instance.

Recall that $\{0,1,2,4\}$ is a $(7,4,2)$ set. Suppose we remove the 0-th, 1-st, 2-nd and 4-th rows and columns. Consider a particular non-zero element. It appears 7 times in the original table. When 4 rows and 4 columns are eliminated, it disappears 8 times. However, 2 of those disappearances are duplicates because $\lambda=2$. Thus in the remaining table, it appears $7-8+2 = 1$ time. This justifies that the set of non-squares $\{3,5,6\}$ is a $(7,3,1)$-set.

In general, the complement of a difference set is also a difference set. Thus in a (v, k, λ)-set, we may take $k < \frac{v}{2}$. We cannot have $k = \frac{v}{2}$ as otherwise $v - 1$ and k are relatively prime. In order for $\lambda(v - 1) = k(k - 1)$, k must divide λ so that $k \leq \lambda$. This is impossible. We can now begin a systematic search for (v, k, λ)-sets with $\lambda < k < \frac{v}{2}$ and $k \leq 10$.

k	$k(k-1)$	v	λ	k	$k(k-1)$	v	λ
3	6	7	1	8	56	57	1
4	12	13	1			29	2
5	20	21	1	9	72	73	1
		11	2			37	2
6	30	31	1			25	3
		16	2			19	4
7	42	43	1	10	90	91	1
		22	2			46	2
		15	3			31	3

The set $\{1,2,4\}$ catches our attention in another way. If we multiply each of its elements by 2, we get back the same set. So we start with each element, multiply it by 2, and multiplying the product by 2 until we reach some number that has already occurred. With 0, we just get $\{0\}$. With 1, 2 or 4, we get $\{1,2,4\}$, and with 3, 5 and 6, we get $\{3,5,6\}$. This is an induced partition of the 7 elements into three subsets. Such subsets are called **cyclotomic cosets**, with respect to the prime 2 in this case. Suppose we use the prime 3 instead of 2. With 0, we of course still just get $\{0\}$. With any non-zero element, we get $\{1,2,3,4,5,6\}$.

A number is said to be a **multiplier** of a set if the same set is obtained after multiplication by this number. Thus any number is a multiplier for the set $\{0\}$, 1 is always a multiplier of any set, and both 2 and 4 are multipliers for the set $\{1,2,4\}$ in $\mathbf{Z}_7$.

The following has been attributed to **H. Ryser**.

The Multiplier Conjecture.

If there exists a (v, k, λ)-set and p is a prime divisor of $k - \lambda$, then p is a multiplier of this difference set.

The conjecture is known to be true if $p > \lambda$, and we will refer to this as the **Multiplier Theorem**. Its proof is beyond the scope of this book. Its main consequence is that if $p > \lambda$ is a divisor of $k - \lambda$, then a (v, k, λ)-set must be a union of cyclotomic cosets induced by p.

It is generally believed that the condition $p > \lambda$ is superfluous. In the chart below, the prime divisors of $k - \lambda$ which fail this condition are in bold face.

v	k	λ	$k - \lambda$	p	v	k	λ	$k - \lambda$	p
7	3	1	2	2	29	8	2	6	**2,3**
13	4	1	3	3	57	8	1	7	7
11	5	2	3	3	19	9	4	5	5
21	5	1	4	2	25	9	3	6	**2,3**
16	6	2	4	**2**	37	9	2	7	7
31	6	1	5	5	73	9	1	8	2
15	7	3	4	**2**	31	10	3	7	7
22	7	2	5	5	46	10	2	8	**2**
43	7	1	6	2,3	91	10	1	9	3

Consider first the four cases for which no prime divisors of $k - \lambda$ satisfy the condition $p > \lambda$. In the cases $(16,6,2)$ and $(46,10,2)$, the prime $p = 2$ does not even induce cyclotomic cosets. In the case $(25,9,3)$, the prime $p = 2$ induces the cyclotomic cosets (0), $(5,10,20,15)$ and

$$(1, 2, 4, 8, 16, 7, 14, 3, 6, 12, 24, 23, 21, 17, 9, 18, 11, 22, 19, 13),$$

whereas the prime $p = 3$ induces the cyclotomic cosets (0), $(5,15,20,10)$ and

$$(1, 3, 9, 2, 6, 18, 4, 12, 11, 8, 24, 22, 16, 23, 19, 7, 21, 13, 14, 17).$$

No union of cyclotomic cosets has size 9. It follows that if the Multiplier Conjecture is true, then there do not exist $(16,6,2)$, $(46,10,2)$ and $(25,9,3)$-sets.

In the case $(15,7,3)$, the prime $p = 2$ induces the cyclotomic cosets

$$(0), (5, 10), (1, 2, 4, 8), (3, 6, 12, 9) \text{ and } (7, 14, 13, 11).$$

The following tables show that both $\{0,1,2,4,5,8,10\}$ and $\{0,5,7,10,11,13,14\}$ are $(15,7,3)$-sets, even though $\{0,3,5,6,9,10,12\}$ is not. This lends weight to the validity of the conjecture.

$-$	0	1	2	4	5	8	10
0	$-$	14	13	11	10	7	5
1	1	$-$	14	12	11	8	6
2	2	1	$-$	13	12	9	7
4	4	3	2	$-$	14	11	9
5	5	4	3	1	$-$	12	10
8	8	7	6	4	3	$-$	13
10	10	9	8	6	5	2	$-$

$-$	0	5	7	10	11	13	14
0	$-$	10	8	5	4	2	1
5	5	$-$	13	10	9	7	6
7	7	2	$-$	12	11	9	8
10	10	5	3	$-$	14	12	11
11	11	6	4	1	$-$	13	12
13	13	8	6	3	2	$-$	14
14	14	9	7	4	3	1	$-$

Suppose a difference set exists in $\mathbf{Z}_v$. The Multiplier Conjecture can point us the way to find it. Once found, we do not need the conjecture any more as we only have to verify that what we have found is indeed a difference set.

In the case (13,4,1), $k - \lambda = 4 - 1 = 3$ is divisible by the prime $p = 3$. The cyclotomic cosets induced by this prime are

$$(0), (1, 3, 9), (2, 6, 5), (4, 12, 10) \text{ and } (7, 8, 11).$$

The following tables show that the union of $\{0\}$ with any of the cosets of size 3 is a (13,4,1)-set.

$-$	0	1	3	9
0	$-$	12	10	4
1	1	$-$	11	5
3	3	2	$-$	7
9	9	8	6	$-$

$-$	0	2	5	6
0	$-$	11	8	7
2	2	$-$	10	9
5	5	3	$-$	12
6	6	4	1	$-$

$-$	0	4	10	12
0	$-$	9	3	1
4	4	$-$	7	5
10	10	6	$-$	11
12	12	8	2	$-$

$-$	0	7	8	11
0	$-$	6	5	2
7	7	$-$	12	9
8	8	1	$-$	10
11	11	4	3	$-$

In the case (21,5,1), $k - \lambda = 5 - 1 = 4$ is divisible by the prime $p = 2$. The cyclotomic cosets induced by this prime are

$$(0), (7, 14), (3, 6, 12), (9, 18, 15), (1, 2, 4, 8, 16, 11) \text{ and } (5, 10, 20, 19, 17, 13).$$

The following tables show that the union of $\{7,14\}$ with either of the cosets of size 3 is a (21,5,1)-set.

$-$	3	6	7	12	14
3	$-$	18	17	12	10
6	3	$-$	20	15	13
7	4	1	$-$	16	14
12	9	6	5	$-$	19
14	11	8	7	2	$-$

$-$	7	9	14	15	18
7	$-$	19	14	13	10
9	2	$-$	16	15	12
14	7	5	$-$	20	17
15	8	6	1	$-$	18
18	11	9	4	3	$-$

In the case (57,8,1), $k - \lambda = 8 - 1 = 7$ is divisible by the prime $p = 7$. The cyclotomic cosets induced by this prime are

$$(0), (19), (38), (1, 7, 49), (2, 14, 41), (3, 21, 33),$$

$$(4, 28, 25), (5, 35, 17), (6, 42, 9), (8, 56, 50), (10, 13, 34),$$

$$(11, 20, 26), (12, 27, 18), (15, 48, 51), 16, 55, 43), (22, 40, 52),$$

$$(23, 47, 44), (24, 54, 36), (29, 32, 53), (30, 39, 45) \text{ and } (31, 46, 37).$$

A (57,8,1)-set must be the union of two of $\{0\}$, $\{19\}$ and $\{38\}$ with two of the cosets of size 3. The table below shows that $\{0,1,5,7,17,35,38,49\}$ is indeed a difference set.

$-$	0	1	5	7	17	35	38	49
0	$-$	56	52	50	40	22	19	8
1	1	$-$	53	51	41	23	20	9
5	5	4	$-$	55	45	27	24	13
7	7	6	2	$-$	47	29	26	15
17	17	16	12	10	$-$	39	36	25
35	35	34	30	38	18	$-$	54	43
38	38	37	33	31	21	3	$-$	46
49	49	48	44	42	32	14	11	$-$

Suppose a difference set does not exist in $\mathbf{Z}_v$. The Multiplier Theorem becomes a most useful tool in establishing non-existence. In the case (31,10,3), $k - \lambda = 10 - 3 = 7$ is divisible by the prime $p = 7$ which satisfies $p > \lambda$. The cyclotomic cosets induced by this prime are

$$(0), (1, 7, 18, 2, 14, 5, 4, 28, 10, 8, 25, 20, 16, 19, 9)$$

$$\text{and } (3, 21, 23, 6, 11, 15, 12, 22, 30, 24, 13, 29, 17, 26, 27).$$

No union of them has size 10. By the Multiplier Theorem, a (31,10,3)-set cannot exist.

In the case $(31,6,1)$, $k - \lambda = 6 - 1 = 5$ is divisible by the prime $p = 5$ which satisfies $p > \lambda$. The cyclotomic cosets induced by this prime are

$$(0), (1, 5, 25), (3, 15, 13), (9, 14, 8), (27, 11, 24), (19, 2, 10),$$

$$(26, 6, 30), (16, 18, 28), (17, 23, 22), (20, 7, 4) \text{ and } (29, 21, 20).$$

If a $(31,6,1)$-set exists, it must be a union of two of the cosets of size 3. Apparently, we have to examine $\binom{10}{2} = 45$ combinations. However, we observe that the first ten powers of 3, namely 1, 3, 9, 27, 19, 26, 16, 17, 20 and 29, generate different cosets. Call them C_i, $0 \le i \le 9$. We only have to consider five non-equivalent cases.

Case 1. $C_0 \cup C_1, C_1 \cup C_2, \ldots, C_8 \cup C_9, C_9 \cup C_0$.
In $C_0 \cup C_1 = \{1, 3, 5, 13, 15, 25\}$, 1 does not appear as a difference.
Case 2. $C_0 \cup C_2, C_1 \cup C_3, \ldots, C_8 \cup C_0, C_9 \cup C_1$.
In $C_0 \cup C_2 = \{1, 5, 8, 9, 14, 25\}$, 2 does not appear as a difference.
Case 3. $C_0 \cup C_3, C_1 \cup C_4, \ldots, C_8 \cup C_1, C_9 \cup C_2$.
In $C_0 \cup C_3 = \{1, 5, 11, 14, 25, 27\}$, 1 does not appear as a difference.
Case 4. $C_0 \cup C_4, C_1 \cup C_5, \ldots, C_8 \cup C_2, C_9 \cup C_3$.
In $C_0 \cup C_4 = \{1, 2, 5, 10, 19, 25\}$, 2 does not appear as a difference.
Case 5. $C_0 \cup C_5, C_1 \cup C_6, C_2 \cup C_7, C_3 \cup C_8, C_4 \cup C_9$.
In $C_0 \cup C_5 = \{1, 5, 6, 25, 26, 30\}$, 2 does not appear as a difference.

It follows that a $(31,6,1)$-set does not exist.

Examples

Example 6.4.1.

We have seen in Example 6.1.1 that the finite field $\mathbf{GF}(9)$ consists of the elements 0, 1, 2, t, $t+1$, $t+2$, $2t$, $2t+1$ and $2t+2$, where the coefficients are elements of $\mathbf{Z}_3$ and t is a root of the irreducible polynomial t^2+1.

(a) Use the non-zero squares and the non-squares to construct two supplementary difference sets.

(b) Construct a (18,9,8,4,3)-design from these supplementary difference sets.

Solution:

(a) The non-zero squares are $1 = 1^2 = 2^2$, $2 = (t+2)^2 = (2t+1)^2$, $t+2 = (t+1)^2 = (2t+2)^2$ and $2t+1 = t^2 = (2t)^2$ in $\mathbf{GF}(9)$. The non-squares are t, $t+1$, $2t$ and $2t+2$. The difference charts for both sets are shown below. While neither is a difference set, between the two charts, every non-zero element of $\mathbf{GF}(9)$ appears exactly three times.

$-$	1	2	$t+2$	$2t+1$
1	0	2	$2t+2$	t
2	1	0	$2t$	$t+1$
$t+2$	$t+1$	t	0	$2t+1$
$2t+1$	$2t$	$2t+2$	$t+2$	0

$-$	t	$t+1$	$2t$	$2t+2$
t	0	2	$2t$	$2t+1$
$t+1$	1	0	$2t+1$	$2t+2$
$2t$	t	$t+2$	0	1
$2t+2$	$t+2$	$t+1$	2	0

(b) We can construct a (18,9,8,4,3)-design using these two sets as starter blocks and adding to each block every element of $\mathbf{GF}(9)$.

$(1,2,t+2,2t+1)$ $(t,t+1,2t,2t+2)$
$(2,0,t,2t+2)$ $(t+1,t+2,2t+1,2t)$
$(0,1,t+1,2t)$ $(t+2,t,2t+2,2t+1)$
$(t+1,t+2,2t+2,1)$ $(2t,2t+1,0,2)$
$(t+2,t,2t,2)$ $(2t+1,2t+2,1,0)$
$(t,t+1,2t+1,0)$ $(2t+2,2t,2,1)$
$(2t+1,2t+2,2,t+1)$ $(0,1,t,t+2)$
$(2t+2,2t,0,t+2)$ $(1,2,t+1,t)$
$(2t,2t+1,1,t)$ $(2,0,t+2,t+1)$

Example 6.4.2.

Construct a (91,10,1)-set.

Solution:

To construct a (91,10,1)-set, consider the cyclotomic cosets induced by the prime divisor $p = 3$ of $k - \lambda = 10 - 1 = 9$. They are

$$(0), (7, 21, 63), (14, 42, 35), (28, 84, 70), (49, 56, 77),$$

$$(1, 3, 9, 27, 81, 61), (2, 6, 18, 54, 71, 31), (4, 12, 36, 17, 51, 62),$$

$$(5, 15, 45, 44, 41, 32), (8, 24, 72, 34, 11, 33), (10, 30, 90, 88, 82, 64),$$

$$(16, 48, 53, 68, 22, 66), (19, 57, 80, 58, 83, 67), (20, 60, 89, 85, 73, 37),$$

$$(23, 69, 25, 75, 43, 38), (29, 87, 79, 55, 74, 40) \text{ and } (46, 47, 50, 59, 86, 76).$$

A (91,10,1)-set must be the union of $\{0\}$ with a coset of size 3 and a coset of size 6. The table below shows that $\{0,1,3,9,27,49,56,61,77,81\}$ is indeed a difference set.

$-$	0	1	3	9	27	49	56	61	77	81
0	$-$	90	88	82	64	42	35	30	14	10
1	1	$-$	89	83	65	43	36	31	15	11
3	3	2	$-$	85	67	45	38	33	17	13
9	9	8	6	$-$	73	51	44	39	23	19
27	27	26	24	18	$-$	69	62	57	41	37
49	49	48	46	40	22	$-$	84	79	63	59
56	56	55	53	47	29	7	$-$	86	70	66
61	61	69	58	52	34	12	5	$-$	75	71
77	77	76	74	68	50	28	21	16	$-$	87
81	81	80	78	72	54	32	25	20	4	$-$

Example 6.4.3.

Prove that a (22,7,2)-set does not exist.

Solution:

If a (22,7,2)-set exists, then the prime divisor $p = 5$ of $k - \lambda = 7 - 2 = 5$ satisfies $p > \lambda$. By the Multiplier Theorem, the difference set is a union of cyclotomic cosets induced by $p = 5$. These are

$$(0), (11), (1, 5, 3, 15, 9), (2, 10, 6, 8, 18), (4, 20, 12, 16, 14), (7, 13, 21, 17, 19).$$

A (22,7,2)-set must be the union of $\{0\}$ and $\{11\}$ with a coset of size 5. Such a set does not exist since

$$
\begin{array}{ccccccccl}
2 & = & 3 - 1 & = & 5 - 3 & = & 11 - 9 & \text{in} & \{0,1,3,5,9,11,15\}, \\
2 & = & 2 - 0 & = & 8 - 6 & = & 10 - 8 & \text{in} & \{0,2,6,8,10,11,18\}, \\
4 & = & 4 - 0 & = & 16 - 12 & = & 20 - 16 & \text{in} & \{0,4,11,12,14,16,20\}, \\
2 & = & 13 - 11 & = & 19 - 17 & = & 21 - 19 & \text{in} & \{0,7,11,13,17,19,21\}.
\end{array}
$$

Exercises

1. Use the non-zero squares in $\mathbf{Z}_{11}$ to construct a $(11,5,2)$-set.

2. Use the non-zero squares and the non-squares in $\mathbf{Z}_{13}$ to construct two supplementary difference sets.

3. Construct a $(37,9,2)$-set.

4. Construct a $(73,9,1)$-set.

5. Prove that a $(29,8,2)$-set does not exist.

6. Prove that a $(43,7,1)$-set does not exist.

Section 6.5. Triple Systems

A **triple system** is a block design with $k = 3$. By taking all possible triples (0,1,2), (0,1,3), (0,2,3) and (1,2,3) of a 4-element set $\{0,1,2,3\}$, we have a (4,4,3,3,2)-design. There are other possibilities, and we now consider an important subclass.

A **Steiner** triple system is a block-design with $k = 3$ and $\lambda = 1$. A Steiner triple system with v elements is said to be of order v. From the equation $\lambda(v - 1) = r(k - 1)$, we have $r = \frac{v-1}{2}$. From the equation $bk = vr$, we have $b = \frac{v(v-1)}{6}$. In order for r and b to be integers, we must have $v \equiv 1$ or 3 (mod 6).

This necessary condition also turns out to be sufficient if we allow the degenerate (0,1,0,3,1) and (1,3,1,3,1)-designs. We shall establish this via an explicit construction for each of the two cases.

First Construction. $v \equiv 3 \pmod 6$.

Since the order $v = 3(2t + 1)$ is a multiple of 3, we choose as elements a_i, b_i and c_i, $0 \leq i \leq 2t$. Note that $r = 3t + 1$ and $b = (2t + 1)(3t + 1)$.

For all $i < j$, a_i and a_j must meet. Hence we form all $t(2t + 1)$ such pairs (a_i, a_j). Since we are constructing a triple system, we throw in b_p where $p = i \oplus j$ for some binary operation $\oplus$. Similarly, we form the triples (b_i, b_j, c_p) and (c_i, c_j, a_p).

Let us write down the necessary properties of this operation $\oplus$. First, it must be commutative. Otherwise, we may have $i \oplus j = p$ while $j \oplus i = q \neq p$, so that (a_i, a_j, b_p) and (a_i, a_j, b_q) are triples in the system. However, this contradicts $\lambda = 1$ as a_i meets a_j at least twice.

Second, $\oplus$ must satisfy the cancellation law, in that if $i \oplus j = i \oplus \ell$, then $j = \ell$. Otherwise, we may have $i \oplus j = p$ and $i \oplus \ell = p$ with $j \neq \ell$, so that (a_i, a_j, b_p) and (a_i, a_ℓ, b_p) are triples in the system. However, this also contradicts $\lambda = 1$ as a_i meets b_p at least twice.

It would appear that addition modulo $2t + 1$ is a possible choice. We illustrate with the case $t = 3$.

+	0	1	2	3	4	5	6
0	0	1	2	3	4	5	6
1	1	2	3	4	5	6	0
2	2	3	4	5	6	0	1
3	3	4	5	6	0	1	2
4	4	5	6	0	1	2	3
5	5	6	0	1	2	3	4
6	6	0	1	2	3	4	5

Note that each a_i meets every b_p except for $p \equiv 2i \pmod{2t+1}$. This is because (a_i, a_i, b_{2i}) is not taken as a triple. Similarly, b_i has not met c_{2i} and c_i has not met a_{2i}. So far, we have used up $3t(2t+1)$ triples. Those pairs of elements which have not met must then meet in the remaining $2t+1$ triples. One of them could be (a_0, b_0, c_0) because no two of them have met earlier. Now a_1 has not met b_2, which in turn has not met c_4, which in turn has not met a_8. However, adding the triple (a_1, b_2, c_4) will only work if $1 \equiv 8 \pmod{2t+1}$ or $t = 3$.

We have a Steiner triple system of order 21 by taking the following blocks:

$$
\begin{array}{llll}
(a_0, a_1, b_1) & (b_0, b_1, c_1) & (c_0, c_1, a_1) & (a_0, b_0, c_0) \\
(a_0, a_2, b_2) & (b_0, b_2, c_2) & (c_0, c_2, a_2) & (a_1, b_2, c_4) \\
(a_0, a_3, b_3) & (b_0, b_3, c_3) & (c_0, c_3, a_3) & (b_1, c_2, a_4) \\
(a_0, a_4, b_4) & (b_0, b_4, c_4) & (c_0, c_4, a_4) & (c_1, a_2, b_4) \\
(a_0, a_5, b_5) & (b_0, b_5, c_5) & (c_0, c_5, a_5) & (a_3, b_6, c_5) \\
(a_0, a_6, b_6) & (b_0, b_6, c_6) & (c_0, c_6, a_6) & (b_3, c_6, a_5) \\
(a_1, a_2, b_3) & (b_1, b_2, c_3) & (c_1, c_2, a_3) & (c_3, a_6, b_5) \\
(a_1, a_3, b_4) & (b_1, b_3, c_4) & (c_1, c_3, a_4) & \\
(a_1, a_4, b_5) & (b_1, b_4, c_5) & (c_1, c_4, a_5) & \\
(a_1, a_5, b_6) & (b_1, b_5, c_6) & (c_1, c_5, a_6) & \\
(a_1, a_6, b_0) & (b_1, b_6, c_0) & (c_1, c_6, a_0) & \\
(a_2, a_3, b_5) & (b_2, b_3, c_5) & (c_2, c_3, a_5) & \\
(a_2, a_4, b_6) & (b_2, b_4, c_6) & (c_2, c_4, a_6) & \\
(a_2, a_5, b_0) & (b_2, b_5, c_0) & (c_2, c_5, a_0) & \\
(a_2, a_6, b_1) & (b_2, b_6, c_1) & (c_2, c_6, a_1) & \\
(a_3, a_4, b_0) & (b_3, b_4, c_0) & (c_3, c_4, a_0) & \\
(a_3, a_5, b_1) & (b_3, b_5, c_1) & (c_3, c_5, a_1) & \\
(a_3, a_6, b_2) & (b_3, b_6, c_2) & (c_3, c_6, a_2) & \\
(a_4, a_5, b_2) & (b_4, b_5, c_2) & (c_4, c_5, a_2) & \\
(a_4, a_6, b_3) & (b_4, b_6, c_3) & (c_4, c_6, a_3) & \\
(a_5, a_6, b_4) & (b_5, b_6, c_4) & (c_5, c_6, a_4) & \\
\end{array}
$$

Although addition modulo $2t + 1$ works only in the case $t = 3$, all that is required is a little modification. We have the ideal situation when none of a_0, b_0 and c_0 meet in the first phase of the construction. If this is also true for a_i, b_i and c_i for all $i > 0$, we can simply add the triples (a_i, b_i, c_i) for $0 \leq i \leq 2t$.

What this means is the $\oplus$ must have a third property, namely, $i \oplus i = i$ for all i. Such an operation is said to be **idempotent**, and it is not as rare as one may think. The prototype is set-theoretic union or intersection, that is, $S \cup S = S$ and $S \cap S = S$ for any set S. Numerical operations such as max (maximum) and gcd (greatest common divisor) are also idempotent. Unfortunately, although both are commutative, neither satisfies the cancellation law.

Where can one find an operation which is idempotent, commutative and satisfies the cancellation law? The diagram below shows the addition table modulo 7, except that the index rows and columns have been rearranged to agree with the main diagonal. This new operation $\oplus$ has all the desired properties.

$\oplus$	0	2	4	6	1	3	5
0	0	1	2	3	4	5	6
2	1	2	3	4	5	6	0
4	2	3	4	5	6	0	1
6	3	4	5	6	0	1	2
1	4	5	6	0	1	2	3
3	5	6	0	1	2	3	4
5	6	0	1	2	3	4	5

We have another Steiner triple system of order 21 by taking the following blocks:

(a_0, a_1, b_4)	(b_0, b_1, c_4)	(c_0, c_1, a_4)	(a_0, b_0, c_0)
(a_0, a_2, b_1)	(b_0, b_2, c_1)	(c_0, c_2, a_1)	(a_1, b_1, c_1)
(a_0, a_3, b_5)	(b_0, b_3, c_5)	(c_0, c_3, a_5)	(a_2, b_2, c_2)
(a_0, a_4, b_2)	(b_0, b_4, c_2)	(c_0, c_4, a_2)	(a_3, b_3, c_3)
(a_0, a_5, b_6)	(b_0, b_5, c_6)	(c_0, c_5, a_6)	(a_4, b_4, c_4)
(a_0, a_6, b_3)	(b_0, b_6, c_3)	(c_0, c_6, a_3)	(a_5, b_5, c_5)
(a_1, a_2, b_5)	(b_1, b_2, c_5)	(c_1, c_2, a_5)	(a_6, b_6, c_6)
(a_1, a_3, b_2)	(b_1, b_3, c_2)	(c_1, c_3, a_2)	
(a_1, a_4, b_6)	(b_1, b_4, c_6)	(c_1, c_4, a_6)	
(a_1, a_5, b_3)	(b_1, b_5, c_3)	(c_1, c_5, a_3)	
(a_1, a_6, b_0)	(b_1, b_6, c_0)	(c_1, c_6, a_0)	
(a_2, a_3, b_6)	(b_2, b_3, c_6)	(c_2, c_3, a_6)	
(a_2, a_4, b_3)	(b_2, b_4, c_3)	(c_2, c_4, a_3)	
(a_2, a_5, b_0)	(b_2, b_5, c_0)	(c_2, c_5, a_0)	
(a_2, a_6, b_4)	(b_2, b_6, c_4)	(c_2, c_6, a_4)	
(a_3, a_4, b_0)	(b_3, b_4, c_0)	(c_3, c_4, a_0)	
(a_3, a_5, b_4)	(b_3, b_5, c_4)	(c_3, c_5, a_4)	
(a_3, a_6, b_1)	(b_3, b_6, c_1)	(c_3, c_6, a_1)	
(a_4, a_5, b_1)	(b_4, b_5, c_1)	(c_4, c_5, a_1)	
(a_4, a_6, b_5)	(b_4, b_6, c_5)	(c_4, c_6, a_5)	
(a_5, a_6, b_2)	(b_5, b_6, c_2)	(c_5, c_6, a_2)	

Actually, this is isomorphic to the earlier triple system, and may be obtained directly from that system by replacing $b_1, b_2, b_3, b_4, b_5, b_6, c_1, c_2, c_3, c_4,$ c_5, c_6 with $b_4, b_1, b_5, b_2, b_6, b_3, c_2, c_4, c_6, c_1, c_3, c_5$ respectively. Later, we will encounter Steiner triple systems of the same order that are not isomorphic.

One remaining issue requires attention. Can we always rearrange the index rows and columns of the addition table modulo $2t + 1$ so that they agree with the main diagonal?

The answer is affirmative provided that the elements on the main diagonal are distinct, and we now show that this is indeed the case. Since addition modulo $2t + 1$ satisfies the cancellation law, each element appears exactly once in each of the $2t + 1$ rows. Hence it appears an odd number of times. Since addition modulo $2t + 1$ is commutative, each element appears an even number of times off the main diagonal. It follows that each element appears at least once on the main diagonal. Since there is just enough room for all of them, each appears exactly once on the main diagonal. In other words, the elements on the main diagonal are distinct, as desired.

Second Construction. $v \equiv 1 \pmod 6$.

The order $v = 3(2t) + 1$ is 1 more than a multiple of 3. However, hoping to ride on the success in the First Construction, we take as elements a_i, b_i and c_i, $0 \le i \le 2t - 1$, as well as d. Note that $r = 3t$ and $b = t(6t + 1)$.

For all $i < j$, a_i and a_j must meet. Hence we form all $t(2t - 1)$ such pairs (a_i, a_j). Since we are constructing a triple system, we throw in b_p where $p = i \oplus j$ for some commutative operation $\oplus$ which satisfies the cancellation law. Similarly, we form the triples (b_i, b_j, c_p) and (c_i, c_j, a_p).

What about the condition that $\oplus$ be idempotent? Unfortunately, this cannot be forced since we will be starting from addition modulo $2t$. Since $2t$ is even, the elements on the main diagonal of the addition table modulo $2t$ are not distinct. In fact, the odd numbers will not appear while each even number appears twice.

Well, half a loaf is better than none. So we proceed to rearrange the index row and column with all the even numbers appearing in order, followed by all the odd numbers in order. Half of the elements, namely the even ones, satisfy the idempotent law. So in the second stage of the construction, we add the blocks (a_{2i}, b_{2i}, c_{2i}) for $0 \le i \le t - 1$.

$\oplus$	0	2	4	1	3	5
0	0	1	2	3	4	5
2	1	2	3	4	5	0
4	2	3	4	5	0	1
1	3	4	5	0	1	2
3	4	5	0	1	2	3
5	5	0	1	2	3	4

For $0 \le i \le t - 1$, a_{2i+1} has still to meet b_{2i}, which must be in a new block. We need a third member to make the triple, and this is where the element d comes in. So the other blocks we add in the second stage are (d, a_{2i+1}, b_{2i}), (d, b_{2i+1}, c_{2i}) and (d, c_{2i+1}, a_{2i}).

We illustrate the general method by constructing a Steiner triple system of order 19. Although the process is slightly more complicated, the idea behind it is essentially the same. The elements are a_i, b_i and c_i, $0 \le i \le 5$, and d.

As in the First Construction, 45 of the blocks are divided into three groups and assembled using an operation $\oplus$ modified from addition modulo 6. Its table is in the diagram above. The remaining 12 blocks are below the last horizontal line in the chart below.

(a_0, a_1, b_3)	(a_0, a_2, b_1)	(a_0, a_3, b_4)
(a_0, a_4, b_2)	(a_0, a_5, b_5)	(a_1, a_2, b_4)
(a_1, a_3, b_1)	(a_1, a_4, b_5)	(a_1, a_5, b_2)
(a_2, a_3, b_5)	(a_2, a_4, b_3)	(a_2, a_5, b_0)
(a_3, a_4, b_0)	(a_3, a_5, b_3)	(a_4, a_5, b_1)
(b_0, b_1, b_3)	(b_0, b_2, c_1)	(b_0, b_3, c_4)
(b_0, b_4, c_2)	(b_0, b_5, c_5)	(b_1, b_2, c_4)
(b_1, b_3, c_1)	(b_1, b_4, c_5)	(b_1, b_5, c_2)
(b_2, b_3, c_5)	(b_2, b_4, c_3)	(b_2, b_5, c_0)
(b_3, b_4, c_0)	(b_3, b_5, c_3)	(b_4, b_5, c_1)
(c_0, c_1, a_3)	(c_0, c_2, a_1)	(c_0, c_3, a_4)
(c_0, c_4, a_2)	(c_0, c_5, a_5)	(c_1, c_2, a_4)
(c_1, c_3, a_1)	(c_1, c_4, a_5)	(c_1, c_5, a_2)
(c_2, c_3, a_5)	(c_2, c_4, a_3)	(c_2, c_5, a_0)
(c_3, c_4, a_0)	(c_3, c_5, a_3)	(c_4, c_5, a_1)
(a_0, b_0, c_0)	(a_2, b_2, c_2)	(a_4, b_4, c_4)
(d, a_1, b_0)	(d, b_1, c_0)	(d, c_1, a_0)
(d, a_3, b_2)	(d, b_3, c_2)	(d, c_3, a_2)
(d, a_5, b_4)	(d, b_5, c_4)	(d, c_5, a_4)

An important subclass of a Steiner triple system is a **Kirkman** triple system, which is defined as a resolvable Steiner triple system. There is a classic problem which is known somewhat awkwardly as Kirkman's Schoolgirl Problem. The problem goes as follows. Fifteen schoolgirls are taking a walk each day in five groups of three. Construct a schedule so that no two girls are in the same group during the week.

A Kirkman triple system with v elements is said to be of order v. The affine plane of order 3 is a Kirkman triple system of order 9, and the classic problem seeks a Kirkman triple system of order 15. It is clear that if a Kirkman triple system exists, then we must have $v \equiv 3 \pmod{6}$. The converse of this result is also true, but a complete proof is beyond the scope of this book. So we concentrate on solving the classic problem by constructing a Kirkman triple system of order 15.

The affine space of order 2 has 8 points that form the vertices of the unit cube $ABDC-EFHG$. They may be given coordinates $A(0,0,0)$, $B(1,0,0)$, $C(0,1,0)$, $D(1,1,0)$, $E(0,0,1)$, $F(1,0,1)$, $G(0,1,1)$ and $H(1,1,1)$. It has 28 lines divided into 7 groups of 4 mutually parallel lines. Each line passes through 2 points and is the intersection of 3 planes. Together, they constitute a resolvable (28,8,7,2,1)-design.

Lines	Points
$y=0, z=0, y+z=0$	AB
$y=1, z=0, y+z=1$	CD
$y=0, z=1, y+z=1$	EF
$y=1, z=1, y+z=0$	GH
$z=0, x=0, z+x=0$	AC
$z=1, x=0, z+x=1$	BD
$z=0, x=1, z+x=1$	EG
$z=1, x=1, z+x=0$	FH
$x=0, y=0, x+y=0$	AE
$x=1, y=0, x+y=1$	BF
$y=0, y=1, x+y=1$	CG
$y=1, y=1, x+y=0$	DH
$x=0, y+z=0, x+y+z=0$	AG
$x=0, y+z=1, x+y+z=1$	CE
$x=1, y+z=0, x+y+z=1$	BH
$x=1, y+z=1, x+y+z=0$	DF
$y=0, z+x=0, x+y+z=0$	AF
$y=0, z+x=1, x+y+z=1$	BE
$y=1, z+x=0, x+y+z=1$	CH
$y=1, z+x=1, x+y+z=0$	DG
$z=0, x+y=0, x+y+z=0$	AD
$z=0, x+y=1, x+y+z=1$	BC
$z=1, x+y=0, x+y+z=1$	EH
$z=1, x+y=1, x+y+z=0$	FG
$y+z=0, z+x=0, x+y=0$	AH
$y+z=0, z+x=1, x+y=1$	BG
$y+z=1, z+x=0, x+y=1$	CF
$y+z=1, z+x=1, x+y=0$	DE

To obtain a solution to Kirkman's Schoolgirl Problem, we extend the affine space of order 2 to the projective space of order 2. We add 7 ideal points I, J, K, L, M, N and O to the respective parallel classes of lines. The diagram below shows the affine space embedded in the projective space.

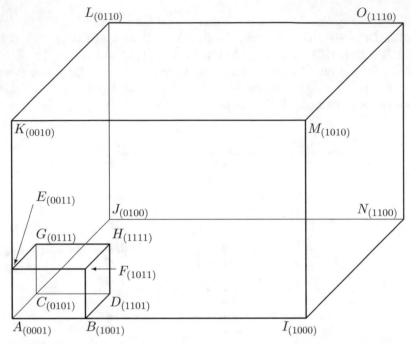

Each point is given coordinates (x, y, z, w) as follows. For the ordinary points, the x, y and z coordinates have the usual meaning while the w coordinate is always 1. For an ideal point P, its w coordinate is 0, and its x, y and z coordinates are the same as the ordinary point Q such that P is the ideal point of the line joining Q to A. Note that there is no point with coordinates $(0,0,0,0)$.

The projective space of order 2 has 7 ideal lines, shown in the first row of the following table. Each subsequent row consists of 4 mutually parallel ordinary lines. Each column consists of 5 mutually non-intersecting lines, and the 7 groups of blocks show that this $(35,15,7,3,1)$-design is indeed resolvable. As such, it is not isomorphic to the Steiner triple system of order 15 obtained from the general construction given earlier, because that triple system is not resolvable.

JKL	KIM	IJN	ILO	JMO	KNO	LMN
IAB				IGH	ICD	IEF
	JAC		JEG		JFH	JBD
		KAE	KDH	KBF		KCG
	LDF	LBH		LCE	LAG	
MCH		MDG	MAF		MBE	
NFG	NEH		NBC	NAD		
ODE	OBG	OCF				OAH

We can also explore the relationship between the planes and the points of either space. The affine space of order 2 has 14 planes each containing 4 points. Together, they constitute a resolvable (14,8,7,4,3)-design. The projective space of order 2 has 15 points and 15 planes, each plane containing 7 points. Together, they constitute a symmetric (15,15,7,7,3)-design.

Affine Space		Projective Space	
Planes	| Points		Planes
$x = 0$	$ACEG$	JKL	$x = 0$
$x = 1$	$BDFH$	JKL	$x + w = 0$
$y = 0$	$ABEF$	IKM	$y = 0$
$y = 1$	$CDGH$	IKM	$y + w = 0$
$z = 0$	$ABCD$	IJN	$z = 0$
$z = 1$	$EFGH$	IJN	$z + w = 0$
$y + z = 0$	$ABGH$	ILO	$y + z = 0$
$y + z = 1$	$CDEF$	ILO	$y + z + w = 0$
$z + x = 0$	$ACFH$	JMO	$z + x = 0$
$z + x = 1$	$BDEG$	JMO	$z + x + w = 0$
$x + y - 0$	$ADEH$	KNO	$x + y = 0$
$x + y = 1$	$BCFG$	KNO	$x + y + w = 0$
$x + y + z = 0$	$ADFG$	LMN	$x + y + z = 0$
$x + y + z = 1$	$BCEH$	LMN	$x + y + z + w = 0$
		$IJKLMNO$	$w = 0$

Examples

Example 6.5.1.
Modify addition modulo 3 to construct a Steiner triple system of order 9.

Solution:
The addition table modulo 3 is shown in the left of the diagram below, and the modified table in the right.

+	0	1	2
0	0	1	2
1	1	2	0
2	2	0	1

$\oplus$	0	2	1
0	0	1	2
2	1	2	0
1	2	0	1

The elements are $a_0, a_1, a_2, b_0, b_1, b_2, c_0, c_1$ and c_2. By the First Construction, the blocks are

$$(a_0, a_1, b_2), \quad (a_0, a_2, b_1), \quad (a_1, a_2, b_0),$$
$$(b_0, b_1, c_2), \quad (b_0, b_2, c_1), \quad (b_1, b_2, c_0),$$
$$(c_0, c_1, a_2), \quad (c_0, c_2, a_1), \quad (c_1, c_2, a_0),$$
$$(a_0, b_0, c_0), \quad (a_1, b_1, c_1), \quad (a_2, b_2, c_2).$$

Example 6.5.2.
Use addition modulo 2 to construct a Steiner triple system of order 7.

Solution:
The addition table modulo 2 is shown in the diagram below.

+	0	1
0	0	1
1	1	0

The elements are $a_0, a_1, b_0, b_1, c_0, c_1$ and d. By the Second Construction, the blocks are (a_0, a_1, b_1), (b_0, b_1, c_1), (c_0, c_1, a_1), (b_0, a_1, d), (c_0, b_1, d), (a_0, c_1, d) and (a_0, b_0, c_0).

Example 6.5.3.
The integers from 0 to 13 are divided into four triples and a pair such that in the five difference tables below, each non-zero difference appears exactly twice. Use these to construct a Kirkman triple system of order 15 on the elements a_i, $0 \le i \le 13$ and b.

$-$	0	4	6
0	0	10	8
4	4	0	12
6	6	2	0

$-$	2	5	13
2	0	11	3
5	3	0	6
13	11	8	0

$-$	3	7	12
3	0	10	5
7	4	0	9
12	9	5	0

$-$	9	10	11
9	0	13	12
10	1	0	13
11	2	1	0

$-$	1	8
1	0	7
8	7	0

Solution:
The starter blocks are (a_0, a_4, a_6), (a_2, a_5, a_{13}), (a_3, a_7, a_{12}), (a_9, a_{10}, a_{11}) and (a_1, a_8, b). The remaining blocks are generated from the starter blocks by adding 2 to the subscript of each element in the preceding block, modulo 14.

(a_0, a_4, a_6)	(a_2, a_5, a_{13})	(a_3, a_7, a_{12})	(a_9, a_{10}, a_{11})	(a_1, a_8, b)
(a_2, a_6, a_8)	(a_4, a_7, a_1)	(a_5, a_9, a_0)	(a_{11}, a_{12}, a_{13})	(a_3, a_{10}, b)
(a_4, a_8, a_{10})	(a_6, a_9, a_3)	(a_7, a_{11}, a_2)	(a_{13}, a_0, a_1)	(a_5, a_{12}, b)
(a_6, a_{10}, a_{12})	(a_8, a_{11}, a_5)	(a_9, a_{13}, a_4)	(a_1, a_2, a_3)	(a_7, a_0, b)
(a_8, a_{12}, a_0)	(a_{10}, a_{13}, a_7)	(a_{11}, a_1, a_6)	(a_3, a_4, a_5)	(a_9, a_2, b)
(a_{10}, a_0, a_2)	(a_{12}, a_1, a_9)	(a_{13}, a_3, a_8)	(a_5, a_6, a_7)	(a_{11}, a_4, b)
(a_{12}, a_2, a_4)	(a_0, a_3, a_{11})	(a_1, a_5, a_{10})	(a_7, a_8, a_9)	(a_{13}, a_6, b)

Exercises

1. Modify addition modulo 5 to construct a Steiner triple system of order 15.

2. Modify addition modulo 4 to construct a Steiner triple system of order 13.

3. The integers from 0 to 13 are divided into four triples and a pair such that in the five difference tables below, each non-zero difference appears exactly twice. Use these to construct a Kirkman triple system of order 15 on the elements a_i, $0 \leq i \leq 13$ and b.

$-$	0	8	12
0	0	6	2
8	8	0	10
12	12	4	0

$-$	3	7	10
3	0	10	7
7	4	0	11
10	7	3	0

$-$	4	5	13
4	0	13	5
5	1	0	6
13	9	8	0

$-$	6	9	11
6	0	11	9
9	3	0	12
11	5	2	0

$-$	1	2
1	0	13
2	1	0

4. (a) The following diagram shows a pair of orthogonal Latin squares of order 4. Use them to construct 16 triples of the form (a_i, b_j, c_k), where $0 \leq i, j, k \leq 3$, so that no two elements appear together in more than one triple.

0	1	2	3
1	0	3	2
2	3	0	1
3	2	1	0

0	2	3	1
3	1	0	2
1	3	2	0
2	0	1	3

(b) Add the triple (d, e, f) and 18 others to the 16 triples in (a) to form a Kirkman triple system of order 15.

5. (a) The idempotent operation $\oplus$ is modified from the addition modulo 7 by reordering the index rows and columns in the operation table. Use it to construct 21 triples of the form (a_i, a_j, b_k), where $0 \leq i \leq j \leq 6$, $0 \leq k \leq 6$ and $i \oplus j = k$.

(b) Add the triples (c, a_i, b_i) for $0 \leq i \leq 6$ and 7 others to the 21 triples in (a) to form a Kirkman triple system of order 15.

6. For each point in the Fano Plane (a, b, c), (a, d, e), (a, f, g), (b, d, f), (b, e, g), (c, d, g) and (c, e, f), there are exactly 4 lines which do not pass through that point. Use this fact to construct a Kirkman triple system of order 15 on the elements $a_0, a_1, b_0, b_1, c_0, c_1, d_0, d_1, e_0, e_1, f_0, f_1, g_0, g_1$ and h.

Section 6.6. Convenient Buildings

A building is said to be *convenient* if for any two floors, there is at least one elevator which stops on both of them. Suppose the building has m elevators each of which stops on n floors. There are no restrictions on the choice of these floors. They do not have to be consecutive, and need not include the ground floor. What is the maximum number $f(m, n)$ of floors in this convenient building?

To establish the answer to this or any extremal problem, we need to do two things. First, we must show by an explicit construction that the answer can be attained. What we have covered in this chapter so far would be useful here. Second, we must prove by a general argument that the answer cannot be improved. What we have covered back in Chapter 1 would be useful here.

We first prove three useful preliminary results.

Observation 1. $f(m + 1, n) \geq f(m, n)$.

Proof:
Having an extra elevator never hurts, though it may not help.

Observation 2. $f(m, n + 1) \geq f(m, n) + 1$.

Proof:
The extra stop for each elevator can all be on a new floor.

Observation 3. $f(m, kn) \geq kf(m, n)$.

Proof:
Pile k copies of a convenient building with $f(m, n)$ floors on top of one another to form a building with $kf(m, n)$ floors and connect the corresponding elevators in each copy so that each stops on kn floors. The same elevator which links the i-th and j-th floors in each copy will link the i-th floor of any copy to the j-th floor of any other copy. Thus the new building is convenient, and we have $f(m, kn) \geq kf(m, n)$.

We now study the function $f(m, n)$ by keeping m constant.

For $m = 1$, we have $f(1, n) = n$. The building can certainly have n floors. If it has more, the elevator will not stop on some floor. No elevator will stop on both this floor and any other floor.

For $m = 2$, we still have $f(2, n) = n$. By Observation 1, we have $f(2, n) \geq f(1, n) = n$. If the building has more floors, each elevator will not stop on some floor. If they skip different floors, no elevator will stop on both. If they skip the same floor, no elevator will stop on both this floor and any other floor.

The first interesting case is $m = 3$. Here we make use of the block design with parameters $(3,3,2,2,1)$, consisting of the blocks $(1,2)$, $(1,3)$ and $(2,3)$. Each element is a floor and each block is an elevator stopping on the floors represented by elements in the block. Thus we have the lower bound $f(3,2) \geq 3$.

We next prove that $f(3,2k) = 3k$. By $f(3,2) \geq 3$ and Observation 3, $f(3,2k) \geq 3k$. The total number of stops is $6k$. If each floor is served by at least 2 elevators, then the number of floors is at most $3k$. If some floor is served by at most 1 elevator, it can be linked to at most $2k - 1$ other floors. Counting this floor, the building can have at most $2k$ floors. It follows that $f(3,2k) = 3k$.

We now prove that $f(3,2k+1) = 3k+1$. By Observation 2,

$$f(3,2k+1) \geq f(3,2k) + 1 = 3k + 1.$$

The total number of stops is $6k + 3$. If each floor is served by at least 2 elevators, then the number of floors is at most $3k + 1$. If some floor is served by at most 1 elevator, it can be linked to at most $2k$ other floors. Counting this floor, the building can have at most $2k + 1$ floors. It follows that $f(3,2k+1) = 3k + 1$.

The cases $m = 4$ and $m = 5$ are slightly more difficult because of the absence of perfect buildings. The next perfect building occurs at $m = 6$. Here we have the block design with parameters $(6,4,3,2,1)$, consisting of the blocks $(1,2)$, $(1,3)$, $(1,4)$, $(2,3)$, $(2,4)$ and $(3,4)$. This yields the lower bound $f(6,2) \geq 4$.

We first prove that $f(6,2k) = 4k$. By $f(6,2) \geq 4$ and Observation 3, $f(6,2k) \geq 4k$. The total number of stops is $12k$. If each floor is served by at least 3 elevators, then the number of floors is at most $4k$. If some floor is served by at most 2 elevators, it can be linked to at most $4k - 2$ other floors. Counting this floor, the building can have at most $4k - 1$ floors. It follows that $f(6,2k) = 4k$.

We now prove that $f(6,2k+1) \leq 4k+2$. Observe that the total number of stops is $12k + 6$. If each floor is served by at least 3 elevators, then the number of floors is at most $4k + 2$. If some floor is served by at most 2 elevators, it can be linked to at most $4k$ other floors. Counting this floor, the building can have at most $4k + 1$ floors.

Finally, we give a general construction to show that $f(6,2k+1) \geq 4k+2$. Let the floors be $a_1, a_2, \ldots, a_k$, $b_1, b_2, \ldots, b_k$, $c_1, c_2, \ldots, c_k$, $d_1, d_2, \ldots, d_k$, e and f. Let the first elevator stop at all the a's and b's, the second at all the a's and c's, the third at all the a's and d's, the fourth at all the b's and c's, the fifth at all the b's and d's, and the sixth at all the c's and d's. Then these $4k$ floors are all linked.

If we add e as the last stop of the first and sixth elevator and f as the last stop of the second and fifth elevator, they are also linked to the other $4k$ floors. However, e and f are not linked. So we replace d_k by f in the sixth elevator. This destroys the links between d_k on the one hand and e and the c's on the other. The remedy is to add e as the last stop of the third elevator and d_k as the last stop of the fourth elevator. It follows that $f(6, 2k+1) = 4k+2$.

The next perfect building where $n = 2$ occurs at $m = 10$, but Observation 3 is no longer sharp. We can get just one isolated value, namely, $f(10, 4) = 10$. For $k \geq 3$, we only have $5k \leq f(10, 2k) \leq 6k - 2$.

The first perfect building where $n = 3$ is the Fano plane, the block design with parameters $(7,7,3,3,1)$. The blocks are $(1,2,3)$, $(1,4,5)$, $(1,6,7)$, $(2,4,6)$, $(2,5,7)$, $(3,4,7)$ and $(3,5,6)$. This yields $f(7, 3) \geq 7$. By this and Observation 3, $f(7, 3k) \geq 7k$. The total number of stops is $21k$. If each floor is served by at least 3 elevators, then the number of floors is at most $7k$. If some floor is served by at most 2 elevators, it can be linked to at most $6k - 2$ other floors. Counting this floor, the building can have at most $6k - 1$ floors. It follows that $f(7, 3k) = 7k$.

It follows from this result and Observation 2 that

$$f(7, 3k+1) \geq f(7, 3k) + 1 = 7k + 1.$$

To prove that $f(7, 3k+1) \leq 7k+2$, observe that the total number of stops is $21k + 7$. If each floor is served by at least 3 elevators, then the number of floors is at most $7k + 2$. If some floor is served by at most 2 elevators, it can be linked to at most $6k$ other floors. Counting this floor, the building can have at most $6k + 1$ floors. It is conjectured that the lower bound is the correct value.

The results for $m \geq 8$ are rather sporadic. For $8 \leq m \leq 11$, we have the following lower bounds.

$$f(8, 3k+1) \geq 7k + 2$$
$$f(8, 3k+2) \geq 7k + 4$$
$$f(8, 3k) \geq 7k$$
$$f(9, 4k) \geq 10k$$
$$f(9, 5k) \geq 12k$$
$$f(9, 7k) \geq 17k$$
$$f(10, 5k) \geq 13k$$
$$f(10, 6k) \geq 16k$$
$$f(10, 6k+1) \geq 17k$$
$$f(11, 4k) \geq 11k$$
$$f(11, 7k) \geq 19k$$
$$f(11, 10k) \geq 28k$$

For $12 \leq m \leq 13$, we can prove that $f(12, 3k) = 9k$, $f(12, 4k) = 12k$ and $f(13, 4k) = 13k$. The following table summarizes the results we know thus far, for $m, n \leq 13$. Numbers in italic are only lower bounds. Other numbers are exact values. Those in boldface indicate perfect buildings. Blanks indicate that we have no information.

$n =$		1	2	3	4	5	6	7	8	9	10	11	12	13
	1	**1**	**2**	**3**	**4**	**5**	**6**	**7**	**8**	**9**	**10**	**11**	**12**	**13**
	2	1	2	3	4	5	6	7	8	9	10	11	12	13
	3	1	**3**	4	6	7	9	10	12	13	15	16	18	19
	4	1	3	5	6	8	10	11	13	15	16	18	20	21
	5	1	3	5	7	9	10	12	14	16	18	19	21	23
	6	1	**4**	6	8	10	12	14	16	18	20	22	24	26
$m =$	7	1	4	**7**	8	11	14	*15*	18	21	*22*	25	28	*29*
	8	1	4	7	9	*11*	*14*	*16*	*18*	*21*	*23*	*25*	*28*	30
	9	1	4	7	10	*12*	*14*	*17*	20	21	*24*	*27*	30	*31*
	10	1	**5**	7	10	*13*	*16*	18	20	23	*26*	28	32	*34*
	11	1	5	8	11	*13*	*16*	19	*22*	*24*	*28*	*30*	*33*	*35*
	12	1	5	**9**	12	–	18	–	24	27	–	–	36	–
	13	1	5	9	**13**	–	–	–	26	–	–	–	39	–

Examples

Example 6.6.1.
Prove that $f(7, 3k + 2) = 7k + 4$.

Solution:
To prove that $f(7, 3k + 2) \leq 7k + 4$, observe that the total number of stops is $21k + 14$. If each floor is served by at least 3 elevators, then the number of floors is at most $7k + 4$. If some floor is served by at most 2 elevators, it can be linked to at most $6k + 2$ other floors. Counting this floor, the building can have at most $6k + 3$ floors. We now give a general construction to show that $f(7, 3k + 2) \geq 7k + 4$. Let the floors be $a_1, a_2, \ldots, a_{k+1}$, $b_1, b_2, \ldots, b_{k+1}$, $c_1, c_2, \ldots, c_{k+1}$, $d_1, d_2, \ldots, d_{k+1}$, $e_1, e_2, \ldots, e_k$, $f_1, f_2, \ldots, f_k$ and $g_1, g_2, \ldots, g_k$. Let the first elevator stop at all the a's, b's and e's, the second at all the a's, c's and f's, the third at all the a's, d's and g's, the fourth at all the b's, c's and g's, the fifth at all the b's, d's and f's, the sixth at all the c's, d's and e's, and the seventh at all the e's, f's and g's. Then all the floors are linked, with two wasted stops in the seventh elevator. It follows that $f(7, 3k + 2) = 7k + 4$.

Example 6.6.2.
Prove that $f(4, 3k) = 5k$.

Solution:
Suppose in a 5-floor building, the first elevator stops on floors 0, 1 and 4, the second on 0, 2 and 4, the third on 0, 3 and 4, and the fourth on 1, 2 and 3. Then every two floors are linked directly by at at least one elevator. Hence $f(4, 3) \geq 5$. By Observation 3, $f(4, 3k) \geq 5k$. Suppose the building has $5k + 1$ floors. The total number of elevator doors is $4(3k) < 3(5k + 1)$. Hence some floor is served by at most two elevators, and it cannot be just one. Let these be elevators A and B. Between them, they must serve every floor. Hence there are exactly $2(3k) - (5k + 1) = k - 1$ floors served by both of them. Denote by A the set of floors served by elevator A, and by B the set of floors served by elevator B. Then $|A - B| = |B - A| = 2k + 1$. Each floor in $A - B$ must be linked to each floor in $B - A$ by elevator C or elevator D. Since $2(3k) < 2(4k + 2)$, some floor in $(A - B) \cup (B - A)$ is served by exactly one of elevator C and elevator D. We may assume that it is in $A - B$ and served by elevator C. Then every floor in $B - A$ must be served by elevator C. Now some other floor in $A - B$ is served by elevator D but not elevator C. Again, every floor in $B - A$ must be served by elevator D. Hence at most $2k - 2$ floors in $A - B$ can be served by elevator C or elevator D, which is a contradiction.

Example 6.6.3.
Prove that $f(5, 5k) = 9k$.

Solution:
Suppose in a 9-floor building, the first elevator stops on floors 0, 1, 3, 5 and 7, the second on 0, 2, 4, 6 and 8, the third on 1, 2, 4, 6 and 8, the fourth on 2, 3, 4, 5 and 7, and the fifth on 3, 5, 6, 7 and 8. Then every two floors are linked directly by at at least one elevator. Hence $f(5, 5) \geq 9$. By Observation 3, $f(5, 5k) \geq 9k$. Suppose the building has $9k + 1$ floors. None of them may be served by only one elevator. Let x be the number of floors served by exactly two elevators. Then the total number of elevator doors is $5(5k) \geq 2x + 3(9k + 1 - x)$. Hence $x \geq 2k + 3$. Consider a floor served by exactly two elevators. Call them A and B. Between them, they must serve every floor. Hence there are exactly $2(5k) - (9k + 1) = k - 1$ floors served by both of them. Denote by A the set of floors served by elevator A, and by B the set of floors served by elevator B. Then $|A - B| = |B - A| = 4k + 1$. Consider a floor not in $A \cap B$ which is served by exactly two elevators. By symmetry, we may assume that it is in $A - B$. It must be linked to all floors in $B - A$ by say elevator C. Then elevator C serves all $4k + 1$ floors in $B - A$ and $k - 1$ floors in $A - B$. Now the remaining $3k + 2$ floors in $A - B$ must be linked to all floors in $B - A$ by the remaining two elevators. However, each of them can link to only $2k$ floors in $B - A$, but the set has $4k + 1$ floors.

Exercises

1. Prove that $f(4, 3k + 1) = 5k + 1$.

2. Prove that $f(4, 3k + 2) = 5k + 3$.

3. Prove that $f(5, 5k + 1) = 9k + 1$.

4. Prove that $f(5, 5k + 2) = 9k + 3$.

5. Prove that $f(5, 5k + 3) = 9k + 5$.

6. Prove that $f(5, 5k + 4) = 9k + 7$.

Additional Exercises

Section A.0.

1. A restaurant offers choices of five soups, six salads, three appetizers, two main courses and four desserts.

   (a) Freda is going to have a complete dinner consisting of one of each item. How many choices are there for her?

   (b) Fred can only afford to have a main course and one other item. How many choices are there for him?

2. There are three roads between towns A and B, five roads between towns B and C and six roads between towns A and C.

   (a) In how many ways can Eric go directly to town B or C from town A?

   (b) In how many ways can Erica go, directly or indirectly, to town C from town A?

3. Ethel has four close friends, Anita, Betty, Celia and Daisy. During one month, she has dinner with each of them 12 times, with every two of them 6 times, with every three of them 4 times, and with all four of them 3 times. How many times during that month does she have dinner with at least one of them?

4. The six most active students in a class formed 30 different committees, every two of which had at least one member in common. Prove that it is possible to form one more committee having a common member with each of these 30 committees.

5. Ten students solve a total of 35 problems in a test. Each problem is solved by exactly one student. There is at least one student who solves exactly one problem, at least one student who solves exactly two problems, and at least one student who solves exactly three problems. Prove that there is also at least one student who solves at least five problems.

6. Ten boys and ten girls are standing in a row. Each boy counts the number of children to his left, and each girl counts the number of children to her right. Prove that the total number counted by the boys is equal to the total number counted by the girls.

7. Prove by mathematical induction that for all $n \geq 1$,
$$(1 + 2 + \cdots + n)^2 = 1^3 + 2^3 + \cdots + n^3.$$

8. Prove by mathematical induction that for all $n \geq 1$,

$$\frac{1}{1 \times 4} + \frac{1}{4 \times 7} + \cdots + \frac{1}{(3n - 2)(3n + 1)} = \frac{n}{3n + 1}.$$

9. On every planet of some planetary system sits an astronomer observing the closest planet.. All distances between planets are different. Prove that if the number of planets is odd, then at least one of them is not observed.

10. In a parliament, each member slapped the face of exactly three other members of this prominent body. Several parliamentary committees are to be organized so that each member of the parliament works in exactly one of them. To avoid conflicts, it is necessary to fill it with members who have never slapped one another in the face. What is the minimum number of committees for which this condition can always be met?

11. A boy and a girl are sitting at opposite ends of a long bench. One at a time, twenty other children take seats on the bench. A boy is said to be brave if at the moment he takes his seat, the next child in each direction, not necessarily in an adjacent seat, is a girl. Similarly, a girl is said to be brave if she takes a seat between two boys. At the end, the boys and girls are sitting alternately. What is the number of brave children?

12. Sisyphus is given a new job: he must move stones, one at a time, from one of three piles to another. If he moves a stone from a pile with a stones to a pile with b stones, then Zeus gives him $b - a + 1$ coins. If this number is negative, it means that Zeus is taking $a - b - 1$ coins from Sisyphus instead. If he is short of money, magnanimous Zeus allows him to continue moving the stones on credit. At some moment, each pile has the same number of stones as it does initially. What is the greatest amount of money that Sisyphus may have made up to that point?

Section A.1.

1. In each of the letter triangles in the diagram below, start with a letter L and move from letter to adjacent letter either to the right or down.

   (a) How many ways are there of tracing out the word LARDER in the letter triangle on the left?

   (b) Starting from each letter L in the letter triangle on the right, express the number of ways of tracing out the word LADDER in terms of a binomial coefficient.

(c) How many ways are there of tracing out the word LADDER in the letter triangle on the right?

(d) Derive an identity from the results in (b) and (c).

```
L   A   R   D   E   R                           L
A   R   D   E   R                         L   A
R   D   E   R                         L   A   D
D   E   R                         L   A   D   D
E   R                         L   A   D   D   E
R                         L   A   D   D   E   R
```

2. In each of the letter triangles in the diagram below, start with a letter C and move from letter to adjacent letter either to the right or down.

   (a) How many ways are there of tracing out the word COFFIN in the letter triangle on the left?

   (b) Starting from each letter C in the letter triangle on the right, express the number of ways of tracing out the word COFFEE in terms of a binomial coefficient .

   (c) How many ways are there of tracing out the word COFFEE in the letter triangle on the right?

   (d) Derive an identity from the results in (b) and (c).

```
C   O   F   F   I   N                           C
O   F   F   I   N                           C   O
F   F   I   N                           C   O   F
F   I   N                           C   O   F   F
I   N                           C   O   F   F   E
N                           C   O   F   F   E   E
```

3. Use a combinatorial argument to prove that $\binom{m+n}{2} - \binom{m}{2} - \binom{n}{2} = mn$ for all positive integers m and n.

4. Use a combinatorial argument to prove that $\binom{n}{k}\binom{k}{m} = \binom{n}{m}\binom{n-m}{k-m}$ for all positive integers $n \geq k \geq m$.

5. Use $S_1 = 1+2+\cdots+n = \frac{n(n+1)}{2}$ and the Binomial Theorem to derive the closed form for $S_2 = 1^2 + 2^2 + \cdots + n^2$.

6. Use $S_1 = 1+2+\cdots+n = \frac{n(n+1)}{2}$, $S_2 = 1^2+2^2+\cdots+n^2 = \frac{n(n+1)(2n+1)}{6}$ and the Binomial Theorem to derive the close form for

$$S_3 = 1^3 + 2^3 + \cdots + n^3.$$

7. (a) Prove that $\binom{100}{0} + \binom{100}{4} + \cdots + \binom{100}{100} = 2^{98} - 2^{49}$.

    (b) Prove that $\binom{100}{1} + \binom{100}{5} + \cdots + \binom{100}{97} = 2^{98}$.

    (c) Prove that $\binom{100}{2} + \binom{100}{6} + \cdots + \binom{100}{98} = 2^{98} + 2^{49}$.

    (d) Prove that $\binom{100}{3} + \binom{100}{7} + \cdots + \binom{100}{99} = 2^{98}$.

8. (a) Prove that $\binom{100}{0} + \binom{100}{3} + \cdots + \binom{100}{99} = \frac{1}{3}(2^{100} - 1)$.

    (b) Prove that $\binom{100}{1} + \binom{100}{4} + \cdots + \binom{100}{100} = \frac{1}{3}(2^{100} - 1)$.

    (c) Prove that $\binom{100}{2} + \binom{100}{5} + \cdots + \binom{100}{98} = \frac{1}{3}(2^{100} + 2)$.

9. A building has six floors and two elevators which operate independently. They move at the same constant speed from floor to floor, and stoppage time is negligible. A person on the floor just below the top floor is waiting for an elevator. What is the probability that the first elevator to arrive is coming from above?

10. A movie theater has $n + k$ seats; their numbers are indicated on the tickets. The first n people, including Howard, who come to a show take n seats without paying attention to the seat numbers. But the remaining k ticket holders are sticklers. If any of them finds her or his assigned seat occupied, the person sitting there is evicted; that person then looks for his or her proper seat and evicts the usurper; and so forth. This migration ends with the spectator whose assigned seat is unoccupied. Find the probability that Howard will not have to change his seat.

11. The game Bridge is played among four players identified as South, West, North and East. South and North are partners, as are West and East. A standard deck of cards is used, and thirteen are dealt to each player. A round of bidding determines which player gets to play the hand. Suppose South is playing with Spades as the special suit called trumps. Then North lays down her or his cards for all to see. Usually, South and North will have more than half of the trumps and the distribution of the missing trumps between West and East is important to South.

    (a) If South and North have 7 Spades between them, which is the most likely distribution of the other 6, 3-3, 4-2, 5-1 or 6-0?

    (b) If South and North have 8 Spades between them, what is the most likely distribution of the other 5, 3-2, 4-1 or 5-0?

    (c) If South and North have 9 Spades between them, what is the most likely distribution of the other 4, 2-2, 3-1 or 4-0?

    (d) If South and North have 10 Spades between them, what is the more likely distribution of the other 3, 2-1 or 3-0?

(e) If South and North have 11 Spades between them, what is the more likely distribution of the other 3, 1-1 or 2-0?

12. In the game Keno, a player bets \$1 on five of the numbers from 1 to 80 inclusive. A machine draws at random twenty of the numbers from 1 to 80 inclusive. Between the player's numbers and the machine's numbers, if the number of matches is 2 or less, the player loses his bet. Otherwise, the player gets back the \$1 and receives an additional \$2, \$25 or \$331 if the number of matches is 3, 4 or 5, respectively. On the average, does the player win or lose money?

Section A.2.

1. What sequence is generated by the function $\frac{1}{\sqrt{1-4x}}$?

2. Find the generating function for the sequence $\{a_n\}$ where $a_n = n$ for all non-negative integers n.

3. Find the number of ways of distributing n apples among four children with each getting at least 3,

   (a) using generating functions;

   (b) without using generating functions.

4. Find the number of ways of distributing n apples among four children with each getting an even number,

   (a) using generating functions;

   (b) without using generating functions.

5. Sylvester has caught n mice which he arranges in a circle and numbers them $1, 2, \ldots, n$ in clockwise order. Starting with mouse number 1, Sylvester goes around the circle in clockwise order, skipping over one mouse and eating the next one. The circle of mice closes in to fill the gap created by Sylvester's appetite. He goes round and round by the same rule, until only one mouse is left. This lucky mouse is then set free. Denote by $f(n)$ the number assigned to the lucky mouse initially. Then $f(1) = 1$, $f(2) = 1$ and $(3) = 3$.

   (a) Express $f(2n)$ and $f(2n + 1)$ in terms of $f(n)$.

   (b) Determine $f(100)$.

6. Let $f(n)$ denote the number of way in which n can be expressed as a sum of powers of 2, if each distinct power may be used up to three times. Then $f(0) = f(1) = 1$ and $f(2) = f(3) = 2$.

   (a) Prove that $f(2n+1) = f(2n) = f(n-1) + f(n)$.

   (b) Determine $f(100)$.

7. Solve the recurrence relation $a_n - 2a_{n-1} + a_{n-2} = 0$ with initial conditions $a_0 = 1$ and $a_1 = 2$, using

   (a) the method of characteristic equations;

   (b) the method of generating functions.

8. Solve the recurrence relation $a_n = a_{n-2} + 2 \cdot 3^n$ with initial conditions $a_0 = i_1 = 0$, using

   (a) the method of characteristic equations;

   (b) the method of generating functions.

9. In the puzzle called the New Tower of Hanoi, there are four pegs in the playing board. There are n disks of different sizes, all stacked on the first peg, in ascending order of size from the top. The objective is to transfer this tower to the fourth peg. The rule is that we may only move a disk on top of a peg to the top of another peg, and a disk may not be placed on top of a smaller disk. What is the minimum number of moves required for n disks?

10. In the puzzle called the Twin Towers of Hanoi, there are three pegs in the playing board. There are n disks, one each of sizes $1, 2, 3, \ldots, n$. Those of odd sizes are stacked on the first peg, and those of even sizes are stacked on the third peg. On both pegs, the disks are in ascending order of size from the top. The object is to merge these two towers on the second peg. The rule is that we may only move a disk on top of a peg to the top of another peg, and a disk may not be placed on top of a smaller disk. What is the minimum number of moves required for n disks?

11. Consider the sequence $\{a_n\}$ where $a_n = 3^n + 2 \cdot 3^{n-1} + \cdots + 2^{n-1} \cdot 3 + 2^n$.

   (a) Express its generating function $A(x)$ as a product of the generating functions of two other sequences.

   (b) Determine a_n from $A(x)$.

12. Consider the sequence $\{a_n\}$ where $a_n = 1 \cdot 3^n + 2 \cdot 3^{n-1} + \cdots + n \cdot 3 + (n+1)$.

   (a) Express its generating function $A(x)$ as a product of the generating functions of two other sequences.

   (b) Determine a_n from $A(x)$.

Section A.3.

1. Seven children are to be seated in a circle.

   (a) How many different cyclic arrangements are there if Johnny and Jackie must be together?

   (b) Derive from (a) the number of different cyclic arrangements if Johnny and Jackie must be separated.

2. Seven children are to be seated in a circle.

   (a) How many different cyclic arrangements are there if Johnny and Jackie must be separated?

   (b) Derive from (a) the number of different cyclic arrangements if Johnny and Jackie must be together.

3. Find the numbers of letter strings of lengths 1, 2, 3 and 4 that can be spelled with the letters of MADAM.

4. Find the numbers of letter strings of lengths 1, 2, 3 and 4 that can be spelled with the letters of MELEE.

5. What sequence is generated exponentially by $\frac{e^{2x} - e^{-2x}}{4x}$?

6. Find the exponential generating function for the sequence $\{a_n\}$ where $a_n = n$ for all non-negative integers n.

7. Determine the number of binary sequences of length n with an even number of 0s

   (a) using the method of exponential generating functions;

   (b) without using the method of exponential generating functions.

8. Determine the number of binary sequences of length n with an odd number of 0s

   (a) using the method of exponential generating functions;

   (b) without using the method of exponential generating functions.

9. Find the total number of letter strings that can be spelled with the letters of MADAM using the method of exponential generating functions.

10. Find the total number of letter strings that can be spelled with the letters of MELEE using the method of exponential generating functions.

11. A random permutation of the integers from 1 to 100 inclusive is decomposed into cycles. What is the probability that the element 13 appears in a cycle of length 13?

12. In a certain city, only pairwise exchanges of apartments are permitted. If two families swap their apartments, they cannot take part in any other exchanges on the same day. Prove that any permutation of the apartments can be achieved in two days of exchanges.

Section A.4.

1. (a) A cinema has n seats in a row, occupied by n boys. At the intermission, they all go to the bathroom. When they come back, each either occupies his own seat or a seat next to his own. Express the number of ways this can be done in terms of the Fibonacci numbers.

   (b) A restaurant has n seats at a round table, occupied by n girls. Just before the main course is served, they all get up to lose some calories. When they come back, each either occupies her own seat or a seat next to her own. Express the number of ways this can be done in terms of the Fibonacci numbers.

2. A counter starts at the bottom left corner of a $2 \times n$ chessboard. In each move, it may move to an adjacent square in the same row or the same column. It may pass through each square at most once, and may not pass through all four squares which form a 2×2 subboard. Let a_n denote the number of different paths which ends on any square of the rightmost column.

   (a) Determine a_n for $n \leq 3$.

   (b) Express a_n in terms of the Fibonacci numbers.

3. Prove that every positive integer has a unique expression as a sum of non-adjacent Fibonacci numbers.

4. Alice and Brian play a game which starts with a pile of n markers. The two of them alternately take markers from the diminishing pile. Alice goes first, and may not take all n markers. In each turn, the number of markers taken must be at least one but at most twice the number of markers taken by the opponent in the preceding turn. The winner is the one who takes the last marker. Which player has a winning strategy if

   (a) $n = 20$;

   (b) $n = 21$?

5. Prove that for any positive integer k, there is a Fibonacci number ending with k zeros.

newpage

6. Compute $\sum_{n=0}^{\infty} \dfrac{F_n}{2^n}$, where F_n are the Fibonacci numbers.

7. Each of four girls sitting at a square table is holding exactly one ribbon. A girl may hold a ribbon by herself, or she may hold one between herself and another girl, not necessarily sitting next to her. We may also have 3 or 4 girls holding the same ribbon, forming a triangle or a square. Draw all possible configurations of how this can be done.

8. There are $4^4 = 256$ quadruples where each number may be 1, 2, 3 and 4. Each quadruple represents a shuffling of a deck of four cards (A,K,Q,J) . The shuffling comes in four steps. In each step, the top card goes to the position indicated by the next number in the quadruple. For instance, with the quadruple (2,2,3,1), first the Ace goes to position 2, resulting in (K,A,Q,J). Then the King goes to position 2, restoring (A,K,Q,J). Next the Ace goes to position 3, resulting in (K,Q,A,J). Finally, the King stays put. Of the 256 shuffles, only 15 of them restore the original order (A,K,Q,J), 15 being the fourth Bell number. Find these shuffles.

9. What is the minimum number of people in a party such that either there exist two disjoint pairs of people who know each other or there exist two disjoint pairs of people who do not know each other.

10. What is the minimum number of people in a party such that either there exists a person knowing three other people, or a person not knowing three other people.

11. What is the minimum number of people in a party such that there exist four people A, B, C and D, either with A knowing B, B knowing C and C knowing D, or with A not knowing B, B not knowing C and C not knowing D?

12. What is the minimum number of people in a party such that there exist four people A, B, C and D, either with A knowing B, B knowing C, C knowing D and D knowing A, or with A not knowing B, B not knowing C, C not knowing D and D not knowing A?

Section A.5.

1. Thirteen dots are marked on a square as shown in the diagram below. They appear on both sides, and each must be painted the same way as on the other side. Determine the number of different ways of painting these dots in red, yellow and blue.

2. Sixteen segments are marked on a square as shown in the diagram below. They appear on both sides, and each must be painted the same way as on the other side. Determine the number of different ways of painting these segments in green and gold.

3. (a) Determine the rotational symmetries of a solid with two parallel equilateral triangular bases and three square lateral faces.

   (b) Determine the cycle index if the faces of this solid are to be painted.

4. (a) Determine the rotational symmetries of a solid obtained by gluing together two regular tetrahedra on a common equilateral triangular face.

   (b) Determine the cycle index if the faces of this solid are to be painted.

5. An order n prism is obtained by placing two horizontal regular n-gons directly over each other and joining the corresponding pairs of vertices with vertical edges of length equal to the edges of the n-gons. A planar representation of an order 8 prism is shown in the diagram below.

(a) Determine the rotational symmetries of an order 8 prism.

(b) Determine the cycle index if the faces of this solid are to be painted.

6. An order n antiprism is obtained by placing two horizontal regular n-gons directly over each other, rotating one of them so that each of its vertices is equidistant from two adjacent vertices of the other n-gon, and joining these pairs of vertices with edges of length equal to the edges of the n-gons. A planar representation of an order 8 antiprism is shown in the diagram below.

(a) Determine the rotational symmetries of an order 8 antiprism.

(b) Determine the cycle index if the faces of this solid are to be painted.

7. Around each vertex of a cube, we cut off a tetrahedron so that the resulting solid has six faces that are regular octagons and eight faces that are equilateral triangles.

(a) Draw a planar representation of this solid.

(b) Determine the cycle index if the faces of this solid are to be painted.

8. Around each vertex of a regular octahedron, we cut off a tetrahedron so that the resulting solid has eight faces that are regular hexagons and six faces that are squares.

(a) Draw a planar representation of this solid.

(b) Determine the cycle index if the faces of this solid are to be painted.

9. Each face of a cube is to be painted in one of four colors. Determine the number of coloring patterns in which each color is used at least once.

10. Each vertex of a cube is to be painted in one of four colors. Determine the number of coloring patterns in which each color is used exactly twice.

11. Determine the number of distributions of three identical red objects and one blue object into three identical boxes.

12. Determine the number of distributions of three identical objects into two identical black boxes and two identical white boxes.

Section A.6.

1. A certain community of lions and ponies is defined by the following postulates.

    (1) There are at least two lions.

    (2) Each lion has bitten at least three ponies.

    (3) For any pair of lions, there is exactly one pony that both have bitten.

    (4) For any pair of ponies, there is at least one lion that has bitten both of them.

    Let A be a lion and P be a pony such that A has not bitten P. Then the number of ponies that A has bitten is equal to the number of lions that have bitten P.

2. A certain community of lions and ponies is defined by the following postulates.

    (1) There are at least two lions.

    (2) Each lion has bitten at least three ponies.

    (3) For any pair of lions, there is exactly one pony that both have bitten.

    (4) For any pair of ponies, there is at least one lion that has bitten both of them.

    (a) Prove that each lion has bitten the same number of ponies.

    (b) Prove that each pony has been bitten by the same number of lions.

3. From the pair of orthogonal 5×5 Latin squares, construct a 5×5 magic square in which the five numbers in each row and column add up to the same magic constant.

$$
\begin{bmatrix}
0 & 1 & 2 & 3 & 4 \\
1 & 2 & 3 & 4 & 0 \\
2 & 3 & 4 & 0 & 1 \\
3 & 4 & 0 & 1 & 2 \\
4 & 0 & 1 & 2 & 3
\end{bmatrix}
\begin{bmatrix}
0 & 1 & 2 & 3 & 4 \\
3 & 4 & 0 & 1 & 2 \\
1 & 2 & 3 & 4 & 0 \\
4 & 0 & 1 & 2 & 3 \\
2 & 3 & 4 & 0 & 1
\end{bmatrix}
$$

4. From the pair of orthogonal 5×5 Latin squares, construct a 5×5 magic square in which the five numbers in each row and column add up to the same magic constant.

$$
\begin{bmatrix}
0 & 1 & 2 & 3 & 4 \\
2 & 3 & 4 & 0 & 1 \\
4 & 0 & 1 & 2 & 3 \\
1 & 2 & 3 & 4 & 0 \\
3 & 4 & 0 & 1 & 2
\end{bmatrix}
\begin{bmatrix}
0 & 1 & 2 & 3 & 4 \\
4 & 0 & 1 & 2 & 3 \\
3 & 4 & 0 & 1 & 2 \\
2 & 3 & 4 & 0 & 1 \\
1 & 2 & 3 & 4 & 0
\end{bmatrix}
$$

5. A possible magic constant for the polyomino magic square in the diagram below is the 3×5 rectangle. Construct it using the three polyominoes in each row, column and diagonal.

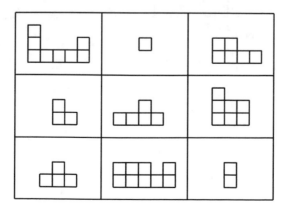

6. A possible magic constant for the polyomino magic square in the diagram below is the 3×5 rectangle. Construct it using the three polyominoes in each row, column and diagonal.

7. Magic squares may be generalized into magic configurations consisting of sets of collinear points, such as the one shown in the diagram below.

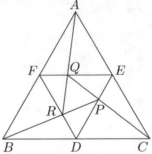

Each point is assigned a generalized pentomino, as shown in the diagram below. Some cells (the shaded ones) may be missing from certain pieces. If a pieces is not connected, each part stays in the same rigid position relative to the other parts. Show that a 4 × 4 square with an edge cell missing may be constructed with the three pieces along each of the lines *AFB*, *BDC*, *CEA*, *AQR*, *BRP*, *CPQ*, *DPE*, *EQF* and *FRD*. Rotations and reflections of the pieces are allowed.

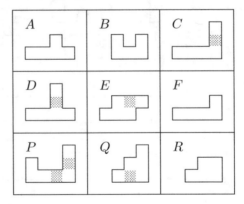

8. Magic squares may be generalized into magic configurations consisting of sets of points in special positions relative to each other, such as the one shown in the diagram below.

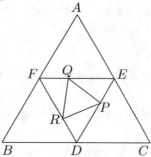

Each point is assigned a generalized pentomino, as shown in the diagram below. Some cells (the shaded ones) may be missing from certain pieces. If a piece is not connected, each part stays in the same rigid position relative to the other parts. Show that a 4×4 square with an edge cell missing may be constructed with the three pieces at the vertices of the equilateral triangles ABC, DEF, PQR, AEF, BFD and CDE. Rotations and reflections of the pieces are allowed.

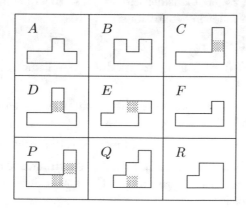

9. Let $2s - 1$ and $2s + 1$ be twin primes. Give each of the numbers 0 to $(2s - 1)(2s + 1) - 1$ a pair of coordinates (m, n), where m and n are the remainders obtained when the number is divided respectively by $2s - 1$ and $2s + 1$. Consider the set consisting of the numbers satisfying one of the following conditions:

   (1) $n = 0$;

   (2) $mn \neq 0$, m is a square modulo $2s - 1$ and n is a square modulo $2s + 1$;

   (3) $mn \neq 0$, m is a non-square modulo $2s - 1$ and n is a non-square modulo $2s + l$.

   (a) Construct this set for the case $s = 3$.

   (b) Prove that in general, if exactly one coordinate of a number is 0, then this number appears $s^2 - 1$ times as a difference of two numbers in this set.

10. Let $2s - 1$ and $2s + 1$ be twin primes. Give each of the numbers 0 to $(2s - 1)(2s + 1) - 1$ a pair of coordinates (m, n), where m and n are the remainders obtained when the number is divided respectively by $2s - 1$ and $2s + 1$. Consider the set consisting of the numbers satisfying one of the following conditions:

   (1) $n = 0$;

   (2) $mn \neq 0$, m is a square modulo $2s - 1$ and n is a square module $2s + 1$;

   (3) $mn \neq 0$, m is a non-square modulo $2s - 1$ and n is a non-square modulo $2s + l$.

   (a) Verify that this set is a difference set of order 3 for the case $s = 2$.

   (b) Prove that in general, if neither coordinate of a number is 0, then this number appears $s^2 - 1$ times as a difference of two numbers in this set.

11. Eight students were solving eight problems. Every problem was solved by exactly k students. What is the minimum value of k to guarantee the existence of two students such that every problem was solved by at least one of them?

12. Eleven students are solving k problems. Each problem is solved by exactly four students. What is the minimum value of k for which it is possible that no three students solve all problems among them?

Appendix B

Additional Examples

Section B.0.

Example B.0.1.

On an east-west shipping lane are ten ships sailing individually. The first five from the west are sailing eastwards while the other five ships are sailing westwards. They sail at the same constant speed at all times. Whenever two ships meets, each turns around and sails in the opposite direction. When all ships have returned to port, how many meetings of two ships have taken place?

Solution:
Let us consider what happens when two ships meet. Each continues where the other would have gone. Since we are interested in the total number of meetings rather than the numbers of meetings for individual ships, we may pretend that the ships just sail on. Since there are 5 ships from each side, the total number of meetings is $5 \times 5 = 25$.

Example B.0.2.

Integers are placed in each square of a 10×10 table, with no two integers in neighbouring squares differing by more than 5 (two squares are considered neighbors if they share a common edge). Prove that two of the integers must be equal.

Solution:
Let m be the smallest number in the table and n be the largest. Starting from the square containing m, we move horizontally to the column where the square containing n is, and then move vertically to that square. Note that all movement are between neighboring squares. Since the value changes by at most 5 in each step, $n - m \leq 19 \times 5 = 95$. It follows that there are at most 96 different numbers among the 100 on the board, so that two of them must be equal.

Example B.0.3.

Two relatively prime positive integers m and n and the number 0 are given. A calculator can execute only one operation: to calculate the arithmetic mean of two given positive integers if they are both even or both odd. Prove that using this calculator you can obtain all the positive integers 1 through n, if you can enter into the calculator only the three numbers initially given or results of previous calculations.

Solution:
We use induction on n. The base case $n = 1$ is trivial. Suppose the result holds for $n = k$. Choose among m, $k + 1$ and 0 two numbers of the same parity, and calculate their arithmetic mean x. The larger of the two chosen numbers is different from 0. Replace this non-zero number by x and repeat the operation for the new trio of numbers, until one of the positive numbers in the trio becomes less than $k+1$. This will eventually happen, because the sum of two positive numbers in the trio decreases after each operation. Now, let g be the largest positive integer less than $k + 1$ which can be obtained using this calculator. Suppose that $g \leq k$. By the assumption, all positive integers 1 through g can be obtained using this calculator, If 1 and $k+1$ are of the same parity, then we can calculate the arithmetic mean y of g and $k + 1$, which contradicts the definition of g since $g < y < k + 1$. If g and $k + 1$ are of opposite parities, we can calculate the arithmetic mean of $g - 1$ and $k + 1$ to get the same contradiction.

Example B.0.4.
Standing in a circle are 99 girls, each with a candy. In each move, each girl gives her candy to either neighbor. If a girl receives two candies in the same move, she eats one of them. What is the minimum number of moves after which only one candy remains?

Solution:
Let the girls be labeled 1 to 99 in clockwise order. We first show that the task can be accomplished in 98 moves. In each of the first 49 moves, if she still has a candy, the k-th girl gives hers to the $(k-1)$-st girl for $2 \leq k \leq 50$ and to the $(k + 1)$-st girl for $51 \leq k \leq 99$. The 1-st girl, who will always have a candy, gives hers to the 99-th girl. After 49 moves, only the 1-st and the 99-th girl have candies. These two candies can be passed, in opposite directions, to the 50-th girl in another 49 moves. We now show that the task cannot be accomplished in less than 98 moves. We will not allow the girls to eat the candies, but each must pass all she has to the same neighbor. Our target is to have all candies in the hands of one girl. Consider what happens to a candy in two consecutive moves. It either returns to the girl who has it initially, or is passed to a girl two places away. Suppose the candies all end up with the 50-th girl in at most 98 moves. The number of moves is not enough for the candy initially with the 50-th girl to go once around the circle before returning to her. It follows that the number of moves must be even. However, in order for the candy initially with the 49-th girl to end up in the hands of the 50-th girl in an even number of moves, it must go once around the circle, and that takes 98 moves.

Section B.1.

Example B.1.1.

Let $\binom{n}{k}$ be the number of ways of choosing a subset of k objects from a set of n objects. Prove that if k and ℓ are positive integers less than n, then $\binom{n}{k}$ and $\binom{n}{\ell}$ have a common divisor greater than 1.

Solution:

Let $0 < k < \ell < n$. Then $\binom{\ell}{k} < \binom{n}{k}$. Suppose we have n players from which we wish to choose a team of size ℓ, and to choose k captains among the team players. The team can be chosen in $\binom{n}{\ell}$ ways and the captains can be chosen in $\binom{\ell}{k}$ ways. On the other hand, if we choose the captains first among all the players, the number of ways is $\binom{n}{k}$. From the remaining $n - k$ players, there are $\binom{n-k}{\ell-k}$ ways of choosing the $\ell - k$ non-captain players. Hence $\binom{n}{\ell}\binom{\ell}{k} = \binom{n}{k}\binom{n-k}{\ell-k}$. Now $\binom{n}{k}$ divides $\binom{n}{\ell}\binom{\ell}{k}$. If it is relatively prime to $\binom{n}{\ell}$, then it must divide $\binom{\ell}{k}$. This is a contradiction since $\binom{\ell}{k} < \binom{n}{k}$.

Example B.1.2.

Find closed forms for the following:

(a) $$\binom{n}{0} + \frac{1}{2}\binom{n}{1} + \cdots + \frac{1}{n+1}\binom{n}{n};$$

(b) $$\binom{n}{0} - \frac{1}{2}\binom{n}{1} + \cdots + (-1)^n \frac{1}{n+1}\binom{n}{n}.$$

Solution:

We use the identity that $\frac{1}{k+1}\binom{n}{k} = \frac{1}{n+1}\binom{n+1}{k+1}$.

(a) We have

$$\binom{n}{0} + \frac{1}{2}\binom{n}{1} + \cdots + \frac{1}{n+1}\binom{n}{n}$$
$$= \frac{1}{n+1}\binom{n+1}{1} + \frac{1}{n+1}\binom{n+1}{2} + \cdots + \frac{1}{n+1}\binom{n+1}{n+1}$$
$$= \frac{1}{n+1}(2^{n+1} - 1).$$

(b) We have

$$\binom{n}{0} - \frac{1}{2}\binom{n}{1} + \cdots + (-1)^n\frac{1}{n+1}\binom{n}{n}$$
$$= \frac{1}{n+1}\binom{n+1}{1} - \frac{1}{n+1}\binom{n+1}{2} + \cdots + (-1)^n\frac{1}{n+1}\binom{n+1}{n+1}$$
$$= \frac{1}{n+1}.$$

Example B.1.3.

The owner of a new factory decides that each worker will get one day off per week. The offical day off is determined at random upon hiring, and this is done independently for each worker. In a momentary loss of sanity, the owner also stipulated that on any day when some workers are off officially, all other workers get an unofficial day off. When he comes to his senses, the rules have been approved by the Labor Board, and he is committed to them. What is the optimal number of workers he should hire to maximize the expected productivity?

Solution:

Let the number of workers be n. The probability that the first day is a working day is $(\frac{6}{7})^n$ since the official day off of each worker must be one of the other six days. The expected productivity for that day is therefore $n(\frac{6}{7})^n$ man-days. By symmetry, this is the same for each of the seven days, so that the overall expected productivity is $7n(\frac{6}{7})^n = \frac{n6^n}{7^{n-1}}$ man-days. In going from n workers to $n+1$ workers, the ratio of the new expected productivity to the old is $\frac{6(n+1)}{7n}$. When $n = 6$, this ratio is 1. When $n < 6$, it is greater than 1. When $n > 6$, it is less than 1. Hence the maximum value of the expected productivity occurs at $n = 6$ and $n = 7$. Since there is no point in paying an unnecessary extra salary, the owner should hire 6 workers.

Example B.1.4.

In the classic cartoon strip "Pogo" by Walt Kelly, there are three card-playing bats named Bewitched, Bothered and Bemildred. Their house rules are "Deuces are wild; one-eyed Jacks are Kings; and all Hearts are Aces." Thus the four Twos can stand for any cards, the Jack of Spades becomes a second King of Spades, and all Hearts are Aces of Hearts, including the Jack which first becomes a King and then an Ace. How many five-card poker hands are straight flushes under these rules?

Solution:

We consider five cases.

Case 1. The hand contains all 4 Twos.

Whatever the fifth card is, the hand is a straight flush. There are 48 such hands.

Case 2. The hand contains 3 Twos.

There are no straight flushes for Hearts. For each of Diamonds and Clubs, there are 10 possible straight flushes without deuces. From (A,K,Q,J,10), we may have any of the $\binom{5}{2} = 10$ subset of 2 cards. From (K,Q,J,10,9), in order to avoid duplication, we must choose 9 plus 1 of the other 4 cards in $\binom{4}{1} = 4$ ways. This applies to remaining possible straight flushes, bringing the total to $10 + 9 \times 4 = 46$. For Spades. we must eliminate the pair (K,J). It follows that the grand total in this case is $2 \times 46 + 45 = 137$.

Case 3. The hand contains 2 Twos.

There are no straight flushes for Hearts. For each of Diamonds and Clubs, there are 10 possible straight flushes without deuces. From (A,K,Q,J,10), we may have any of the $\binom{5}{3} = 10$ subsets of 3 cards. From (K,Q,J,10,9), in order to avoid duplication, we must choose 9 plus 2 of the other 4 cards in $\binom{4}{2} = 6$ ways. This applies to the remaining possible straight flushes, bringing the total to $10 + 9 \times 6 = 64$. For Spades, we must eliminate the triples containing both K and J, namely (A,K,J), (K,Q,J), (K,J,10) and (K,J,9). It follows that the grand total in this case is $2 \times 64 + 60 = 188$.

Case 4. The hand contains only 1 Two.

There are no straight flushes for Hearts. For the other three suits, the other four cards of the hand are either consecutive or with a single gap in between. For each of Diamonds and Clubs, there are 11 sets of 4 consecutive cards and 10 sets of 5 consecutive cards. Any of the middle 3 cards in 5 consecutive cards can be the gap. Hence the total number of straight flushes in Diamonds or Clubs is $11 + 10 \times 3 = 41$. For Spades, there are 7 sets of four consecutive cards and 6 sets of 5 consecutive cards, yielding $7 + 6 \times 3 = 25$. To these, we must add the 4-card sets (A,K,Q,10), (A,Q,J,10), (K,Q,10,9), (Q,J,10,9) and (Q,10,9,8). It follows that the grand total in this case is 112.

Case 5. The hand contains 0 Twos.

The numbers of straight flushes are 6 for Spades, 0 for Hearts and 10 for each of Diamonds and Clubs, for a total of 26.

The final answer is $26\binom{4}{0} + 112\binom{4}{1} + 188\binom{4}{2} + 137\binom{4}{3} + 48\binom{4}{4} = 2198$ since the 4 Twos are distinct.

Section B.2.

Example B.2.1.

Find the generating function for the sequence $\{n^2\}$.

Solution:

The generating function is $A(x) = x + 4x^2 + 9x^3 + 16x^4 + \cdots$. Then we have $xA(x) = x^2 + 4x^3 + 9x^4 + \cdots$ so that $(1-x)A(x) = x + 3x^2 + 5x^3 + 7x^4 + \cdots$. Now $x(1-x)A(x) = x^2 + 3x^3 + 5x^4 + \cdots$ so that

$$(1-x)^2 A(x) = x + 2x^2 + 2x^3 + 2x^4 + \cdots = \frac{2}{1-x} - 2 - x.$$

It follows that $A(x) = \frac{2-(2+x)(1-x)}{(1-x)^3} = \frac{x(1+x)}{(1-x)^3}$.

Example B.2.2.

Find the sequence generated by the function $\frac{x(1+4x+x^2)}{(1-x)^4}$.

Solution:

Note that $\binom{-4}{n} = \frac{(-4)(-5)\cdots(-(n+3))}{n!} = (-1)^n\binom{n+3}{3}$. By Newton's Binomial

Theorem, $\frac{1}{(1-x)^4} = \sum_{n=0}^{\infty}\binom{-4}{n}(-x)^n = \sum_{n=0}^{\infty}\binom{n+3}{3}x^n$. It follows that

$$
\begin{aligned}
\frac{x(1+4x^2+x^3)}{(1-x)^4} &= \sum_{n=1}^{\infty}\binom{n+2}{3}x^n + 4\sum_{n=2}^{\infty}\binom{n+1}{3}x^2 + \sum_{n=3}^{\infty}\binom{n}{3}x^n \\
&= x+4x^2+4x^2 + \sum_{n=3}^{\infty}\left(\binom{n+2}{3}+4\binom{n+1}{3}+\binom{n}{3}\right)x^n \\
&= \sum_{n=0}^{\infty}n^3x^n.
\end{aligned}
$$

Thus the sequence generated is $\{n^3\}$.

Example B.2.3.

Let a_n be the number of binary sequences of length n which do not contain 101. Find a recurrence relation for $\{a_n\}$.

Solution:

Clearly, $a_0 = 1$, $a_1 = 2$ and $a_2 = 4$. For $n \geq 3$, the number of such sequences ending in 00 is a_{n-2} because the first $n-2$ terms form a sequence counted in a_{n-2}. If such a sequence ends in 01, it must end in 001, and there are a_{n-3} such sequences. Consider now sequences ending in 10 or 11. The first $n-1$ terms form a sequence counted in a_{n-1} ending in 1. The number of sequences of length $n-1$ and ending in 0 is a_{n-2}. Hence the number of such sequences is $2(a_{n-1} - a_{n-2})$. Combining these preliminary results, we have
$a_n = a_{n-2} + a_{n-3} + 2(a_{n-1} - a_{n-2}) = 2a_{n-1} - a_{n-2} + a_{n-3}$.

Example B.2.4.

Solve the recurrence relation $a_n - a_{n-1} - a_{n-3} + a_{n-4} = 0$ with initial conditions $a_0 = 0$, $a_1 = 2$, $a_2 = 1$ and $a_3 = 3$.

Solution:

The characteristic equation is $0 = x^4 - x^3 - x + 1 = (x-1)^2(x^2+x+1)$. Hence the character roots are 1, 1 and $\frac{1\pm\sqrt{3}}{2}$. It follows that the general solutions is $a_n = C_1 + C_2 n + C_3\cos\frac{2n\pi}{3} + C_4\sin\frac{2n\pi}{3}$. From the initial conditions,

$$
\begin{aligned}
0 &= C_1 && &+& &C_3 && &(1) \\
2 &= C_1 &+& C_2 &-& \tfrac{1}{2}C_3 &+& \tfrac{\sqrt{3}}{2}C_4 & &(2) \\
1 &= C_1 &+& 2C_2 &-& \tfrac{1}{2}C_3 &-& \tfrac{\sqrt{3}}{2}C_4 & &(3) \\
3 &= C_1 &+& 3C_2 &+& C_3 && & &(4)
\end{aligned}
$$

Adding (2) and (3), we have $3 = 2C_1 + 3C_2 - C_3$. Comparing this with (4), we have $C_1 = 2C_3$. Combining this with (1), we have $C_1 = C_3 = 0$. Hence $C_2 = 1$ and $C_4 = \frac{2}{\sqrt{3}}$, so that $a_n = n + \frac{2}{\sqrt{3}}\sin\frac{2n\pi}{3}$.

Section B.3.

Example B.3.1.
Find the exponential generating function for the sequence $\{1 \cdot 4 \cdots (3n+1)\}$.

Solution:
The exponential generating function is

$$\sum_{n=0}^{\infty} 1 \cdot 4 \cdots (3n+1) \frac{x^n}{n!}$$

$$= 1 + \sum_{n=1}^{\infty} \frac{\left(-\frac{4}{3}\right)\left(-\frac{4}{3}-1\right)\cdots\left(-\frac{4}{3}-n+1\right)}{n!}(-3x)^n$$

$$= (1-3x)^{-\frac{4}{3}}.$$

Example B.3.2.
Find the sequence generated exponentially by the function $\frac{1}{1-x}$.

Solution:
The function $displaystyle\frac{1}{1-x} = \sum_{n=0}^{\infty} x^n = \sum_{n=0}^{\infty} n!\frac{x^n}{n!}$ generates exponentially the sequence $\{n!\}$.

Example B.3.3.
Find the numbers of letter strings of length 1, 2, 3 and 4 that can be spelled with the letters of METAMORPHIC.

Solution:
The exponential generating function for M $1 + x + \frac{1}{2}x^2$, and that for each of the other letters is $1 + x$. Thus the overall exponential generating function is

$$(1+x)^9 \left(1 + x + \frac{1}{2}x^2\right)$$

$$= (1 + 9x + 36x^2 + 84x^3 + 126x^4 + \cdots)\left(1 + x + \frac{1}{2}x^2\right)$$

$$= 1 + 10x + \frac{91}{2}x^2 + \frac{249}{2}x^3 + 228x^4 + \cdots$$

$$= 1 + 10x + 91\frac{x^2}{2!} + 747\frac{x^3}{3!} + 5472\frac{x^4}{4!} + \cdots.$$

The respective numbers are 10, 91, 747 and 5472.

Example B.3.4.
A deck of 16 cards consists of each of the numbers 1, 2, 3 and 4 in red, tan, blue and grey. The deck is shuffled and, as they are dealt one at a time face up on top of a pile, the dealer counts 3 for the first card, 2 for the second card and 1 for the third card. As soon as a number called matches the number on the card dealt, a new pile is started and the process is repeated.

If there are no matches after the number 1 is called, the next card is placed face down on top of the pile. Altogether four piles are formed, with possibly some cards unused.

(a) Prove that the sum of the face-up cards on top of the piles is equal to the number of unused cards.

(b) The result in (a) is trivial if no matches occur at all — there are no face-up cards and all cards are used. Prove that the probability of such an anti-climax is just over $\frac{3}{100}$.

Solution:

(a) If no matches occur in forming a pile, 4 cards are used up and there is no face-up card. If a match occurs at the number k, $1 \le k \le 3$, then the face-up card of the pile is k and only $4 - k$ cards are used up. Thus the face-up card gives the correct number of unused cards in forming this pile. Since this is true for all four piles, the sum of the numbers on the face-up cards is indeed equal to the number of unused cards.

(b) A diagram of the forbidden positions is shown below.

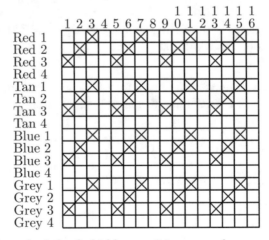

The chessboard of the forbidden positions may be represented as the disjoint union of three 4×4 chessboards. On each chessboard, the number of ways of placing k rooks, $0 \le k \le 4$, is equal to $\binom{4}{k}^2 k!$. Hence the overall rook polynomial is

$$(1 + 16x + 72x^2 + 96x^3 + 24x^4)^3$$
$$= 1 + 48x + 984x^2 + 11296x^3 + 80136x^4 + 366336x^5 + 1093248x^6$$
$$+ 2115072x^7 + 2586816x^8 + 1907712x^9 + 787968x^{10}$$
$$+ 165888x^{11} + 13824x^{12}.$$

It follows from the Principle of Inclusion-Exclusion that the number of permutations of the 16 cards without producing any matches is

$$16! - 48 \times 15! + 984 \times 14! - 11296 \times 13! + 80136 \times 12!$$
$$-366336 \times 11! + 1093248 \times 10! - 2115072 \times 9! + 2586816 \times 8!$$
$$-1907712 \times 7! + 787968 \times 6! - 165888 \times 5! + 13824 \times 4!$$
$$= 20922789888000 - 62768369664000 + 85783438540800$$
$$-70340426956800 + 38385272217600 - 14622960844800$$
$$+3967178342400 - 767517327360 + 104300421120$$
$$-9614868480 + 567336960 - 19906560 + 331776$$
$$= 654637510656.$$

Hence the probability of an anti-climax is $\frac{654637510656}{20922789888000} \approx 0.031$, which is just over $\frac{3}{100}$.

Section B.4.

Example B.4.1.
Prove that the number of sequences of 1s, 2s and 3s with sum n, such that no 2 immediately follows a 1 is given by the Fibonacci sequence.

Solution:
Let a_n be the desired number. Then $a_0 = 1$ for the empty sequence, $a_1 = 1$ for the sequence $\{1\}$, and $a_2 = 2$ for the sequences $\{1,1\}$ and $\{2\}$. For $n \geq 3$, the number of sequences with sum n and starting with 2 is a_{n-2}, since any of the a_{n-2} sequences with sum $n-2$ can follow a 2 to yield a sequence with sum n. Similarly, the number of sequences with sum n and starting with 3 is a_{n-3}. For those starting with 1, it can be followed by any of the a_{n-1} sequence except those which starts with 2, and there are $a_{(n-1)-2} = a_{n-1}$ of them. It follows that $a_n = a_{n-2} + a_{n-3} + (a_{n-1} - a_{n-3}) = a_{n-1} + a_{n-2}$. The initial values indicate that $\{a_n\}$ is the Fibonacci sequence itself.

Example B.4.2.
Prove that $F_1 + 2F_2 + \cdots + nF_n = (n+1)F_{n+2} - F_{n+4} + 3$.

Solution:
Using $F_1 + F_2 + \cdots + F_n = F_{n+2} - 2$, we have

$$3 + (n+1)F_{n+2} - (F_1 + 2F_2 + \cdots + nF_n)$$
$$= (F_1 + F_2) + (n+1)(F_1 + F_2 + \cdots + F_n + 2) - (F_1 + 2F_2 + \cdots + nF_n)$$
$$= (F_1 + F_2) + (nF_1 + (n-1)F_2 + \cdots + F_n) + 2n + 2$$
$$= F_1 + F_2 + (F_1+2) + (F_1+F_2+2) + \cdots + (F_1+F_2+\cdots+F_n+2) + 2$$
$$= F_1 + F_2 + F_3 + F_4 + \cdots + F_{n+2} + 2$$
$$= F_{n+4}.$$

Example B.4.3.
Find a five word English phrase such that, according to how the phrase is bracketed, the number of different interpretations is the Catalan number $C_4 = 14$.

Solution:
A possible phrase is *pretty little brown doll house*. The fourteen interpretations are:

1. *pretty ((light brown) (doll house))*, meaning
   an attractive light-brown house for a doll.

2. *pretty (light ((brown doll) house))*, meaning
   an attractive and light-weight house for a brown doll.

3. *pretty ((light (brown doll)) house)*, meaning
   an attractive house for a light-weight brown doll.

4. *pretty (light (brown (doll house)))*, meaning
   an attractive light-weight brown house for a doll.

5. *pretty (((light brown) doll) house)*, meaning
   an attractive house for a light-brown doll.

6. *((pretty light) (brown doll)) house*, meaning
   a house for a very light-weight brown doll.

7. *(pretty ((light brown) doll)) house*, meaning
   a house for an attractive light-brown doll.

8. *((pretty (light brown)) doll) house*, meaning
   a house for a doll in an attractive light-brown color.

9. *(pretty (light (brown doll))) house*, meaning
   a house for an attractive light-weight brown doll.

10. *(((pretty light) brown) doll) house*, meaning
    a house for a very light-brown doll.

11. *((pretty light) brown) (doll house)*, meaning
    a very light-brown house for a doll.

12. *(pretty (light brown)) (doll house)*, meaning
    a house in an attractive light-brown color for a doll.

13. *(pretty light) ((brown doll) house)*, meaning
    a very light-weight house for a brown doll.

14. *(pretty light) (brown (doll house))*, meaning
    a very light-weight brown house for a doll.

Example B.4.4.
What is the minimum number of people in a party such that there exist four people A, B, C and D,

(a) either with A knowing B, B knowing C, C knowing A and D knowing A, or with A not knowing B, B not knowing C, C not knowing A, D not knowing A;

(b) either with A knowing B, B knowing C, C knowing D, D knowing A and A knowing C, or with A not knowing B, B not knowing C, C not knowing D, D not knowing A and A not knowing C?

Solution:

(a) We may have six people P_1, P_2, P_3, Q_1, Q_2 and Q_3, with two people knowing each other if and only if they share the same letter. There are no triples A, B and C satisfying A not knowing B, B not knowing C and C not knowing A. The only triples A, B and C satisfying A knowing B, B knowing C and C knowing A are those sharing the same letter, but none of them knows any of the other three people. Hence six people are not enough. Suppose we have seven people P, Q, R, S, T, U and V. Since $R(3,3) = 6$, we may assume by symmetry that there exist three people, say P, Q and R, who know one another. If any of them knows any of the other four, adding the new person yields a desired quartet. Hence none of P, Q and R knows any of S, T, U and V. If S, T, U and V all know one another, they form a desired quartet. If not, we may assume by symmetry and S does not know T. Then any quartet including S, T and at least one of P, Q and R has the desired property.

(b) We may have nine people P_1, P_2, P_3, Q_1, Q_2, Q_3, R_1, R_2 and R_3, with two people knowing each other if and only if they share the same letter or the same subscript. Then a typical quartet satisfying A knowing B, B knowing C, C knowing D and D knowing A consists of P_1, P_2, Q_2 and Q_1, but P_1 does not know Q_2 and P_2 does not know Q_1. There are two typical quartets satisfying A not knowing B, B not knowing C, C not knowing D and D not knowing A. However, in each case, we have A knowing C and B knowing D. It follows that 9 people are not enough. Suppose we have ten people. Consider any of them, say P. We may assume by symmetry that P knows five or more of the other nine people. Call them Q, R, S, T and U. If one of them knows two or the other four, say Q knowing R and S, and P, Q, R and S is a desired quartet. If not, there are at most two pairs of acquaintances among the five, say Q knows R and S knows T. Then any quartet including U will have the desired property.

Section B.5.

Example B.5.1.
A 2×3 rectangle is divided into 6 unit squares, each of which is painted black or white. Find the number of patterns under all symmetries.

Solution:
The symmetries are the identity I, the half-turn R, the reflection H about the horizontal axis and the reflection V about the vertical axis. Under I, each of the 6 cells forms a cycle of length 1. Under R or V, the 6 cells form 3 cycles of length 2. Under H, the 2 cells on the axis of reflection form 2 cycles of length 1 while the other cells form 2 cycles of length 2. The cycle index is $\frac{x_1^6 + 2x_2^3 + x_1^2 x_2^2}{4}$. The number of coloring patterns is $\frac{2^6 + 2 \times 2^3 + 2^4}{4} = 24$ when we set $x_1 = x_2 = 2$.

Example B.5.2.
The truncated dodecahedron is a semi-regular solid in which every vertex is surrounded by one equilateral triangle and two regular decagons. A planar representation is shown in the diagram below.

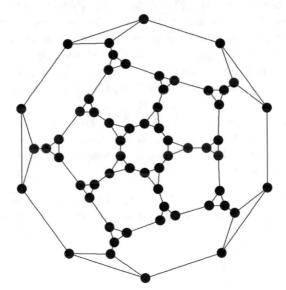

Determine the cycle index if its faces are to be painted.

Solution:
Each of its decagonal faces may be mapped onto any of the twelve decagonal faces in five different ways, yielding a total of $12 \times 5 = 60$ symmetries. The breakdown is as follows.

(1) The identity symmetry.

(2) Twelve $72°$ rotational symmetries about axes joining the centers of opposite decagonal faces.

(3) Twelve $144°$ rotational symmetries about axes joining the centers of opposite decagonal faces.

(4) Fifteen $180°$ rotational symmetries about axes joining the midpoints of opposite sides between two decagonal faces.

(5) Twenty $120°$ rotational symmetries about axes joining opposite triangular faces.

The cycle index is $\frac{x_1^{32}+24x_1^2x_5^6+15x_2^{16}+20x_3^{10}}{60}$.

Example B.5.3.

The truncated icosahedron is a semi-regular solid in which every vertex is surrounded by one regular pentagon and two regular hexagons. A planar representation is shown in the diagram below.

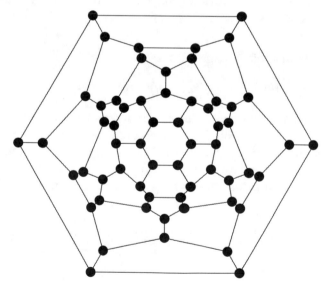

Determine the cycle index if its faces are to be painted.

Solution:

Each of its hexagonal faces may be mapped onto any of the twenty hexagonal faces in three different ways, yielding a total of $20 \times 3 = 60$ symmetries. The breakdown is as follows.

(1) The identity symmetry.

(2) Twenty 120° rotational symmetries about axes joining the centers of opposite hexagonal faces.

(3) Fifteen 180° rotational symmetries about axes joining the midpoints of opposite sides between two hexagonal faces.

(4) Twelve 72° rotational symmetries about axes joining opposite pentagonal faces.

(5) Twelve 144° rotational symmetries about axes joining opposite pentagonal faces.

The cycle index is $\frac{x_1^{32}+20x_1^2x_3^{10}+15x_2^{16}+24x_1^2x_5^6}{60}$.

Example B.5.4.
A square is divided into four square cells. Each cell is painted red, yellow, blue and green. A color-blind person cannot distinguish between red and yellow, and cannot distinguish between blue and green, though he can tell that they are different. Up to rotations of the square, how many different patterns can this person distinguish?

Solution:
There are four relevant color permutations:

$$(r)(y)(b)(g), (r)(y)(bg), (ry)(b)(g), (ry)(bg).$$

The first means all colors stay put. The second means red and yellow are interchanged. The third means blue and green are interchanged. The fourth means that both interchanges take place. Combining with the four rotational symmetries of the square I, R, A and C, we have a total of sixteen symmetries. We calculate the total number of invariants in the following chart.

Invariants	$(r)(y)(b)(g)$	$(r)(y)(bg)$	$(ry)(b)(g)$	$(ry)(bg)$
I	4^{16}	4^2	4^2	4^1
R	4^2	4^1	4^1	4^1
A	4^1	4^1	4^1	4^1
C	4^1	4^1	4^1	4^1

The total number of distinct patterns is given by

$$\frac{256 + 16 + 16 + 16 + 4 + 4 + 4 + 4 + 4 + 4 + 4 + 4 + 4 + 4 + 4 + 4}{16} = 22.$$

Section B.6.

Example B.6.1.

Ten copies of each of the numbers 0, 1, 2, ..., 9 are used to fill in a 10×10 square to form a Latin square. Is is always possible to complete the task if someone else has already placed

(a) 98;

(b) 10

numbers without having two equal numbers in the same row or the same column?

Solution:

(a) The table below cannot be completed into a Latin square.

1		2	3	4	5	6	7	8	9
	1	3	4	5	6	7	8	9	0
2	3	4	5	6	7	8	9	0	1
3	4	5	6	7	8	9	0	1	2
4	5	6	7	8	9	0	1	2	3
5	6	7	8	9	0	1	2	3	4
6	7	8	9	0	1	2	3	4	5
7	8	9	0	1	2	3	4	5	6
8	9	0	1	2	3	4	5	6	7
9	0	1	2	3	4	5	6	7	8

(b) The table below cannot be completed into a Latin square.

0									
	1								
		1							
			1						
				1					
					1				
						1			
							1		
								1	
									1

Example B.6.2.
In a (b, v, r, k, λ)-design where every k subset of the v elements of size k is a block, determine b, r and λ in terms of v and k.

Solution:
Clearly, $b = \binom{v}{k}$. Consider any particular element. The other $k-1$ elements appearing in the same block may be chosen in $\binom{v-1}{k-1}$ ways. Since all k-subsets are blocks, $r = \binom{v-1}{k-1}$. Consider any particular pair of elements. The other $k-2$ elements appearing in the same block may be chosen in $\binom{v-2}{k-2}$ ways. Since all k-subsets are blocks, $\lambda = \binom{v-2}{k-2}$.

Example B.6.3.
Give a simple proof of Fisher's Inequality for any $(b, v, r, k, 1)$-design.

Solution:
Consider a particular block. No two of its k elements can appear together elsewhere. Since each must appear another $r - 1$ times, there are at least $k(r - 1)$ blocks other than the chosen one. It follows that

$$b \geq 1 + k(r - 1).$$

From

$$v - 1 = r(k - 1),$$

we have

$$r + r^2(k - 1) = vr = bk \geq k + k^2(r - 1).$$

This may be rewritten as

$$(r - k)(r - 1)(k - 1) \geq 0.$$

Since $r > 1$ and $k > 1$, we have $r \geq k$ so that $b \geq v$.

Example B.6.4.
Fourteen evenly spaced points on a circle are labeled $1, 2, \ldots, 14$ in clockwise order. The center of the circle is labeled 15. The length of a chord is measured by the number of arcs between adjacent points which are subtended by the chord. Thus the possible chord lengths are 1, 2, 3, 4, 5 and 6, plus 7 for a diameter. Chords other than diameters are classified as odd-chords and even-chords, according to the parity of the number on the first endpoint of the chord in clockwise order. The diagram below shows an odd-chord and an even-chord of length 3, as well as an odd-chord and an even chord of length 6.

(a) Use all 15 points, 13 chords with distinct lengths and parity, as well as two radii, to construct five triangles.

(b) Use the five triangles in (a) to construct a Kirkman triple system.

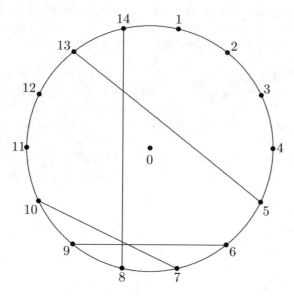

Solution:

(a) A construction is shown in the diagram below. The five triangles are
(0,2,3), (1,9,13), (4,8,11), (5,6,14) and (7,10,12). The triangle with the
diameter (4,11) as a side is shaded.

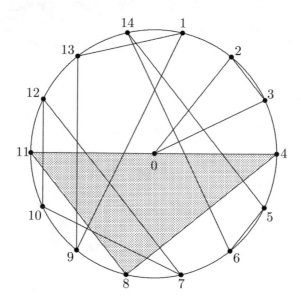

(b) Using the five triangles in (a), we construct the following Kirkman triple system by rotating the triangles through an angle subtended by two adjacent arcs each joining two adjacet points.

(0,2,3)	(1,9,13)	(4,8,11)	(5,6,14)	(7,10,12)
(0,4,5)	(3,11,1)	(6,10,13)	(7,8,2)	(9,12,14)
(0,6,7)	(5,13,3)	(8,12,1)	(9,10,4)	(11,14,2)
(0,8,9)	(7,1,5)	(10,14,3)	(11,12,6)	(13,2,4)
(0,10,11)	(9,3,7)	(12,2,5)	(13,14,8)	(1,4,6)
(0,12,13)	(11,5,9)	(14,4,7)	(1,2,10)	(3,6,8)
(0,14,1)	(13,7,11)	(2,6,9)	(3,4,12)	(5,8,10)

Appendix C

Solutions to Odd-numbered Exercises

Section 0.1.

1. There are obviously 5 odd-looking number with only one digit, namely, 1, 3, 5, 7 and 9. There are $5^2 = 25$ odd-looking numbers with exactly two digits since there are 5 choices for each digit. Similarly, there are $5^3 = 125$ odd-looking numbers with exactly three digits, and $5^4 = 625$ odd-looking numbers with exactly four digits. Hence the total count is $5 + 25 + 125 + 625 = 780$.

3. Let S denote the set of students in that class. Let C, L and A denote the subsets of S consisting of those passing calculus, linear algebra and abstract algebra respectively. Then $|S|=12$, $|C|=8$, $|L|=6$, $|A|=5$, $|C \cap L| = 5$, $|L \cap A| = 3$, $|A \cap C| = 4$ and $|C \cap L \cap A| = 3$. By the Principle of Inclusion-Exclusion, the number of students who failed all three courses is

$$
\begin{aligned}
|\overline{C} \cap \overline{L} \cap \overline{A}| &= |S| - |C| - |L| - |A| \\
&\quad + |C \cap L| + |L \cap A| + |A \cap C| - |C \cap L \cap A| \\
&= 12 - 8 - 6 - 5 + 5 + 3 + 4 - 3 \\
&= 2.
\end{aligned}
$$

5. Let S be the set of positive integers up to 210 that are divisible by 2 and 7. Let A and B be the subsets of S consisting of those divisible by 3 and 5, respectively. Then $|S| = \frac{210}{2 \times 7} = 15$, $|A| = \frac{15}{3} = 5$, $|B| = \frac{15}{5} = 3$ and $|A \cap B| = \frac{15}{3 \times 5} = 1$. By the Principle of Inclusion-Exclusion, $|\overline{A} \cap \overline{B}| = 15 - 5 - 3 + 1 = 8$.

Section 0.2.

1. We consider all partitions of the parliament into two committees. For each partition, we count the total number m of enemies each member has within the committee. We claim that the partition for which m is minimum has the required property. Suppose to the contrary some member has at least two enemies within the committee. Then he would have at most one enemy in the other committee. By switching him to the other committee, m is reduced by at least 1, which is a contradiction.

3. Consider a member with a maximal number m of friends in the society. We claim that all her friends have different numbers of friends greater than 0 but less than or equal to m. There are m possibilities: $1, \ldots, m$ friends. Hence all possibilities are realized. In particular, there exists a person with exactly one friend.

5. Suppose there are positive integer solutions. Then there exists one in which z is minimum. Note that x must be divisible by 3, say $x = 3r$ for some integer r. Then $(3r)^3 = 3y^3 + 9z^3$ or $9r^3 = y^3 + 3z^3$. Now $y = 3s$ for some integer s, and similarly $z = 3t$ for some positive integer t. Now $r^3 = 3s^3 + 9t^3$ and (r, s, t) is another solution with $t < z$, a contradiction.

Section 0.3.

1. We first show that 87 is not enough. We may have in the bag 75 green, 12 red, 12 white and 12 blue balls. The total number of balls of any three colors is at most 99. If 100 are drawn at random, there will be 4 balls of different colors. Hence the requirement is satisfied. Now if we draw only 87 balls, we may end up with 75 green and 12 white balls. We now show that 88 is enough. By symmetry, we may assume that the numbers of green, red, white and blue balls is non-increasing. We must have at least 12 blue balls as otherwise we may not have a blue one when we draw 100 balls. Hence there are at least 24 white and blue balls, meaning that the total number of balls of any two colors is at most $111 - 24 = 87$. The desired result follows immediately.

3. We have $(8-3+1)^2 = 36$ different 3×3 subboards and $(8-4+1)^2 = 25$ different 4×4 subboards. Starting with a board containing all 0s, we can generate at most $10^{36+25} = 10^{61}$ differently labeled 8×8 boards. However, there are altogether 10^{64} differently labeled 8×8 boards. Since the process is reversible, if we take as the starting position one of these labeled 8×8 boards we cannot generate from the board of all 0s, then the task is impossible.

5. The maximum is 32. If we paint every other row green, then there are 32 green squares. Since the shape must cover two squares in the same column, exactly one of them is not green, Suppose there are 33 or more green squares. Divide the board into sixteen 2×2 subboards. Then the average number of green squares in a subboard is $\frac{33}{16} > 2$. Hence 3 green squares must be in the same subboard. The given shape can cover these 3 squares, which is a contradiction.

Section 0.4.

1. The holes are the possible numbers of games played by any team so far. The pigeons are the teams. Let there be k teams. Suppose some team has played every other team. Then no team may have played zero games. Hence the number of holes is $k - 1$. On the other hand, if no team has played every other team, then there are also only $k - 1$ holes. It follows that two pigeons must be in the same hole, meaning that two teams must have played the same number of games so far.

3. Let the given integers be a_1, a_2, ..., a_{10}. Consider the ten sums $s_1 = a_1$, $s_2 = a_1 + a_2$, $s_3 = a_1 + a_2 + a_3$, ..., $s_{10} = a_1 + a_2 + \cdots + a_{10}$. If any of these sums is divisible by 10, there is nothing further to prove. Otherwise, all the remainders are different from 0 when these sums are divided by 10. Since there are only 9 such remainders, two of the sums, say s_i and s_j with $i < j$, have the same remainder. The difference between these two sums is divisible by 10. This means that $a_{i+1} + a_{i+2} + \cdots + a_j = s_j - s_i$ is divisible by n.

5. (a) There are 6 such partitions: 7+1, 5+3, 8, 6+2, 5+2+1 and 4+3+1.

   (b) There are 6 such partitions: 7+1, 5+3, 1+1+1+1+1+1+1+1, 3+3+1+1, 5+1+1+1 and 3+1+1+1+1+1.

   (c) We shall exhibit a one-to-one correspondence between the two types of partitions. Starting with one with distinct terms, if any term is even, split it into 2 equal parts. Continue until no even terms are left. What we have is a partition into only odd terms. For instance,
   8=4+4=2+2+2+2=1+1+1+1+1+1+1+1,
   6+2=3+3+1+1,
   5+2+1=5+1+1+1 and
   4+3+1=2+2+3+1=3+2+2+1=3+1+1+1+1+1.
   Conversely, starting with a partition into only odd terms, combine 2 equal terms into one. Continue until no equal terms are left. This reversal of the above process yields a partition into distinct terms. For instance,
   1+1+1+1+1+1+1+1=2+2+2+2=4+4=8,
   3+3+1+1=6+2,
   5+1+1+1=5+2+1 and
   3+1+1+1+1+1=3+2+2+1=3+4+1=4+3+1.

Section 0.5.

1. More generally, we prove that the game can be won when the deck has n cards for any positive integer n. The basis $n = 1$ is trivial as we win immediately with only one card. Suppose the game can be won for some $n \geq 1$. Consider the game with $n + 1$ cards. Suppose card number $n + 1$ starts at the bottom. Then it will never leave its position, and we are really playing with a deck of n cards. By the induction hypothesis, the game will be won.

   Suppose card number j starts at the bottom for some $j \leq n$. It will stay there unless it is bumped up, and only card number ℓ can do so. For the time being, let us pretend that card number $n + 1$ is card number j. It will not be unmasked since the real card number j is hidden safely away at the bottom of the deck, unless the fake card number j surfaces to the top. If that does not happen, card number j is buried at the bottom forever, and once again, we are playing with a deck of n cards. By the induction hypothesis, the game will be won. If card number $n + 1$ comes to the top, it will go to the bottom at the next move, and we are back to the first case which we have already resolved. Note that if $j = 1$, then card number $n + 1$, serving as the fake card number 1, must come to the top by the induction hypothesis. This completes the induction argument.

3. (a) For $n = 1$, the left side is equal to $2^1 = 2$ while the right side is equal to $2^2 - 2 = 2$. Assume that the result holds for some $n \geq 1$. Then $2 + 2^2 + \cdots + 2^n + 2^{n+1} = 2^{n+1} - 2 + 2^{n+1} = 2^{n+2} - 2$. By mathematical induction, the result holds for all $n \geq 1$.

   (b) For $n = 1$, the left side is equal to $1^2 = 1$ while the right side is equal to $\frac{1(1+1)(2\times 1+1)}{6} = 1$. Assume that the result holds for some $n \geq 1$. Then

$$
\begin{aligned}
1^2 &+ 2^2 + \cdots + n^2 + (n + 1)^2 \\
&= \frac{n(n + 1)(2n + 1)}{6} + (n + 1)^2 \\
&= \frac{n + 1}{6}(n(2n + 1) + 6(n + 1)) \\
&= \frac{(n + 1)(n + 2)(2n + 3)}{6}.
\end{aligned}
$$

By mathematical induction, the result holds for all $n \geq 1$.

5. First let us prove the base. If the number given equals 1 or 2, then the existence of the required representation is simple. Now denote the given number by n. and find the largest power of 2 not exceeding n. Let it be 2^m; that is, $2^m \leq n \leq 2^{m+1}$. The difference $d = n - 2^m$ is less than n, and also less than 2^m, since $2^{m+1} = 2^m + 2^m$. By the induction hypothesis, d can be represented as a sum of several different powers of 2, and it is clear that 2^m is too big to be included. Thus, adding 2^m, we get the required expression for n. The induction is complete.

Section 0.6.

1. The situation is simpler at the start, with only one amoeba. After one move, we have two amoebas. Suppose we assign the value 1 to the initial amoeba, x to the one going north and y to the one going east. After the move, the initial amoeba is replaced by the other two. If we want the total value of amoebas to remain 1, we must have $x + y = 1$. By symmetry, we may take $y = x$. The value of an amoeba is determined by its location. So we may assign values to the cells themselves, as shown in the diagram below.

$\frac{1}{16}$	$\frac{1}{32}$	$\frac{1}{64}$	$\frac{1}{128}$	$\frac{1}{256}$
$\frac{1}{8}$	$\frac{1}{16}$	$\frac{1}{32}$	$\frac{1}{64}$	$\frac{1}{128}$
$\frac{1}{4}$	$\frac{1}{8}$	$\frac{1}{16}$	$\frac{1}{32}$	$\frac{1}{64}$
$\frac{1}{2}$	$\frac{1}{4}$	$\frac{1}{8}$	$\frac{1}{16}$	$\frac{1}{32}$
1	$\frac{1}{2}$	$\frac{1}{4}$	$\frac{1}{8}$	$\frac{1}{16}$

The total value of the cells in the first row is $S = 1 + \frac{1}{2} + \frac{1}{4} + \frac{1}{8} + \cdots$. Then $2S = 2 + 1 + \frac{1}{2} + \frac{1}{4} + \frac{1}{8} + \cdots$. Subtracting the previous equation from this one, we have $S = 2$. Since each cell in the second row is half in value of the corresponding cell in the first row, the total value of the cells in the second row is 1. Similarly, the total values of the cells in the remaining rows are $\frac{1}{2}$, $\frac{1}{4}$, $\frac{1}{8}$, ... Hence the total value of the cells in the entire quadrant is 4. Note that the total value of the six prison cells is $2\frac{3}{4}$. Remember that the total value of the amoebas is the invariant 1. If the Great Escape is to be successful, the amoebas must fit into the non-prison cells with total value $1\frac{1}{4}$. Now each of the first row and the first column holds exactly one amoeba at any time. If the amoeba on the first row is outside the prison, its value is at most $\frac{1}{8}$. The remaining space with total value $\frac{1}{16} + \frac{1}{32} + \frac{1}{64} + \cdots = \frac{1}{8}$ must be wasted.

Similarly, we have to leave vacant cells in the first column with total value at least $\frac{1}{8}$. Since $1\frac{1}{4} - 2 \times \frac{1}{8} = 1$, we have no room to play at all. In order for the Great Escape to be successful, all cells outside the prison and not on the first row or first column must be occupied. However, this requires that the number of moves be infinite. Hence the Great Escape cannot be achieved in a finite number of moves.

3. Let the critical measure be the current number S of heads of the dragon.

   (a) With Prince Ivan acting alone, S either decreases by 21 or increases by 1981. Both are multiples of 7. Hence the remainder when S is divided by 7 is invariant. Since 100 and 0 leave different remainders when divided by 7, the task is impossible.

   (b) With Prince Igor acting alone, S either decreases by 21 or increases by 2007. Both are multiples of 3. Hence the remainder when S is divided by 3 is invariant. Since 100 and 0 leave different remainders when divided by 3, the task is also impossible.

   (c) The task is now possible. First, Prince Igor uses his sword once to change S to 2107. This can be reduced to 7 by using the King's sword. Now Prince Ivan uses his sword twice to change S to 3969. This can be be reduced to 0 using the King's sword.

5. If each of the 10 cells along a diagonal is infected, then the infection can spread on both sides one diagonal at a time, until the whole board is infected. We now prove that this is impossible with less than 10 cells infected initially. The diagram below shows that when a cell is infected by two neighbors, the perimeter of the infected areas is unchanged, whereas if it is infected by three or four neighbors, the perimeter actually goes down. So we take the total perimeter of all infected areas as our critical measure. It is initially at most $4 \times 9 = 36$ since we have at most 9 infected cells. If the whole board is infected, the value of the critical measure will be 40. This is impossible.

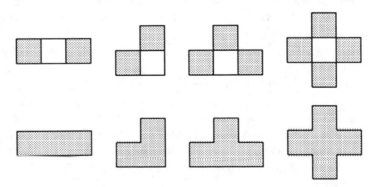

Section 1.1.

1. We have $\binom{n}{2} + \binom{n+1}{2} = \frac{n(n-1)}{2} + \frac{n(n+1)}{2} = \frac{n}{2}((n-1)+(n+1)) = n^2$.

3. Let $A = \{1, 4, \ldots, 1000\}$, $B = \{2, 5, \ldots, 998\}$ and $C = \{3, 6, \ldots, 999\}$. We can either choose three integers from the same set or one from each set. The total number of choices is $\binom{334}{3} + 2\binom{333}{3} + 334 \times 333^2$.

5. Initially, the plane is $\binom{n}{0} = 1$ region. When the first line is drawn, a new region is carved out. When the second line is drawn, it would have carved out a new region too, except that it intersects the first line, and the point of intersection divides the second line into two segments. Each segment divides an existing region into two. So two regions are gained, one because of the introduction of a line, and the other because of the introduction of a point of intersection. Since there are $\binom{n}{1}$ lines and $\binom{n}{2}$ points of intersection, the total number of regions is $\binom{n}{0} + \binom{n}{1} + \binom{n}{2} = \frac{n^2+n+2}{2}$.

Section 1.2.

1. There is a one-to-one correspondence between the solutions to $p+q+r = 52$ and the binary sequences of length 54, consisting of 52 0s and 2 1s. The number of 0s before the first 1 is the value of p, the number of 0s between the two ones is the value of q, and the number of 0s after the second 1 is the value of r. Clearly, the number of desired binary sequence is $\binom{52+2}{2} = 1431$.

3. Since $\binom{7}{3} = 35$, we use all 35 labels of length up to 5. The total point loss is

$$1 \times \binom{2}{2} + 2 \times \binom{3}{2} + 3 \times \binom{4}{2} + 4 \times \binom{5}{2} + 5 \times \binom{6}{2} = 140.$$

5. Since each player must contribute 1 to the total, we may replace the cards with the number 10 by cards with the number 0, and adjust the desired total to 23. Since at most two players can exceed their quotas, the desired number is given by $\binom{26}{3} - 4\binom{16}{3} + 6\binom{6}{3} = 480$.

Section 1.3.

1. From a pool of n players, we choose a team of arbitrary size and a captain from among the team members. There are $\binom{n}{k}$ ways of choosing a team of size k, and k ways to choose the captain from within the team.

    (a) Alternatively, we can choose the captain first, in n ways, and then the rest of the team, in 2^{n-1} ways.

(b) We claim that the number of teams of odd sizes is equal to the number of teams of even sizes. Consider all teams with the same captain. Since $n \geq 2$, we can choose an arbitrary player who is not the captain. For each team which includes this player, there is a corresponding team which excludes this player. Clearly, one team in each corresponding pair has odd size and the other has even size. This one-to-one correspondence establishes the desired result.

3. (a) By Example 1.3.2, we have

$$\binom{n+1}{1} + 2\binom{n+1}{2} + \cdots + m\binom{n+m}{m}$$

$$= \left(\binom{n+1}{1} - \binom{n+1}{0}\right) + 2\left(\binom{n+2}{2} - \binom{n+1}{1}\right) + \cdots$$

$$+ m\left(\binom{n+m+1}{m} - \binom{n+m}{m-1}\right)$$

$$= m\binom{n+m+1}{m} - \binom{n+1}{0} - \binom{n+2}{1} - \cdots - \binom{n+m}{m-1}$$

$$= m\binom{n+m+1}{m} - \binom{n+m+1}{m-1}$$

$$= \frac{m(n+1)}{n+2}\binom{n+m+1}{m}$$

(b) Let there be n boys and $m + 1$ girls numbered from 1 to $m + 1$. We wish to choose a team of size $n + 1$, and a manager from the remaining players provided that if this is a girl, her number must not be higher than that of every girl team member. We consider two different selection processes. In the first, let $k + 1$ be the highest number among the girl team members. Then $1 \leq k \leq m$. For each such k, the remaining n players may be chosen from the boys as well as the girls numbered from 1 to k, and the number of choices is $\binom{n+k}{n}$. The manager can then be chosen in k ways. Hence the total number of ways is the left side of the given expression. In the second process, we choose the team in $\binom{n+m+1}{n+1}$ ways, and the manager in m ways for a total of $m\binom{n+m+1}{m}$ ways. However, we have to eliminate those choices in which the manager is a girl with a higher number than that of any girl team member. We can choose $n + 2$ players and make the girl with the highest number the manager. This can be done in $\binom{n+m+1}{n+2} = \binom{n+m+1}{m-1}$ ways. Applying the Subtraction Rule, the total number of ways simplifies to the right side of the given expression.

5. We have n players from whom we choose a team of size k. Any number of the k team members may be a team captain. Suppose we first choose the captains and then the members. Let there be i captains, $0 \le i \le k$. The captains can be chosen in $\binom{n}{i}$ ways, and the remaining members can be chosen in $\binom{n-i}{k-i}$ ways. Summing from $i = 0$ to $i = k$ yields the left side. On the other hand, we can choose the members first in $\binom{n}{k}$ ways, and since any member can be a captain, the number of choices is 2^k.

Section 1.4.

1. Differentiating $(1+x)^n = \sum_{i=0}^{n} \binom{n}{i} x^i$ with respect to the variable x, we have $n(1+x)^{n-1} = \sum_{i=1}^{n} i \binom{n}{i} x^{i-1}$.

   (a) Setting $x = 1$ yields the desired result.

   (b) Setting $x = -1$ yields the desired result.

3. The given sum is the coefficient of x^n in $S = \sum_{i=1}^{m} i(1+x)^{n+i}$. Then

$$(1+x)S = \sum_{i=1}^{m} i(1+x)^{n+i+1} = \sum_{j=2}^{m+1} (j-1)(1+x)^{n+j}. \text{ We have}$$

$$\begin{aligned}
-xS &= S - (1+x)S \\
&= \sum_{i=1}^{m} (1+x)^{n+i} - m(1+x)^{n+m+1} \\
&= \frac{(1+x)^{n+m+1} - (1+x)^{n+1}}{x} - m(1+x)^{n+m+1}.
\end{aligned}$$

Hence $S = \frac{m}{x}(1+x)^{n+m+1} - \frac{1}{x^2}(1+x)^{n+m+1} + \frac{1}{x^2}(1+x)^{n+1}$. The coefficient of x^n in the first term is m times the coefficient of x^{n+1} in $(1+x)^{n+m+1}$ or $m\binom{n+m+1}{m}$. The coefficient of x^n in the second term is the same as the coefficient of x^{n+2} in $(1+x)^{n+m+1}$ or $\binom{n+m+1}{m-1}$. The coefficient of x^n in the third term is 0. Hence the coefficient of x^n in S is $\frac{m(n+1)}{n+2}\binom{n+m+1}{m}$.

5. We have $(1+2x)^n = (x + (1+x))^n = \sum_{i=0}^{n} \binom{n}{i}(1+x)^i$. The coefficient of the term x^k on the left side is $\binom{n}{k}2^k$ while that on the right side is $\sum_{i=0}^{k} \binom{n}{i}\binom{n}{k-i}$.

Section 1.5.

1. The desired term is $\binom{3}{0,1,1,1}(2x_2)(3x_3)(4x_4)$ and the desired coefficient is $6 \times 24 = 144$.

3. We change each term into a binary sequence as follows. Write down a number of 0s equal to the number of x_1s in the term. Insert a 1 after this block. Then write down a number of 0s equal to the number of x_2s, followed by another 1. Next, write down a number of 0s equal to the number of x_3s, followed by yet another 1. Finally, write down a number of 0s equal to the number of x_4s. Note that each binary sequence consists of 3 1s and 3 0s, and there are $\binom{3+3}{3} = 20$ such binary sequences.

5. The number of desired distributions is

$$
3\binom{6}{4,1,1} + 6\binom{6}{3,2,1} + \binom{6}{2,2,2} = 90 + 360 + 90 = 540.
$$

Section 1.6

1. More generally, suppose your probability of winning a game is p against the master and q against the novice, where $p < q$. There are three ways in which you can win the match. You may win all three games, win the first two only or win the last two only. If you start the match against the novice, your probability of winning the match is

$$
qpq + qp(1 - q) + (1 - q)pq = pq(2 - q).
$$

If you start the match against the master, your probability of winning the match is

$$
pqp + pq(1 - p) + (1 - p)qp = pq(2 - p).
$$

Since $p < q$, $2 - p > 2 - q$, and you are better off starting the match against the master.

3. The answer is negative. Here is a counter-example. In the first apartment building, there is only 1 pet on the lower floor, and it is a cat. There are 2 pets on the upper floor, 1 of which is a cat. Hence there are 3 pets in the building, 2 of which are cats. In the second apartment building, there are 11 pet on the lower floor, 10 of which are cats. There are also 11 pets on the upper floor, 5 of which are cats. Hence there are 22 pets in the building, 15 of which are cats. Now $1 > \frac{10}{11}$, $\frac{1}{2} > \frac{5}{11}$ but $\frac{2}{3} < \frac{15}{22}$.

5. The expected return is independent of the choice of the number. There are $6^3 = 216$ possible outcomes when three dice are rolled. In $5^3 = 125$ cases, the chosen number does not appear. In $\binom{3}{1} \times 5^2 = 75$ of these cases, it appears once. In $\binom{3}{2} \times 5 = 15$ cases, it appears twice. In only 1 case does the chosen number appear three times. Hence the expected gain is $(-1) \times \frac{125}{216} + 1 \times \frac{75}{216} = 2 \times \frac{15}{216} + 3 \times \frac{1}{216} = -\frac{17}{216} \approx -0.08$, and you expect to lose on the average 8 cents each time you play.

Section 2.1.

1. We have

$$\binom{-\frac{1}{2}}{n} = \frac{(-\frac{1}{2})(-\frac{1}{2} - 1) \cdots (-\frac{1}{2} - n + 1)}{n!}$$

$$= \left(-\frac{1}{2}\right)^n \frac{1 \cdot 3 \cdots (2n-1)}{n!}$$

$$= \left(-\frac{1}{2}\right)^n \frac{(2n)!}{(n!)^2 2^n}$$

$$= \left(-\frac{1}{4}\right)^n \binom{2n}{n}.$$

3. (a) We have

$$\frac{3}{(1+x)(1-2x)} = \frac{A}{1+x} + \frac{B}{1-2x}.$$

Clearing fractions, $3 = A(1 - 2x) + B(1 + x)$. Setting $x = -1$, $3 = 3A$ or $A = 1$. Setting $x = \frac{1}{2}$, $3 = \frac{3}{2}B$ or $B = 2$. Thus the sequence generated is $\{(-1)^n + 2^{n+1}\}$.

(b) We have

$$\frac{-8x^2}{(1-x)(1+x)(1-3x)} = \frac{A}{1-x} + \frac{B}{1+x} + \frac{C}{1-3x}.$$

Clearing fractions,

$$-8x^2 = A(1+x)(1-3x) + B(1-x)(1-3x) + C(1-x)(1+x).$$

Setting $x = 1$, $-8 = -4A$ or $A = 2$. Setting $x = -1$, $-8 = 8B$ or $B = -1$. Setting $x = \frac{1}{3}$, $-\frac{8}{9} = \frac{8}{9}C$ so that $C = -1$ Thus the sequence generated is $\{2 - (-1)^n - 3^n\}$.

(c) We have

$$\frac{2 - 3x + 3x^2}{(1+x)(1-x)^2} = \frac{A}{1+x} + \frac{B}{1-x} + \frac{C}{(1-x)^2}.$$

Clearing fractions,

$$2 - 3x + 3x^2 = A(1-x)^2 + B(1+x)(1-x) + C(1+x).$$

Setting $x = -1$, $8 = 4A$ or $A = 2$. Setting $x = 1$, $2 = 2C$ or $C = 1$. Setting $x = 0$, $2 = A + B + C$ so that $B = -1$ Thus the sequence generated is $\{2(-1)^n + n\}$.

5. (a) The sequence is $\{0, \frac{1}{\sqrt{2}}, 1, \frac{1}{\sqrt{2}}, 0, -\frac{1}{\sqrt{2}}, -1, -\frac{1}{\sqrt{2}}\}$, and repeats itself thereafter. Hence the numerator of its generating function is

$$\frac{1}{\sqrt{2}}(x + \sqrt{2}x^2 + x^3 - x^5 - \sqrt{2}x^6 - x^7) = \frac{1}{\sqrt{2}}x(1 + \sqrt{2}x + x^2)(1 - x^4)$$

while its denominator is

$$1 - x^8 = (1 + x^4)(1 - x^4) = (1 + \sqrt{2}x + x^2)(1 - \sqrt{2}x + x^2)(1 - x^4).$$

It follows that the generating function is $\frac{\frac{1}{\sqrt{2}}x}{1 - \sqrt{2}x + x^2}$.

(b) The sequence is $\{1, \frac{1}{\sqrt{2}}, 0, -\frac{1}{\sqrt{2}}, -1, -\frac{1}{\sqrt{2}}, 0, \frac{1}{\sqrt{2}}\}$, and repeats itself thereafter. Hence the numerator of its generating function is

$$\frac{1}{\sqrt{2}}(\sqrt{2} + x - x^3 - \sqrt{2}x^4 - x^5 + x^7) = \frac{1}{\sqrt{2}}(\sqrt{2} - x)(1 + \sqrt{2}x + x^2)(1 - x^4)$$

while its denominator is

$$1 - x^8 = (1 + x^4)(1 - x^4) = (1 + \sqrt{2}x + x^2)(1 - \sqrt{2}x + x^2)(1 - x^4).$$

It follows that the generating function is $\frac{1 - \frac{1}{\sqrt{2}}x}{1 - \sqrt{2}x + x^2}$.

Section 2.2.

1. Define $s = q - 5$ and $t = r + 5$. Then $0 \le s \le 20$ and $0 \le t \le 10$. Each triple (p, q, r) corresponds to a unique triple (p, s, t), and the sum of p, s and t is also 25. The generating functions for p, s and t are $\frac{1-x^6}{1-x}$, $\frac{1-x^{21}}{1-x}$ and $\frac{1-x^{11}}{1-x}$, respectively. The overall generating function is

$$(1 - x^6 - x^{11} - x^{21} + x^{17} + x^{27} + x^{32} - x^{38}) \sum_{n=0}^{\infty} \binom{n+2}{2} x^n.$$

To extract the coefficient of x^{25}, note that only the first five terms in the first factor are relevant. They are complemented by the terms x^{25}, x^{19}, x^{14}, x^4 and x^8 in the second factor. Thus the coefficient of x^{25} is given by $\binom{27}{2} - \binom{21}{2} - \binom{16}{2} - \binom{6}{2} + \binom{10}{2} = 51$. This is the exact expression obtained when we solve this problem by the Principle of Inclusion-Exclusion.

3. Our task is to divide the eight factors $x^2(1+x)^2(1+x+x^2)^2(1-x+x^2)^2$ into two products so that one product represents a 3-sector wheel and the other a 12-sector wheel. Each product must have a copy of x in order that only positive integers are used. Now the former must have one copy of $1+x+x^2$ while the latter must have the other copy of $1+x+x^2$ as well as both copies of $1+x$. There are three cases but only two solutions.

Case 1.

The 3-sector wheel is represented by

$$x(1+x+x^2)(1-x+x^2) = x + x^3 + x^5$$

while the 12-sector wheel is represented by

$$x(1+x)^2(1+x+x^2)(1-x+x^2) = x+2x^2+2x^3+2x^4+2x^5+2x^6+x^7.$$

Thus the labels on the 3-sector wheel are 1, 3 and 5, while the labels on the 12-sector wheel are 1, 2, 2, 3, 3, 4, 4, 5, 5, 6, 6 and 7.

Case 2.

This is impossible as the 3-sector wheel would be represented by

$$x(1+x+x^2)(1-x+x^2)^2 = x - x^2 + 2x^3 - x^4 + 2x^5 - x^6 + x^7.$$

Case 3.

The 3-sector wheel is represented by $x(1+x+x^2) = x+x^2+x^3$ while the 12-sector wheel die is represented by

$$x(1+x)^2(1+x+x^2)(1-x+x^2)^2 = x+x^2+x^3+2x^4+2x^5+2x^6+x^7+x^8+x^9.$$

Thus the labels on the 3-sector wheel are 1, 2 and 3, while the labelss on the 12-sector wheel are 1, 2, 3, 4, 4, 5, 5, 6, 6, 7, 8 and 9.

5. The number of partitions into odd terms is generated by

$$(1 + x + x^2 + \cdots)(1 + x^3 + x^6 + \cdots)(1 + x^5 + x^{10} + \cdots) \cdots$$
$$= \frac{1}{1-x} \cdot \frac{1}{1-x^3} \cdot \frac{1}{1-x^5} \cdots$$
$$= \frac{1-x^2}{1-x} \cdot \frac{1-x^4}{1-x^2} \cdot \frac{1-x^6}{1-x^3} \cdots$$
$$= (1+x)(1+x^2)(1+x^3) \cdots.$$

The last expression is precisely the generating function for the number of partitions into distinct terms. This establishes the desired equality.

Section 2.3.

1. We have

$$
\begin{array}{rcl}
a_n - 2a_{n-1} &=& n2^{n+1}, \\
2a_{n-1} - 2^2 a_{n-2} &=& (n-1)2^{n+1}, \\
2^2 a_{n-2} - 2^3 a_{n-3} &=& (n-2)2^{n+1}, \\
\cdots - \cdots &=& \cdots, \\
2^{n-1} a_1 - 2^n a_0 &=& 2^{n+1}, \\
\hline
a_n - 2^n &=& 2^{n+1} S,
\end{array}
$$

where $S = n + (n-1) + (n-2) + \cdots + 1 = \frac{n(n+1)}{2}$. It follows that $a_n = 2^{n+1} S + 2^n = 2^n(1 + n + n^2)$.

3. Construct a sphere with the center at the center of the block. The intersection of each cut with the sphere is a great circle. Let one of them be the equator. Project the southern hemisphere from the center to the tangent plane at the south pole. Each great half-circle becomes a straight line on the tangent plane. By Example 2.3.2, these $n-1$ lines divide the plane into $\frac{n^2 - n + 2}{2}$ regions. The number of regions on the whole sphere is $n^2 - n + 2$, which is also the number of pieces of cheese.

5. Let the number of sequences with an even number of 0s be a_n, and let the number of those with an odd number of 0s be b_n. Then $a_0 = 0$ and $b_0 = 0$. We can obtain a sequence counted in a_n by adding a 1, a 2 or a 3 to any sequence counted in a_{n-1}, adding a 0 to any sequence counted in b_{n-1}, as well as adding a 3 to any ternary sequence of length $n-1$ with an even number of 0s. By Example 2.3.3, we have $a_n = 3a_{n-1} + b_{n-1} + \frac{1}{2}(3^{n-1} + 1)$ and $b_n = a_{n-1} + 3b_{n-1} + \frac{1}{2}(3^{n-1} - 1)$. Hence

$$
\begin{aligned}
a_n + b_n &= 4(a_{n-1} + b_{n-1}) + 3^{n-1} \\
&= 3^2(a_{n-2} + b_{n-2}) + 4 \cdot 3^{n-2} + 3^{n-1} \\
&= \cdots \\
&= 4^n(a_0 + b_0) + 4^{n-1} + 4^{n-2} \cdot 3 + \cdots + 4 \cdot 3^{n-1} + 3^{n-1} \\
&= 4^n - 3^n.
\end{aligned}
$$

It follows that

$$
\begin{aligned}
a_n &= 2a_{n-1} + 4^{n-1} - 3^{n-1} + \frac{1}{2}(3^{n-1} + 1) \\
&= 2a_{n-1} + 4^{n-1} - \frac{1}{2}3^{n-1} + \frac{1}{2}.
\end{aligned}
$$

Iteration yields

$$
\begin{array}{rlllll}
a_n = & 2a_{n-1}+ & 4^{n-1}- & \tfrac{1}{2}3^{n-1}+ & \tfrac{1}{2}, \\
2a_{n-1} = & 2^2 a_{n-2}+ & 2\cdot 4^{n-2}- & \tfrac{1}{2}(2\cdot 3^{n-2})+ & \tfrac{1}{2}(2), \\
2^2 a_{n-2} = & 2^3 a_{n-3}+ & 2^2 4^{n-3}- & \tfrac{1}{2}(2^2 3^{n-3})+ & \tfrac{1}{2}(2^2), \\
\cdots = & \cdots+ & \cdots- & \cdots+ & \cdots, \\
2^{n-1}a_1 = & 2^n a_0+ & 2^{n-1}- & \tfrac{1}{2}(2^{n-1})+ & \tfrac{1}{2}(2^{n-1}); \\
\hline
a_n = & 2^n a_0+ & \frac{4^n-2^n}{2}- & \frac{3^n-2^n}{2}+ & \frac{2^n-1}{2}.
\end{array}
$$

Hence $a_n = \tfrac{1}{2}(4^n - 3^n + 2^n - 1)$. Alternatively, there are 4^n quaternary sequences of length n, 2^n of them consisting only of 2s and 3s. The remaining $4^n - 2^n$ sequences can be paired up so that the two sequences in each pair differ only in the first digit which is not a 2 and not a 3. Hence one of them has an odd number of 0s while the other has an even number of 0s. Thus the number of quaternary sequences with an even number of 0s is $\frac{4^n+2^n}{2}$. By Example 2.3.3, there are $\frac{3^n+1}{2}$ ternary sequences of length n with an even number of 0s. These may also be considered as quaternary sequences without 3s. Hence the desired total is the difference of these two quantities, namely, $\tfrac{1}{2}(4^n - 3^n + 2^n - 1)$.

Section 2.4.

1. The characteristic equation is $x^3 - 6x^2 + 11x - 6 = 0$. This may be factored as $(x - 1)(x - 2)(x - 3) = 0$. Hence the general solution is $a_n = C_1 + C_2 2^n + C_3 3^n$. From the initial conditions, we have

$$
\begin{array}{rcrcrcl}
C_1 & + & C_2 & + & C_3 & = & 1 \\
C_1 & + & 2C_2 & + & 3C_3 & = & 4 \\
C_1 & + & 4C_2 & + & 9C_3 & = & 12 \,.
\end{array}
$$

Eliminating C_1, we have $C_2 + 2C_3 = 3$ and $C_2 + 3C_3 = 4$. Eliminating C_2, we have $C_3 = 1$. Back substituting yields $C_2 = 1$ and $C_1 = -1$. Hence we have $a_n = -1 + 2^n + 3^n$.

3. We seek a particular solution of the form $a_n = An2^n + Bn^2 2^n$. Then

$$
\begin{aligned}
n2^{n+1} &= a_n - 2a_{n-1} \\
&= An2^n + Bn^2 2^n - 2A(n-1)2^{n-1} - 2B(n^2 - 2n + 1)2^{n-1} \\
&= (A - B)2^n + 2Bn2^n.
\end{aligned}
$$

Hence $A = B = 1$. The general solution is $a_n = C2^n + n2^n + n^2 2^n$. From $a_0 = 1$, we have $C = 1$ so that $a_n = 2^n(1 + n + n^2)$.

5. Let the octagon be $ABCDEFGH$ and let the bug start at A, so that E is the destination. From A, C or G, it takes an even number of moves to reach E. Let a_n denote the number of ways for the bug to get to E for the first time after $2n$ moves. Let b_n denote the same number except that the bug starts from C, or by symmetry from G. We have $a_1 = 0$ and $b_1 = 1$. For $n \geq 2$, we claim that $a_n = 2a_{n-1} + 2b_{n-1}$. This is because the bug's first two moves can only be $A - B - A$, $A - H - A$, $A - B - C$ or $A - H - G$. Similarly, we have $b_n = 2b_{n-1} + a_{n-1}$. Extending these recurrence relations backward to $n = 0$, we have $a_0 = -1$ and $b_0 = 1$. Combining these two recurrence relations, we obtain $\frac{1}{2}(a_n - 2a_{n-1}) = (a_{n-1} - 2a_{n-2}) + a_{n-2}$, which simplifies to $a_n = 4a_{n-1} - 2a_{n-2}$. The characteristic equation is $x^2 - 4x + 2 = 0$, with roots $2 \pm \sqrt{2}$. Hence $a_n = C_1(2 + \sqrt{2})^n + C_2(2 - \sqrt{2})^n$. We have $C_1 + C_2 = a_0 = -1$ and $C_1(2 + \sqrt{2}) + C_2(2 - \sqrt{2}) = a_1 = 0$. Solving this system of linear equations, we have $C_1 = \frac{1}{2}(\sqrt{2} - 1)$ and $C_2 = -\frac{1}{2}(\sqrt{2} + 1)$. Hence $a_n = \frac{1}{\sqrt{2}}((2 + \sqrt{2})^{n-1} - (2 - \sqrt{2})^{n-1})$ for $n \geq 1$, with $a_0 = 0$.

Section 2.5.

1. Let $G(x) = \displaystyle\sum_{n=0}^{\infty} a_n x^n$. We have

$$
\begin{aligned}
0 &= \sum_{n=3}^{\infty} a_n x^n - 6x \sum_{n=3}^{\infty} a_{n-1} x^{n-1} \\
&\quad + 11x^2 \sum_{n=3}^{\infty} a_{n-2} x^{n-2} - 6x^3 \sum_{n=3}^{\infty} a_{n-3} x^{n-3} \\
&= (G(x) - 1 - 4x - 12x^2) - 6x(G(x) - 1 - 4x) \\
&\quad + 11x^2(G(x) - 1) - 6x^3 G(x) \\
&= (1 - 6x + 11x^2 - 6x^3)G(x) - (1 - 2x - x^2).
\end{aligned}
$$

Hence $G(x) = \dfrac{1 - 2x - x^2}{1 - 6x + 11x^2 - 6x^3} = \dfrac{A}{1 - x} + \dfrac{B}{1 - 2x} + \dfrac{C}{1 - 3x}$.
Clearing fractions, we have

$$
1 - 2x - x^2 = A(1 - 2x)(1 - 3x) + B(1 - x)(1 - 3x) + C(1 - x)(1 - 2x).
$$

Setting $x = 1$, we have $-2 = 2A$ so that $A = -1$. Setting $x = \frac{1}{2}$, we have $-\frac{1}{4} = -\frac{1}{4}B$ so that $B = 1$. Finally. setting $x = \frac{1}{3}$, we have $\frac{2}{9} = \frac{2}{9}$ so that $C = 1$. We have $a_n = -1 + 2^n + 3^n$.

3. Note that we have $\sum_{n=1}^{\infty} a_n x^n - 2x \sum_{n=1}^{\infty} a_{n-1} x^{n-1} = \sum_{n=1}^{\infty} n 2^{n+1} x^n$. Let $G(x) = \sum_{n=0}^{\infty} a_n x^n$. Then we have

$$
\begin{aligned}
G(x)(1 - 2x) - 1 &= G(x) - 1 - 2xG(x) \\
&= \frac{4x}{(1 - 2x)^2} \\
&= \frac{2 - 2(1 - 2x)}{(1 - 2x)^2} \\
&= \frac{2}{(1 - 2x)^2} - \frac{2}{(1 - 2x)}.
\end{aligned}
$$

Hence $G(x) = \frac{1}{(1-2x)^3} - \frac{1}{(1-2x)^2} + \frac{1}{1-2x}$. It follows that

$$
a_n = \left(2 \binom{n+2}{2} - 2(n+1) + 1 \right) 2^n = (n^2 + n + 1)2^n.
$$

5. Let the octagon be $ABCDEFGH$ and let the bug start at A, so that E is the destination. From A, C or G, it takes an even number of moves to reach E. Let a_n denote the number of ways for the bug to get to E for the first time after $2n$ moves. Let b_n denote the same number except that the bug starts from C, or by symmetry from G. We have $a_1 = 0$ and $b_1 = 1$. For $n \geq 2$, we claim that $a_n = 2a_{n-1} + 2b_{n-1}$. This is because the bug's first two moves can only be $A - B - A$, $A - H - A$, $A - B - C$ or $A - H - G$. Similarly, we have $b_n = 2b_{n-1} + a_{n-1}$. Extending these recurrence relations backward to $n = 0$, we have $a_0 = -1$ and $b_0 = 1$. Let $G(x)$ be the generating function for $\{a_n\}$ and $H(x)$ be that for $\{b_n\}$. Summing the recurrence relations from $n = 1$ to infinity, we have $G(x) + 1 = 2x(G(x) + 1) + 2x(H(x) - 1)$ and $H(x) - 1 = xG(x) + 2xH(x)$. Solving this system of functional equations, we have

$$
G(x) = \frac{-1 + 4x}{1 - 4x + 2x^2} = \frac{A}{1 - (2 + \sqrt{2})x} + \frac{B}{1 - (2 - \sqrt{2})x}.
$$

Clearing fractions, we have

$$
-1 + 4x = A(1 - (2 - \sqrt{2})x) + B(1 - (2 + \sqrt{2})x).
$$

Setting $x = \frac{2 - \sqrt{2}}{2}$, we have $A = \frac{1}{2}(\sqrt{2} - 1)$. Setting $x = \frac{2 + \sqrt{2}}{2}$, we have $B = -\frac{1}{2}(\sqrt{2} + 1)$. It follows that $a_n = \frac{1}{\sqrt{2}}((2 + \sqrt{2})^{n-1} - (2 - \sqrt{2})^{n-1})$ for $n \geq 1$, with $a_0 = 0$.

Section 2.6.

1. By Dirichlet's Product Formula,

$$\sum_{n=1}^{\infty} \frac{\phi(n)}{n^s} \sum_{m=1}^{\infty} \frac{1}{m^s} = \sum_{r=1}^{\infty} \frac{1}{r^s} \sum_{n|r} \phi(n).$$

Since $\displaystyle\sum_{m=1}^{\infty} \frac{1}{m^s} = \zeta(s)$ while $\displaystyle\sum_{n|r} \phi(n) = r$, we get

$$\zeta(s) \sum_{n=1}^{\infty} \frac{\phi(n)}{n^s} = \sum_{r=1}^{\infty} \frac{r}{r^s} = \zeta(s-1).$$

It follows that $\displaystyle\sum_{n=1}^{\infty} \frac{\phi(n)}{n^s} = \frac{\zeta(s-1)}{\zeta(s)}$.

3. We have $\tau(n) = \displaystyle\sum_{d|n} 1$. By Möbius' Inversion Formula,

$$\sum_{d|n} \mu(d)\tau\left(\frac{n}{d}\right) = 1.$$

5. Suppose $\tau(n) = 2k - 1$ for some positive integer k. Let the positive divisors of n be

$$1 = d_1 < d_2 < \cdots < d_{2k-1} = n.$$

Then $n = d_1 d_{2k-1} = d_2 d_{2k-2} = \cdots = d_{k-1}d_{k+1} = d_k^2$. Conversely, if n is a square, then each of its positive divisors can be paired with another one so that their product is n, except for $\sqrt{n}$ which must be paired with itself. Hence $\tau(n)$ is odd.

Section 3.1.

1. Arrange all five flags on a long pole. The number of arrangements is $\frac{5!}{2!2!1!} = 30$. There are four spaces between the flags. We now cut the long pole in two of these spaces to obtain the three distinct poles. Hence the desired number is $30\binom{4}{2} = 180$.

3. Suppose we have $2n$ objects consisting of 2 of each of n kinds. The total number of ways of arranging them is $\frac{(2n)!}{2!2!\cdots2!} = \frac{(2n)!}{2^n}$. Since this must be an integer, the desired conclusion follows.

5. We consider three cases.

Case 1. Two distint pairs of identical letters.

The letters can be chosen in $\binom{3}{2} = 3$ ways. The chosen letters can be arranged in 6 ways as there are $\binom{4}{2} = 6$ ways of placing one pair of identical letters. This yields a total of 18 ways.

Case 2. One pair of identical letters.

The letters can be chosen in 63 ways since the identical letters can be chosen in 3 ways and the other letters in $\binom{7}{2} = 21$ ways. The chosen letters can be arranged in 12 ways since the two letters not in the identical pair can be placed in 4×3 ways. This yields a total of 756 ways.

Case 3. No identical letters.

The letters can be chosen in $\binom{8}{4} = 70$ way. The chosen letters can be arranged in $4! = 24$ ways. This yields a total of 1680 ways.

The grand total is $18+756+1680=2454$ ways.

Section 3.2.

1. The exponential generating function for each of M, A and T is $1+x+\frac{1}{2}x^2$, and that for each of H, E, I, C and S is $1 + x$. Thus the overall exponential generating function is

$$\left(1 + x + \frac{1}{2}x^2\right)^3 (1+x)^5$$

$$= \left(1 + 2x + 2x^2 + x^3 + \frac{1}{4}x^4\right)\left(1 + x + \frac{1}{2}x^2\right)$$
$$(1 + 5x + 10x^2 + 10x^3 + 5x^4 + \cdots)$$

$$= \left(1 + 3x + \frac{9}{2}x^2 + 4x^3 + \frac{9}{4}x^4 + \cdots\right)$$
$$(1 + 5x + 10x^2 + 10x^3 + 5x^4 + \cdots)$$

$$= 1 + 8x + \frac{59}{2}x^2 + \frac{133}{2}x^3 + \frac{409}{4}x^4 + \cdots$$

$$= 1 + 8x + 59\frac{x^2}{2!} + 399\frac{x^3}{3!} + 2454\frac{x^4}{4!} + \cdots.$$

The respective numbers are 8, 59, 399 and 2454.

3. The exponential generating function for the digit 0 is $\frac{1}{2}(e^x + e^{-x})$. The exponential generating function for the digit 3 is $e^x - 1$ and the exponential generating function for the digit 1 or 2 is e^x.

Hence the overall exponential generating function is

$$\frac{1}{2}(e^x + e^{-x})(e^x - 1)e^{2x}$$

$$= \frac{1}{2}(e^{4x} - e^{3x} + e^{2x} - e^x)$$

$$= \frac{1}{2}\left(\sum_{n=0}^{\infty} 4^n \frac{x^n}{n!} - \sum_{n=0}^{\infty} 3^n \frac{x^n}{n!} + \sum_{n=0}^{\infty} 2^n \frac{x^n}{n!} - \sum_{n=0}^{\infty} \frac{x^n}{n!}\right).$$

It follows that the sequence generated is $a_n = \frac{1}{2}(4^n - 3^n + 2^n - 1)$.

5. The exponential generating function for the digit 0 is $\frac{1}{2}(e^x + e^{-x})$. The exponential generating function for the digit 1 is $e^x - 1 - x$ and the exponential generating function for the digit 2 or 3 is e^x. Hence the overall exponential generating function is

$$\frac{1}{2}(e^x + e^{-x})(e^x - 1 - x)e^{2x}$$

$$= \frac{1}{2}(e^{4x} - e^{3x} + e^{2x} - e^x - xe^{3x} - xe^x)$$

$$= \frac{1}{2}\left(\sum_{n=0}^{\infty} 4^n \frac{x^n}{n!} - \sum_{n=0}^{\infty} 3^n \frac{x^n}{n!} + \sum_{n=0}^{\infty} 2^n \frac{x^n}{n!}\right.$$

$$\left. - \sum_{n=0}^{\infty} \frac{x^n}{n!} - \sum_{n=0}^{\infty} 3^n \frac{x^{n+1}}{n!} - \sum_{n=0}^{\infty} \frac{x^{n+1}}{n!}\right)$$

$$= \frac{1}{2}\left(\sum_{n=0}^{\infty} 4^n \frac{x^n}{n!} - \sum_{n=0}^{\infty} 3^n \frac{x^n}{n!} + \sum_{n=0}^{\infty} 2^n \frac{x^n}{n!}\right.$$

$$\left. - \sum_{n=0}^{\infty} \frac{x^n}{n!} - \sum_{n=1}^{\infty} 3^{n-1} n \frac{x^n}{n!} - \sum_{n=1}^{\infty} n \frac{x^n}{n!}\right).$$

It follows that the sequence generated is $a_0 = \frac{1}{2}(1 - 1 + 1 - 1) = 0$ and for $n \geq 1$, $a_n = \frac{1}{2}(4^n - 3^n + 2^n - 1 - 3^{n-1}n - n)$.

Section 3.3.

1. Let S denote the set of all permutations. For $1 \leq i \leq 9$, let A_i be those in which i is a fixed point. By the Principle of Inclusion-Exclusion, we have

$$|\overline{A_1} \cap \overline{A_3} \cap \overline{A_5} \cap \overline{A_7} \cap \overline{A_9}|$$

$$= |S| - |A_1| - |A_3| - |A_5| - |A_7| - |A_9| + \cdots$$

$$= \binom{5}{0}9! - \binom{5}{1}8! + \binom{5}{2}7! - \binom{5}{3}6! + \binom{5}{4}5! - \binom{5}{5}4!$$

$$= 205056.$$

3. There are $\binom{n}{n-k}D_k$ permutations of $1, 2, \ldots, n$ with $n-k$ fixed points, $0 \leq k \leq n$. Each permutation is counted exactly once. Since there are $n!$ permutations overall, the desired result follows from $\binom{n}{n-k} = \binom{n}{k}$.

5. Let S be the set of all permutations. For $1 \leq k \leq n-1$, let A_k be the set of permutations in which k is followed immediately by $k+1$. By the Principle of Inclusion-Exclusion, the number of desired permutations is given by

$$|\overline{A_1} \cap \overline{A_2} \cap \cdots \cap \overline{A_{n-1}}|$$
$$= |S| - \sum |A_i| + \sum |A_i \cap A_j| - \cdots$$
$$+ (-1)^{n-1}|A_1 \cap A_2 \cap \cdots \cap A_{n-1}|$$
$$= n! + \sum_{k=1}^{n-1}(-1)^k \binom{n-1}{k}(n-k)!$$
$$= n! + \sum_{k=1}^{n-1}(-1)^k \left(\binom{n}{k} - \binom{n-1}{k-1}\right)(n-k)!$$
$$= \sum_{k=0}^{n-1}(-1)^k \binom{n}{k}(n-k)! - \sum_{k=1}^{n-1}(-1)^k \binom{n-1}{k-1}(n-k)!$$
$$= \sum_{k=0}^{n-1}(-1)^k \binom{n}{k}(n-k)! + \sum_{k=0}^{n-2}(-1)^k \binom{n-1}{k}(n-1-k)!$$
$$= \sum_{k=0}^{n}(-1)^k \binom{n}{k}(n-k)! + \sum_{k=0}^{n-1}(-1)^k \binom{n-1}{k}(n-1-k)!$$
$$= D_n + D_{n-1}.$$

Section 3.4.

1. Permuting the rows and columns of the given chessboard, we can convert it to the one shown below. Its rook polynomial is $R(B_2)^2 = (1+2x)^2 = 1 + 4x + 4x^2$.

3. We first compute the rook polynomial of the chessboard B of forbidden positions. We have
$$R(B) = R(B-s) + xR(B/s)$$
$$= R(B_2)R(B_6) + xR(B_1)$$
$$= (1+2x)(1+3x+x^2) + x(1+x)$$
$$= 1 + 6x + 8x^2 + 2x^3.$$

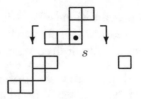

We now return to the problem of forbidden positions. By the Principle of Inclusion-Exclusion, the number of desired permutations is
$$4! - 6 \times 3! + 8 \times 2! - 2 \times 1! = 2.$$

5. Regard the rolls on the first die as numbers, and the rolls on the second die as places. Then we are considering permutations of 1, 2, 3, 4, 5 and 6 which do not violate any of the forbidden combinations. We first compute the rook polynomial of the chessboard B of forbidden positions. We have

$$
\begin{aligned}
R(B) &= R(B - s) + x R(B/s) \\
&= R(B_1) R(B_9) + x R(B_1)^2 \\
&= (1 + x)(1 + 4x + 3x^2) + x(1 + x)^2 \\
&= 1 + 6x + 9x^2 + 4x^3.
\end{aligned}
$$

We now return to the problem of forbidden positions. By the Principle of Inclusion-Exclusion, the number of desired permutations is
$$6! - 6 \times 5! + 9 \times 4! - 4 \times 3! = 192.$$

Section 3.5.

1. (a) We have
$$
\begin{pmatrix} 3 & 6 & 2 & 5 & 7 & 1 & 8 & 4 \\ 1 & 2 & 3 & 4 & 5 & 6 & 7 & 8 \end{pmatrix}, \begin{pmatrix} 3 & 1 & 2 & 8 & 7 & 4 & 5 & 6 \\ 1 & 2 & 3 & 4 & 5 & 6 & 7 & 8 \end{pmatrix}.
$$
Hence
$$
\begin{pmatrix} 1 & 2 & 3 & 4 & 5 & 6 & 7 & 8 \\ 6 & 3 & 1 & 8 & 4 & 2 & 5 & 7 \end{pmatrix}, \begin{pmatrix} 1 & 2 & 3 & 4 & 5 & 6 & 7 & 8 \\ 2 & 3 & 1 & 6 & 7 & 8 & 5 & 4 \end{pmatrix}.
$$
It follows that
$$
\begin{pmatrix} 1 & 2 & 3 & 4 & 5 & 6 & 7 & 8 \\ 6 & 3 & 1 & 8 & 4 & 2 & 5 & 7 \\ 8 & 1 & 2 & 4 & 6 & 3 & 7 & 5 \end{pmatrix}.
$$
The final permutation is $\{2,3,6,4,8,5,7,1\}$.

(b) The factorization of the second permutation is

$$(1, 2, 3)(4, 6, 8)(5, 7) = (1, 2)(1, 3)(4, 6)(4, 8)(5, 7).$$

The computation below leads to {2,3,6,4,8,5,7,1}.

Starting position	{3,6,2,5,7,1,8,4}
Action of (1,2)	{6,3,2,5,7,1,8,4}
Action of (1,3)	{2,3,6,5,7,1,8,4}
Action of (4,6)	{2,3,6,1,7,5,8,4}
Action of (4,8)	{2,3,6,4,7,5,8,1}
Action of (5,7)	{2,3,6,4,8,5,7,1}

3. We may assume that the King smashes box number 1 and box number 2, and refer to the chart in Example 3.5.2. In the first row, each of the 120 permutations leads to a successful scenario here too. In the next three rows, 1 and 2 must appear in different cycles. In the second row, either 1 or 2 must be alone, and this occurs 2 times out of 6. Hence the number of successful permutations is $144 \times \frac{2}{6} = 48$. In the third row, either 1 or 2 must be in the short cycle, and its companion must be one of the other four numbers. This occurs $2 \times 4 = 8$ times out of 15. Hence the number of successful permutations is $90 \times \frac{8}{15} = 48$. In the fourth row, 1 will apear in either cycle, and 2 must be one of the three numbers in the other cycle. This occurs 3 times out of 5. Hence the number of successful permutations is $40 \times \frac{3}{5} = 24$. The grand total is 120+48+48+24=240 and the overall probability is $\frac{240}{6!} = \frac{1}{3}$.

Alternatively, take the permutation {5,2,4,6,1,3}. It consists of three cycles (1,5), (3,4,6) and (2). We rewrite each cycle so that its smallest number appears last, namely, (5,1), (4,6,3) and (2). Finally, we arrange the cycles in ascending order of their last numbers, namely, (5,1), (2), (4,6,3). We now remove the brackets and link the numbers together to form another permutation, namely, {5,1,2,4,6,3}. This is called the associated permutation of {5,2,4,6,1,3}. Note that different permutations have different associated permutations. We claim that a permutation leads to a successful scenario if and only if the last number of its associated permutation is 1 or 2. If it is 1, the permutation is a single cycle. If it is 2, the permutation consists of two cycles with 1 and 2 in different cycles. If it is 3 or more, either the permutation consists of three or more cycles, or it consists of two cycles but with 1 and 2 in the same cycle. This justifies the claim. Now the probability that 1 or 2 is the last number of an associated permutation, or any permutation for that matter, is clearly $\frac{2}{6} = \frac{1}{3}$.

5. Represent a permutation on $\{1, 2, \ldots, n\}$ as a thoroughly shuffled deck of cards numbered from 1 to n. Discard one of them at random. If the longest cycle was unique, then the probability that we have decreased its length by one is $\frac{E(n)}{n}$. For $n \geq 2$, the longest cycle is not always unique, and discarding a card from a cycle which was tied for the longest does not decrease the length of the longest cycle. Hence for $n \geq 2$, the probability of a decrease in the length of the longest cycle is less than $\frac{E(n)}{n}$. Thus, for $n \geq 2$, $E(n-1) > E(n) - \frac{E(n)}{n} = \frac{(n-1)E(n)}{n}$. Dividing by $n - 1$ yields $\frac{E(n-1)}{n-1} > \frac{E(n)}{n}$.

Section 3.6

1. **Initialization:** We have $S_0 = \{1, 2, 3, 4\}$.
   First Iteration. Prefix=1.
   $S_1 = S_0 - \{1\} \cup \{12, 13, 14, 112, 113, 114, \ldots\}$.
   Second Iteration. Prefix=2.
   $S_2 = S_1 - \{2\} \cup \{23, 24, 212, 213, 214, 223, 224 \ldots\}$.
   Third Iteration. Prefix=3.
   $S_3 = S_2 - \{3\} \cup \{34, 312, 313, 314, 323, 324, 334\}$.
   Fourth Iteration. Prefix=4.
   $S_4 = S_3 - \{4\} \cup \{412, 413, 414, 423, 424, 434\}$.
   There are $3 + 5 + 6 + 6 = 20 = \frac{4^3 - 4}{3}$ words of length 3 in S_4. They constitute a maximal comma-free (4,3)-dictionary.

3. **Initialization:** We have $S_0 = \{1, 2, 3, 4\}$.
   First Iteration. Prefix=1.
   $S_1 = S_0 - \{1\} \cup \{12, 13, 112, 113, 1112, 1113, 11112, 11113, \ldots\}$.
   Second Iteration. Prefix=2.
   $S_2 = S_1 - \{2\} \cup \{23, 212, 213, 223, 2112, 2113\}$
   $\cup \{2223, 21112, 21113, 22112, 22113, 22212, 22213, 22223, \ldots\}$.
   Third Iteration. Prefix=3.
   $S_3 = S_2 - \{3\} \cup \{312, 313, 323, 3112, 3113, 3212, 3213, 3223, 3312, 3313\}$
   $\cup \{3323, 31112, 31113, 32112, 32113, 32212, 32213, 32223, 33112, 33113\}$
   $\cup \{33212, 33213, 33223, 33312, 33313, 33323\}$.
   Fourth Iteration. Prefix=112.
   $S_4 = S_3 - \{112\} \cup \{11212, 11213, 11223, \ldots\}$.
   Fifth Iteration. Prefix=113.
   $S_5 = S_4 - \{113\} \cup \{11312, 11313, 11323, \ldots\}$.
   Sixth Iteration. Prefix=212.
   $S_6 = S_5 - \{212\} \cup \{21212, 21213, 21223, \ldots\}$.
   Seventh Iteration. Prefix=213.
   $S_7 = S_6 - \{213\} \cup \{21312, 21313, 21323, \ldots\}$.

Eighth Iteration. Prefix=223.
$S_8 = S_7 - \{223\} \cup \{22312, 22313, 22323, \ldots\}$.
Ninth Iteration. Prefix=312.
$S_9 = S_8 - \{312\} \cup \{31212, 31213, 31223, \ldots\}$.
Tenth Iteration. Prefix=313.
$S_{10} = S_9 - \{313\} \cup \{31312, 31313, 31323, \ldots\}$.
Eleventh Iteration. Prefix=323.
$S_{11} = S_{10} - \{323\} \cup \{32312, 32313, 32323, \ldots\}$.
There are $2 + 7 + 15 + 3 + 3 + 3 + 3 + 3 + 3 + 3 + 3 = 48 = \frac{3^5 - 3}{5}$ words
in S_{11}. They constitute a maximal comma-free (3,5)-dictionary.

5. Take $A = \{1, 2\}$, $B = \{3, 4\}$ and $C = \{5, 6\}$. Then

$$D = \{13, 23, 14, 24, 15, 25, 16, 26, 35, 45, 36, 46\}.$$

Section 4.1.

1. (a) We have

$$a_n = \binom{n}{0} + \sum_{k=1}^{n-1} \left(\binom{n+k-1}{2k-1} + \binom{n+k-1}{2k} \right) + \binom{2n-1}{2n-1}$$

$$= \sum_{k=0}^{n-1} \binom{n+k-1}{2k} + \sum_{k=1}^{n} \binom{n+k-1}{2k-1}$$

$$= a_{n-1} + \sum_{k=0}^{n-1} \binom{n+k}{2k+1}.$$

It follows that $a_n - a_{n-1} = \sum_{k=0}^{n-1} \binom{n+k}{2k+1}$. Similarly, we have
$a_{n-1} - a_{n-2} = \sum_{k=0}^{n-2} \binom{n+k-1}{2k+1}$. Now

$$a_n - 2a_{n-1} = \sum_{k=1}^{n-1} \left(\binom{n+k}{2k+1} - \binom{n+k-1}{2k} \right)$$

$$= \sum_{k=0}^{n-2} \binom{n+k-1}{2k+1}$$

$$= a_{n-1} - a_{n-2}, .$$

so that we indeed have $a_n - 3a_{n-1} + a_{n-2} = 0$.

(b) The characteristic equation of this recurrence relation is $x^2 - 3x + 1 = 0$ and the characteristic roots are $\frac{3 \pm \sqrt{5}}{2}$. It follows that

$$a_n = C_1 \left(\frac{3 + \sqrt{5}}{2} \right)^n + C_2 \left(\frac{3 - \sqrt{5}}{2} \right)^n.$$

Note that we have $a_0 = \binom{0}{0} = 1$ and $a_1 = \binom{1}{0} + \binom{2}{2} = 2$. It follows that $C_1 + C_2 = 1$ and $C_1(\frac{3+\sqrt{5}}{2}) + C_2(\frac{3-\sqrt{5}}{2}) = 2$. The latter simplifies to $C_1 - C_2 = \frac{1}{\sqrt{5}}$. Hence $C_1 = \frac{1+\sqrt{5}}{2\sqrt{5}}$ and $C_2 = -\frac{1-\sqrt{5}}{2\sqrt{5}}$ so that

$$a_n = \frac{1+\sqrt{5}}{2\sqrt{5}}\left(\frac{3+\sqrt{5}}{2}\right)^n - \frac{1-\sqrt{5}}{2\sqrt{5}}\left(\frac{3-\sqrt{5}}{2}\right)^n.$$

Since $(\frac{1\pm\sqrt{5}}{2})^2 = \frac{3\pm\sqrt{5}}{2}$, we have

$$a_n = \frac{1}{\sqrt{5}}\left(\left(\frac{1+\sqrt{5}}{2}\right)^{2n+1} - \left(\frac{1-\sqrt{5}}{2}\right)^{2n+1}\right) = F_{2n}.$$

(c) We have $a_n = \sum_{k=0}^{n}\binom{n+k}{2k} = \sum_{k=0}^{n}\binom{n+k}{n-k}$. Let $j = n - k$. Then

$$a_n = \sum_{j=0}^{n}\binom{2n-j}{j} = F_{2n} \text{ since } \binom{n}{n} \text{ is the last non-zero term.}$$

3. Let a_n be the number of ways. Then $a_0 = a_1 = 1$. In general, if the first space is covered by a unit square, the remaining part of the board may be covered in a_{n-1} ways. If the first space is not covered by a unit square, it must be covered by a domino. The remaining part of the board may be covered in a_{n-2} ways. Hence $a_n = a_{n-1} + a_{n-2}$. From the initial conditions, $a_n = F_{n+1}$.

5. Let b_n denote the number of different paths the bug can take in order to reach E after exactly n minutes, and let a_n denote the same number except that the bug commences from A. Then $b_0 = b_1 = 0$ and $b_2 = 1$, the unique path being $B - A - E$. Suppose $n \geq 2$. Consider the move the bug makes in the first minute. It is either $B - A$ or $B - C$. If the bug ends up at C, the number of different ways to complete its path is b_{n-1} by symmetry. If it ends up at A, the number of different ways to complete its path is a_{n-1}. Now $a_0 = 0$ and $a_1 = 1$, the unique path being $A - E$. Suppose $n \geq 2$. Consider the moves the bug makes in the first minute. It must be $A - B$, and there are b_{n-1} different ways to complete its path. It follows that for $n \geq 3$, $b_n = b_{n-1} + a_{n-1}$ and $a_n = b_{n-1}$. Hence $b_n = b_{n-1} + b_{n-2}$. From the initial conditions, $b_n = F_{n+3}$.

Section 4.2.

1. We use induction on n. For $n = 1$, $1 \times 1 - 0 \times 2 = 1 = (-1)^{1-1}$. Suppose the result holds for some $n \geq 1$. Then

$$
\begin{aligned}
F_{n+1} F_{n+2} - F_n F_{n+3} &= F_{n+1} F_{n+2} - F_n(F_{n+2} + F_{n+1}) \\
&= F_{n+2}(F_{n+1} - F_n) - F_n F_{n+1} \\
&= F_{n-1} F_{n+2} - F_n F_{n+1} \\
&= (-1)^n.
\end{aligned}
$$

3. (a) Let $\begin{aligned}[t] S &= F_0 + F_3 + F_6 + \cdots + F_{3n} \\ &= (F_1 + F_2) + (F_4 + F_5) + \cdots + (F_{3n-2} + F_{3n-1}). \end{aligned}$

Then $2S = F_0 + F_1 + F_2 + \cdots + F_{3n} = F_{3n+2} - 1$ by (7), so that $S = \frac{1}{2}(F_{3n+2} - 1)$.

(b) Let $\begin{aligned}[t] S &= F_1 + F_4 + F_7 + \cdots + F_{3n+1} \\ &= 1 + F_0 + (F_2 + F_3) + (F_5 + F_6) + \cdots + (F_{3n-1} + F_{3n}). \end{aligned}$

Then $2S = 1 + F_0 + F_1 + \cdots + F_{3n+1} = F_{3n+3}$ by (7), so that $S = \frac{1}{2} F_{3n+3}$.

(c) Let $\begin{aligned}[t] S &= F_2 + F_5 + F_8 + \cdots + F_{3n+2} \\ &= (F_0 + F_1) + (F_3 + F_4) + (F_6 + F_7) + \cdots + (F_{3n} + F_{3n+1}). \end{aligned}$

Then $2S = F_0 + F_1 + \cdots + F_{3n+2} = F_{3n+4} - 1$ by (7), so that $S = \frac{1}{2}(F_{3n+4} - 1)$.

5. We use induction on n that every positive integer less than or equal to F_n is a sum of distinct Fibonacci numbers. For $n = 1$, $F_1 = 1$ and there is nothing to prove. Suppose the result holds for some $n \geq 1$. Consider a number k where $F_n < k \leq F_{n+1}$. Then $k \leq F_n + F_{n-1}$ so that $0 < k - F_n \leq F_{n-1} < F_n$. By the induction hypothesis, $k - F_n$ is a sum of distinct Fibonacci numbers all less than F_n. Hence $k = (k - F_n) + F_n$ is also a sum of distinct Fibonacci numbers.

Section 4.3.

1. Suppose we have a working seat plan. Since each student can see the blackboard to the right, they must be seated in ascending order of height from right to left within each row. Have the students report, in ascending order of height, whether they are in row 0 or row 1. Since each student can see the blackboard in front, the running total of 0s must not fall behind the running total of 1s. Hence the number of working seat plans is the Catalan number C_n.

3. Denote by a_n the number of ways $2n - 2$ people can shake hands at a round table without crossing arms. Then $a_1 = a_2 = 1$. Let the people be P_1, P_2,..., P_{2n-2} in cyclic order. P_1 cannot shake hands with P_{2k-1} for any k, as otherwise there will be an odd number of people on each side of their linked arms. Hence P_1 shakes hands with P_{2k} for some k, $1 \leq k \leq n - 1$. There are $2k - 2$ people on one side and $2n - 2k - 2$ people on the other, yielding $a_k a_{n-k}$ handshake patterns. Hence $a_n = a_1 a_{n-1} + a_2 a_{n-2} + \cdots + a_{n-1} a_1$. From the initial conditions, $a_n = C_{n-1}$.

5. Define $y_0 = 0$ and plot the points $(0, y_0), (1, y_1), \ldots, (n, y_n)$ on the coordinate plane. We build a lattice path from $(0,0)$ to (n, n), passing through these points in succession. We go from (k, y_k) to $(k+1, y_{k+1})$. If $y_k = y_{k+1}$, we just make a horizontal move. If $y_k < y_{k+1}$, then we first move vertically to (k, y_{k+1}) and then horizontally to $(k+1, y_{k+1})$. Finally, from (n, y_n), we move vertically to (n, n). If this lattice path ever rises above the line $y = x$, it must occur at a point of the form (k, y_{k+1}). However, this is impossible since $y_{k+1} \leq k$. Since this process is reversible, we have the Catalan number C_n.

Section 4.4.

1. As in Example 4.4.1, we have

$$
\begin{aligned}
S_3(n, 4) &= \sum_{i=1}^{n} \binom{n}{i} S_3(n - i, 3) \\
&= \sum_{i=1}^{n-3} \binom{n}{i} (3^{n-i} - 3 \cdot 2^n + 3) \\
&= \sum_{i=1}^{n-3} \binom{n}{i} 3^{n-i} - 3 \sum_{i=1}^{n-3} \binom{n}{i} 2^n + 3 \sum_{i=1}^{n-3} \binom{n}{i} \\
&= 4^n - \binom{n}{0} 3^n - \binom{n}{n-2} 3^2 - \binom{n}{n-1} 3^1 - \binom{n}{n} 3^0 \\
&\quad - 3 \left(3^n - \binom{n}{0} 2^n - \binom{n}{n-2} 2^2 - \binom{n}{n-1} 2^1 - \binom{n}{n} 2^0 \right) \\
&\quad + 3 \left(2^n - \binom{n}{0} 1^n - \binom{n}{n-2} 1^2 - \binom{n}{n-1} 1^1 - \binom{n}{n} 1^0 \right) \\
&= 4^n - 4 \cdot 3^n + 6 \cdot 2^n - 4 \cdot 1^n - \binom{n}{n-2} S_3(2, 3) \\
&\quad - \binom{n}{n-1} S_3(1, 3) - \binom{n}{n} S_3(0, 3) \\
&= \sum_{i=1}^{4} (-1)^i \binom{4}{i} (4 - i)^n.
\end{aligned}
$$

3. The term $\binom{n}{k}S_2(k,m)$ counts the number of the $S_2(n+1,m+1)$ partitions into $m+1$ subsets of $\{1,2,\ldots,n+1\}$ where $n+1$ is in a subset of size $n-k+1$. The other k elements may be chosen in $\binom{n}{k}$ ways and partitioned into m subsets in $S_2(k,m)$ ways.

5. We prove by induction on n that $S_3(n,n) = n!$. Note that

$$S_3(1,1) = \sum_{i=0}^{1}(-1)^i\binom{1}{i}(1-i) = 1 - 0 = 1.$$

Suppose $S_3(n-1,n-1) = (n-1)!$ for some $n-1 \geq 1$. We have $S_3(n-1,n) = 0$ by Example 4.4.3(b). Hence

$$
\begin{aligned}
S_3(n,n) &= S_3(n,n) - nS_3(n-1,n) \\
&= \sum_{i=0}^{n}(-1)^i\binom{n}{i}(n-i)^n - n\sum_{i=0}^{n}(-1)^i\binom{n}{i}(n-i)^{n-1} \\
&= n\sum_{i=1}^{n}(-1)^i\frac{n-i}{n}\binom{n}{i}(n-i)^{n-1} - n\sum_{i=1}^{n}(-1)^i\binom{n}{i}(n-i)^{n-1} \\
&= n\sum_{i=1}^{n}(-1)^i\binom{n-1}{i}(n-i)^{n-1} - n\sum_{i=1}^{n}(-1)^i\binom{n}{i}(n-i)^{n-1} \\
&= n\sum_{i=1}^{n}(-1)^{i-1}\left(\binom{n}{i} - \binom{n-1}{i}\right)(n-i)^{n-1} \\
&= n\sum_{i=1}^{n}(-1)^{i-1}\binom{n-1}{i-1}(n-i)^{n-1} \\
&= n\sum_{i=0}^{n-1}(-1)^i\binom{n-1}{i}(n-1-i)^{n-1} \\
&= nS_3(n-1,n-1) \\
&= n!.
\end{aligned}
$$

Section 4.5.

1. The term $\binom{k}{m}S_1(n,k)$ counts the number of ways of seating the people 1, 2, $\ldots$, n at exactly k polygonal tables, and converting m of them into round tables. For each of the other polygonal table, list the people there in cyclic order, starting with the person with the smallest number. Seat these people round an $(m+1)$-st round table as follows. Start with the polygonal table whose first person has the highest number among the tables, and seat the rest of the people from at table in order. Then follow with the polygonal table whose first person has the second highest number, and so on. Finally, add the new person $n+1$. This yields a seating of the $n+1$ people at $m+1$ round tables.

We now show that this process is reversible, that is, the table with $n+1$ may be decomposed into its constituent polygonal tables. Remove $n+1$. Start the first polygonal table with the next person, and continue until we reach someone with a lower number than the starter. Then we close that table and start a new one with the next person. This establishes the desired one-to-one correspondence.

3. The term $S_1(k, m)S_2(n + 1, k + 1)$ counts the number of ways of partitioning the people $1, 2, \ldots, n+1$ into $k+1$ subsets, and apart from the subset containing $n + 1$, seat the other k subsets round exactly m tables. The people in the same subset must occupy a block of adjacent seats, but the internal order of seating is immaterial. We now try to establish a one-to-one correspondence between the ways where k is odd and the ways where k is even. Consider the person with the highest number among those not in the same subset as $n+1$. If it is not alone in a subset, detach it from that subset and form a new subset in the same table following the depleted subset. On the other hand, if it is alone in a subset, attach it to the subset in the same table preceding it. These two processes are reversible except when there are exactly m people not in the same subset as $n + 1$, and they are in individual subsets seating alone at the m tables. There are $\binom{n}{m}$ such ways and, taking signs into consideration, the given sum is equal to $(-1)^m\binom{n}{m}$.

5. Note that $S_3(n, m)$ counts the number of distributions of n distinct objects into m distinct boxes with no boxes empty. For each such distribution, we can construct a permutation of the n objects as follows. The objects within each box are arranged in ascending order, and these lists are joined together following a fixed ordering of the boxes. We claim that each permutation counted in $E(n, k)$ is constructed $\binom{k-1}{n-m}$ times. We will insert $m - 1$ dividers in between pairs of objects to indicate how they are to be distributed into the m boxes. Such a permutation has $k - 1$ up-pairs and therefore $n - k$ down-pairs. A divider must be inserted between every down-pair. If $m - 1 \leq n - k$, such a permutation cannot be constructed by our procedure, but then $k - 1 \leq n - m$, and $\binom{k-1}{n-m} = 0$ anyway. Suppose $m - 1 \geq n - k$. We have $m - 1 - n + k$ additional dividers to be inserted between some of the $k - 1$ up-pairs. This can be done in $\binom{k-1}{m-1-n+k} = \binom{k-1}{n-m}$ ways.

Section 4.6.

1. The upper bound follows from the strengthened Greenwood-Gleason Inequality. For the lower bound, let the people A, B, C, D, E, F, G and H stand on the vertices of an octagon in cyclic order. Let two people on adjacent vertices hold a yellow ribbon between them. Let all other pairs hold blue ribbons between them. We have no yellow cliques of size 3 but many blue cliques of size 4. We shall change a few yellow ribbons into blue ones. If A is to be in a blue clique of size 4, the other three members must be among C, D, E, F and G. If we change the ribbon between C and G into a yellow one, then the five of them form a pentagon with yellow edges and blue diagonals. We know that it contains no blue cliques of size 3, and therefore A cannot be in a blue clique of size 4. By symmetry, E cannot be in a blue clique of size 4 either. Changing the ribbon between D and H prevents B and F to be in blue cliques of size 4, and of course C, D, G and H do not form a blue clique of size 4. Hence there are no blue cliques of size 4. It is easy to verify that we have not introduced any yellow cliques of size 3. It follows that we indeed have $R(3, 4) > 8$.

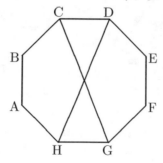

Remark:

Note that we can change more ribbons from blue to yellow, such as the one between B and E, without introducing any yellow trios. If instead, we change the colors of the ribbons between A and E and between B and F, we can present the construction in a symmetric way, like the pentagons used to justify $R(3, 3) = 6$. Now the yellow ribbons are along the stellated polygons with steps 1 and 4 (the latter consisting of four diagonals joining opposite vertices) and the blue ribbons are along the stellated polygons with steps 2 and 3 (the former consisting of two squares on alternating vertices).

3. The upper bound follows from the Greenwood-Gleason Inequality. For
   the lower bound, let the people A, B, C, D, E, F, G, H, A', B', C',
   D', E', F', G', H' and M stand on the vertices of a 17-gon in the order
   shown in the diagram below. Every two of them hold a yellow or a blue
   ribbon between them. A ribbon is yellow if and only if it is along the
   stellated polygons with steps 1, 2, 4 and 8. Suppose there is a yellow
   quartet or a blue quartet. By symmetry, we may assume that M is
   involved. Now A, B, C, D, E, F, G and H hold yellow ribbons with
   M. Their ribbon pattern is essentially the same as in the construction
   which was used to justify $R(3,4) > 8$ (with a yellow ribbon between B
   and E instead). This means that there are no yellow trios among them,
   and therefore no yellow quartets involving M. For those holding blue
   ribbons with M, namely A', B', C', D', E', F', G' and H', their ribbon
   pattern is the same as the symmetric expansion of the construction
   which was used to justify $R(3,4) > 8$ (plus an interchanging of colors).
   This means that there are no blue trios among them, and therefore no
   blue quartet involving M either. It follows that $R(4,4) > 17$.

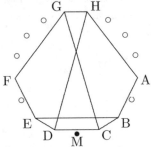 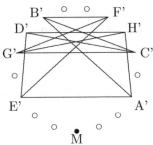

Remark:
Our approach in the last two problems is largely geometric. It is also
possible to describe the constructions in number-theoretic terms. In
Problem 5.6.3, number the 13 people from 0 to 12. Two people i and j
are holding a yellow ribbon between them if and only if $i - j$ is a cube
modulo 13, namely 1, 5, 8 and 12. In Problem 5.6.4, number the 17
people from 0 to 16. Two people i and j are holding a yellow ribbon
between them if and only if $i - j$ is a square modulo 17, namely, 1, 2,
4, 8, 9, 13, 15 and 16.

5. Since $R(3,3) > 5$, we have a yellow and blue clique of size 5 with no
   monochromatic cliques of size 3. Now let a twin show up for each of
   the five people, one at a time. Each pair of twins holds a red ribbon
   between them, and the newcomer holds the same color ribbon with
   every other person as the other twin. Clearly there are no red cliques
   of size 3. A yellow or blue clique of size 3 cannot involve more than
   one of any pair of twins. Since there are no yellow or blue cliques of
   size 3 before the twins show up, this is still the case. Hence there are
   no monochromatic cliques of size 3.

Section 5.1.

1. (a) The Cayley table is shown below.

$+_{2/3}$	(0,0)	(0,1)	(0,2)	(1,0)	(1,1)	(1,2)
(0,0)	(0,0)	(0,1)	(0,2)	(1,0)	(1,1)	(1,2)
(0,1)	(0,1)	(0,2)	(0,0)	(1,1)	(1,2)	(1,0)
(0,2)	(0,2)	(0,0)	(0,1)	(1,2)	(1,0)	(1,1)
(1,0)	(1,0)	(1,1)	(1,2)	(0,0)	(0,1)	(0,2)
(1,1)	(1,1)	(1,2)	(1,0)	(0,1)	(0,2)	(0,0)
(1,2)	(1,2)	(1,0)	(1,1)	(0,2)	(0,0)	(0,1)

(b) This group is Abelian and is therefore not isomorphic to the dihedral group of the equilateral triangle. However, it is isomorphic to Z_6, with the correspondences (0,0) to 0, (0,1) to 4, (0,2) to 2, (1,0) to 3, (1,1) to 1 and (1,2) to 5.

3. (a) The Cayley table is shown below.

$\times_{15}$	1	2	4	7	8	11	13	14
1	1	2	4	7	8	11	13	14
2	2	4	8	14	1	7	11	13
4	4	8	1	13	2	14	7	11
7	7	14	13	4	11	2	1	8
8	8	1	2	11	4	13	14	7
11	11	7	14	2	13	1	8	4
13	13	11	7	1	14	8	4	2
14	14	13	11	8	7	4	2	1

(b) This Abelian group is not isomorphic to the dihedral group of the square, which is non-Abelian. It is also not isomorphic to either Z_8 pr Z_{24}^* because it has 4 self-inverse elements whereas the others have 2 or 8. However, it is isomorhpic to $Z_2 \times Z_4$, with the correspondences 1 to (0,0), 2 to (0,1), 4 to (0,2), 7 to (1,1), 8 to (0,3), 11 to (1,0), 13 to (1,3) and 14 to (1,2).

5. (a) The Cayley table is shown below.

$\circ$	I	R	H	V
I	I	R	H	V
R	R	I	V	H
H	H	V	I	R
V	V	H	R	I

(b) This group is not isomorphic to Z_4 because it has 4 self-inverse elements whereas Z_4 has only 2. However, it is isomorphic to $Z_2 \times Z_2$ with the correspondences I to (0,0), R to (0,1), H to (1,0) and V to (1,1).

Section 5.2.

1. (a) We have $2^3 = 8$ distinct designs because each of the 3 cells can be painted in either of the 2 colors. These are shown in the diagram below.

   (b) There are four patterns altogether, consisting of
      (1) the design with all three cells black;
      (2) the design with all three cells white;
      (3) three designs with two cells black and the other one white;
      (4) three designs with two cells white and the other one black.
      These are marked in the diagram above.

   (c) We have exactly the same four patterns as before.

3. This is shown in the diagram below, where the invariant entries are shaded.

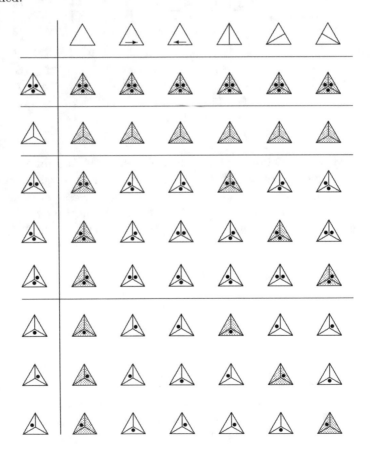

5. We choose arbitrarily one design as the Home Base and call the others
the First Base and the Second Base.

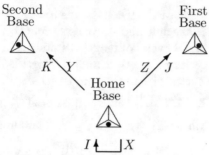

The Home Base is invariant under two symmetries, namely, I and X.
The symmetries J and Z take it to the First Base and the symmetries
K and Y take it to the Second Base. We claim that the number
of symmetries under which another Base is invariant is equal to the
number of symmetries going to it from the Home Base. Since each
symmetry takes the Home Base either back to itself or to another
Base, the total number of invariants within this pattern must be six.
Consider for example the First Base. One of the symmetries which
takes it back to the Home Base is $J^{-1} = K$. Now the symmetry KS
takes the First Base to the same destination as the symmetry S takes
the Home Base. Moreover, the symmetries KI, KJ, KK, KX, KY
and KZ are distinct by the Cancellation Law. As a matter of fact,
they are K, I, J, Z, X and Y respectively. Since the Home Base is
invariant under I and X, the symmetries $KI = K$ and $KX = Z$ take
the First Base to the Home Base. Similarly, the symmetries $KK = J$
and $KY = X$ go from the First Base to the Second Base, and the
First Base is invariant under the symmetries $KJ = I$ and $KZ = Y$,
and only under these two. This can also be verified directly.

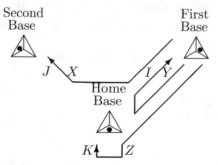

We can analyse the Second Base in the same way. This justifies our
claim that there are exactly six invariant entries within this pattern.

Section 5.3.

1. (a) Under I, each of the 9 cells forms a cycle of length 1. They form
   1 cycle of length 1 and 4 cycles of length 2 under R, 1 cycle of
   length 1 and 2 cycles of length 4 under A or C, 3 cycles of length
   1 and 3 cycles of length 2 under H or V, and 3 cycles of length
   1 and 3 cycles of length 2 under U or D. The cycle index is
   $\dfrac{x_1^9 + x_1 x_2^4 + 2x_1 x_4^2 + 4x_1^3 x_2^3}{8}$. Since each cell is painted in any of
   3 colors, we set $x_1 = x_2 = x_4 = 3$. The number of patterns is

   $$\frac{3^9 + 3^5 + 2 \times 3^3 + 4 \times 3^6}{8} = 2862.$$

   (b) Under only the rotational symmetries, the new cycle index is
   $\dfrac{x_1^9 + x_1 x_2^4 + 2x_1 x_4^2}{4}$. With $x_1 = x_2 = x_4 = 3$, the number of
   patterns is

   $$\frac{3^9 + 3^5 + 2 \times 3^3}{4} = 4995.$$

3. (a) Each of the faces of the regular dodecahedron may be mapped
   onto any of its twelve faces in five different ways, yielding a total
   of $12 \times 5 = 60$ symmetries. The breakdown is as follows.

   (1) The identity symmetry.
   (2) Twelve $72°$ rotational symmetries about axes joining the cen-
       ters of opposite faces.
   (3) Twelve $144°$ rotational symmetries about axes joining the
       centers of opposite faces.
   (4) Fifteen $180°$ rotational symmetries about axes joining the
       midpoints of opposite sides.
   (5) Twenty $120°$ rotational symmetries about axes joining oppo-
       site vertices.

   (b) The cycle index is $\dfrac{x_1^{12}+24x_1^2 x_5^2+15x_2^6+20x_3^4}{60}$.

   Remark:
   The symmetry group of the regular dodecahedron is isomorphic to the
   symmetry group of the regular icosahedron, as well as to the alternating group A_5.

5. The cycle index is $\dfrac{x_1^{18}+9x_1^2 x_2^8+6x_1^2 x_4^4+8x_3^6}{24}$. Setting $x_1 = x_2 = x_3 = x_4 = 3$,
   we have $\dfrac{3^{18}+9\times3^{10}+14\times3^6}{24} = 16165089$ patterns.

Section 5.4.

1. The cycle index is $\frac{x_1^{16}+8x_1x_3^5+3x_2^8}{12}$. Setting $x_1=b+1$, $x_2=b^2+1$, $x_3=b^3+1$ and $x_4=b^4+1$, the inventory function is $\frac{(b+1)^{16}+8(b+1)(b^3+1)^5+3(b^2+1)^8}{12}$. The number of coloring patterns is $\frac{\binom{16}{3}+8\binom{2}{0}\binom{5}{1}}{12}=50$.

3. The cycle index is $\frac{x_1^{12}+3x_2^6+6x_4^3+6x_1^2x_2^5+8x_3^4}{24}$. Setting $x_1=w+2$, $x_2=w^2+2$, $x_3=w^3+2$ and $x_4=w^4+2$, the inventory function is $\frac{(w+2)^{12}+3(w^2+2)^6+6(w^4+2)^3+6(w+2)^2(w^2+2)^5+8(w^3+2)^4}{12}$. The number of coloring patterns is $\frac{\binom{12}{3}2^9+6\binom{2}{1}2^1\binom{5}{1}2^4+8\binom{4}{1}2^3}{24}=4784$.

5. (a) There are $\binom{16}{3}=560$ ways of placing the markers, all of which are invariant under I. Under R, A, C, H or V, there are no invariants. Under D or U, first we may have all three markers in the reflecting diagonal, yielding $\binom{4}{3}=4$ ways. Second, we may have one marker on the reflecting diagonal and one marker on each side of the reflecting diagonal in corresponding positions, yielding $6\binom{4}{1}=24$ ways. It follows that the desired number of ways is $\frac{1}{8}(560+0+0+0+0+0+28+28)=77$.

   (b) The problem is equivalent to painting the 16 squares in black and white with exactly three squares black. The cycle index is $\frac{x_1^{16}+3x_2^8+2x_4^4+2x_1^4x_2^6}{8}$. The inventory function is

   $$\frac{(b+1)^{16}+3(b^2+1)^8+2(b^4+1)^4+2(b+1)^4(b^2+1)^6}{8}.$$

   The final answer is $\frac{\binom{16}{3}+2(\binom{4}{3}\binom{6}{0}+\binom{4}{1}\binom{6}{1})}{8}=77$.

Section 5.5.

1. (a) The total number of patterns is given by

   $$\frac{1}{9}\sum_{d|9}\phi\left(\frac{9}{d}\right)3^d = \frac{\phi(9)3^1+\phi(3)3^3+\phi(1)3^9}{9}$$

   $$= 2195.$$

   (b) The number of patterns with no non-trivial symmetries is given by

   $$\frac{1}{9}\sum_{d|9}\mu\left(\frac{9}{d}\right)3^d = \frac{\mu(9)3^1+\mu(3)3^3+\mu(1)3^9}{9}$$

   $$= 2184.$$

(c) The 11 patterns counted in (a) but not in (b) are sketched below.

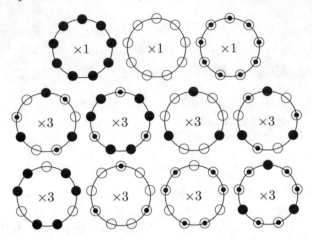

3. (a) The total number of patterns if only rotations are taken into consideration is given by

$$\frac{1}{8} \sum_{d|8} \phi\left(\frac{8}{d}\right) 4^d = \frac{\phi(8)4^1 + \phi(4)4^2 + \phi(2)4^4 + \phi(1)4^8}{8}$$

$$= \frac{16 + 32 + 256 + 65536}{8}$$

$$= 8230.$$

(b) The total number of patterns if reflections are also taken into consideration is given by

$$\frac{1}{16}\left(\sum_{d|8} \phi\left(\frac{8}{d}\right) 4^d + 4(4^4 + 4^5)\right)$$

$$= \frac{\phi(8)4^1 + \phi(4)4^2 + \phi(2)4^4 + \phi(1)4^8 + 4(4^4 + 4^5)}{16}$$

$$= \frac{16 + 32 + 256 + 65536 + 5120}{16}$$

$$= 4435.$$

5. If n is composite, then $n = ab$ for some integers $1 < a, b < n$. Now a divides $(n-1)!$ since it is one of the factors in the product, but a does not divide 1. Hence a does not divide $(n-1)! + 1$, and neither does n.

Section 5.6.

1. Let the colors be denoted by g, y and b. There are six permutations of the colors, $(g)(y)(b)$, $(g)(yb)$, $(y)(bg)$, $(b)(gy)$, (gyb) and (gby). The first means all colors stay put, the next three mean two of the colors are interchanged, while the last two mean the colors are permuted cyclically. Combining with the usual eight symmetries of the square I, R, A, C, H, V, D and U, we have a total of forty-eight symmetries. We calculate the total number of invariants in the following chart.

Invariants	$(g)(y)(b)$	$(g)(yb)$	$(y)(bg)$	$(b)(gy)$	(gyb)	(gby)
I	3^9	3^0	3^0	3^0	0	0
R	3^5	3^4	3^4	3^4	0	0
A	3^3	3^2	3^2	3^2	0	0
C	3^3	3^2	3^2	3^2	0	0
H	3^6	3^3	3^3	3^3	0	0
V	3^6	3^3	3^3	3^3	0	0
D	3^6	3^3	3^3	3^3	0	0
U	3^6	3^3	3^3	3^3	0	0

There are no invariants under cyclic permutations of the three colors. The total number of distinct patterns is $\frac{22896+208+208+208}{48} = 490$.

3. There are 6 symmetries. In the identity, each object is an independent cycle, and the contribution to the cycle index is x_1^4. In the three pairwise switchings of the red objects, the two red objects being switched form a single cycle, and the contribution to the cycle index is $3x_1^2x_2$. In the two cyclic switchings of the red objects, all three red objects are in the same cycle. The contribution to the cycle index is $2x_1x_3$. Hence the cycle index is $\frac{x_1^4+3x_1^2x_2+2x_1x_3}{6}$. Setting $x_1 = x_2 = x_3 = 3$, we have $\frac{81+3\times27+2\times9}{6} = 30$. This is the number of distributions we seek.

5. Let us for now assume that the two red objects are distinct, as are the two black boxes. We have four symmetries, the identity, the switching of the red objects, the switching of the black boxes, and the double switch of the red objects and the black boxes. In the first case, each object is an independent cycle, and the number of invariants is $3^3 = 27$. In the second case, the two red objects form a single cycle, so that the number of invariants is $3^2 = 9$. In the third case, the two black boxes being switched must be empty because the objects are distinct. Hence the number of invariants is $1^3 = 1$. In the last case, the yellow object must be in the white box. The two red objects can both be in the white box, one in a black box and the other in the other black box, or vice versa. Thus the number of invariants is $3^1 = 3$. The number of distributions we seek is $\frac{27+9+1+3}{4} = 10$.

Section 6.1.

1. (a) The number of such polynomials is $\frac{1}{2}(\mu(2)5 + \mu(1)5^2) = 10$.

   (b) The chart below shows all reducible monic quadratic polynomials over $\mathbf{Z}_5$.

$\times$	x	$x+1$	$x+2$	$x+3$	$x+4$
x	x^2	x^2 $+x$	x^2 $+2x$	x^2 $+3x$	x^2 $+4x$
$x+1$	x^2 $+x$	x^2 $+2x+1$	x^2 $+3x+2$	x^2 $+4x+3$	x^2 $+4$
$x+2$	x^2 $+2x$	x^2 $+3x+2$	x^2 $+4x+4$	x^2 $+1$	x^2 $+x+3$
$x+3$	x^2 $+3x$	x^2 $+4x+3$	x^2 $+1$	x^2 $+x+4$	x^2 $+2x+2$
$x+4$	x^2 $+4x$	x^2 $+4$	x^2 $+x+3$	x^2 $+2x+2$	x^2 $+3x+1$

The irreducible monic quadratic polynomials over $\mathbf{Z}_5$ are $x^2 + 2$, x^2+3, x^2+x+1, x^2+x+2, x^2+2x+3, x^2+2x+4, x^2+3x+3, $x^2 + 3x + 4$, $x^2 + 4x + 1$ and $x^2 + 4x + 2$.

   (c) The elements are 0, 1, 2, 3, 4, t, $t+1$, $t+2$, $t+3$, $t+4$, $2t$, $2t+1$, $2t+2$, $2t+3$, $2t+4$, $3t$, $3t+1$, $3t+2$, $3t+3$, $3t+4$, $4t$, $4t+1$, $4t+2$, $4t+3$ and $4t+4$, where $t^2 + 2 = 0$ or $t^2 = 3$.

   (d) The number of minimal polynomials for all these elements is $\frac{1}{2}(\phi(2)5 + \phi(1)5^2) = 15$.

   (e) The minimal polynomials are x for 0, $x + 4$ for 1, $x + 3$ for 2, $x + 2$ for 3, $x + 1$ for 4, $x^2 + 2$ for t and $4t$, $x^2 + 3$ for $2t$ and $3t$, $x^2 + x + 1$ for $t + 3$ and $4t + 3$, $x^2 + x + 2$ for $2t + 3$ and $3t + 3$, $x^2 + 2x + 3$ for $t + 1$ and $4t + 1$, $x^2 + 2x + 4$ for $2t + 1$ and $3t + 1$, $x^2 + 3x + 3$ for $t + 4$ and $4t + 4$, $x^2 + 3x + 4$ for $2t + 4$ and $3t + 4$, $x^2 + 4t + 1$ for $t + 2$ and $4t + 2$, and $x^2 + 4x + 2$ for $2t + 2$ and $3t + 2$.

3. In the chart below, the 56 lines are partitioned into 8 classes of mutually parallel lines. The points on each line are listed. To save space, (i, j) is abbreviated to ij.

$x = 0$	00 01 02 03 04 05 06
$x = 1$	10 11 12 13 14 15 16
$x = 2$	20 21 22 23 24 25 26
$x = 3$	30 31 32 33 34 35 36
$x = 4$	40 41 42 43 44 45 46
$x = 5$	50 51 52 53 54 55 56
$x = 6$	60 61 62 63 64 65 66
$y = 0$	00 10 20 30 40 50 60
$y = 1$	01 11 21 31 41 51 61
$y = 2$	02 12 22 32 42 52 62
$y = 3$	03 13 23 33 43 53 63
$y = 4$	04 14 24 34 44 54 64
$y = 5$	05 15 25 35 45 55 65
$y = 6$	06 16 26 36 46 56 66
$y = x$	00 11 22 33 44 55 66
$y = x + 1$	01 12 23 34 45 56 60
$y = x + 2$	02 13 24 35 46 50 61
$y = x + 3$	03 14 25 36 40 51 62
$y = x + 4$	04 15 26 30 41 52 63
$y = x + 5$	05 16 20 31 42 53 64
$y = x + 6$	06 10 21 32 43 54 65
$y = 2x$	00 12 24 36 41 53 65
$y = 2x + 1$	01 13 25 30 42 54 66
$y = 2x + 2$	02 14 26 31 43 55 60
$y = 2x + 3$	03 15 20 32 44 56 61
$y = 2x + 4$	04 16 21 33 45 50 62
$y = 2x + 5$	05 10 22 34 46 51 63
$y = 2x + 6$	06 11 23 35 40 52 64
$y = 3x$	00 13 26 32 45 51 64
$y = 3x + 1$	01 14 20 33 46 52 65
$y = 3x + 2$	02 15 21 34 40 53 66
$y = 3x + 3$	03 16 22 35 41 54 60
$y = 3x + 4$	04 10 23 36 42 55 61
$y = 3x + 5$	05 11 24 30 43 56 62
$y = 3x + 6$	06 12 25 31 44 50 63
$y = 4x$	00 14 21 35 42 56 63
$y = 4x + 1$	01 15 22 36 43 50 64
$y = 4x + 2$	02 16 23 30 44 51 65
$y = 4x + 3$	03 10 24 31 45 52 66
$y = 4x + 4$	04 11 25 32 46 53 60
$y = 4x + 5$	05 12 26 33 40 54 61
$y = 4x + 6$	06 13 20 34 41 55 62

$$
\begin{array}{ll}
y = 5x & \text{00 15 23 31 46 54 62} \\
y = 5x + 1 & \text{01 16 24 32 40 55 63} \\
y = 5x + 2 & \text{02 10 25 33 41 56 64} \\
y = 5x + 3 & \text{03 11 26 34 42 50 65} \\
y = 5x + 4 & \text{04 12 20 35 43 51 66} \\
y = 5x + 5 & \text{05 13 21 36 44 52 60} \\
y = 5x + 6 & \text{06 14 22 30 45 53 61} \\
\hline
y = 6x & \text{00 16 25 34 43 52 61} \\
y = 6x + 1 & \text{01 10 26 35 44 53 62} \\
y = 6x + 2 & \text{02 11 20 36 45 54 63} \\
y = 6x + 3 & \text{03 12 21 30 46 55 64} \\
y = 6x + 4 & \text{04 13 22 31 40 56 65} \\
y = 6x + 5 & \text{05 14 23 32 41 50 66} \\
y = 6x + 6 & \text{06 15 24 33 42 51 60}
\end{array}
$$

5. (a) Let A and B be any pair of lions. By Postulate 3, there is a pony P that both A and B have bitten. By Example 6.1.3(b), there is a lion C that has not bitten P. By Postulate 2, C has bitten at least three ponies. By Postulate 3, A has bitten at most one of these three and so has B. Hence there is a pony that neither A nor B has bitten.

   (b) Let P and Q be any pair of ponies. By Postulate 4, there is a lion A that has bitten both of them. By Postulate 2, this lion has bitten a third pony R. By Example 6.1.3(a), there is a pony S that A has not bitten. By Postulate 4, there is a lion B that has bitten both R and S. By Postulate 3, B has not bitten either P or Q.

Section 6.2.

1. The completed Latin squares are shown in the diagram below, and the elements of transversals are marked in boldface.

(a)

0	1	2	3	4
1	4	3	0	**2**
2	**3**	4	1	0
3	2	0	**4**	1
4	0	**1**	2	3

(b)

0	1	2	3	4
1	4	0	**2**	3
2	**3**	1	4	0
3	0	**4**	1	2
4	2	3	0	**1**

(c)

0	1	2	3	4
1	0	**4**	2	3
2	**3**	0	4	1
3	4	1	0	**2**
4	2	3	**1**	0

3. Here, $g_1 = 1, g_2 = 2, g_3 = 4, g_4 = 7, g_5 = 8, g_6 = 11, g_7 = 13$ and $g_8 = 14$. From the given transversal, the values of h_i are recorded in the diagram below as the first column. The subsequent columns are obtained by multiplying the first column with the other elements, yielding the f_i values for each. Note that the rows are just permutations of the rows of the Cayley table.

4	8	1	13	2	14	7	11
8	1	2	11	4	13	14	7
11	7	14	2	13	1	8	4
1	2	4	7	8	11	13	14
7	14	13	4	11	2	1	8
13	11	7	1	14	8	4	2
14	13	11	8	7	4	2	1
2	4	8	14	1	7	11	13

The resulting Latin square orthogonal to the Cayley table is shown in the diagram below.

4	2	11	1	13	7	14	8
8	4	7	2	11	14	13	1
1	8	14	4	7	13	11	2
13	14	2	7	1	4	8	11
2	1	13	8	14	11	7	4
14	7	1	11	8	2	4	13
7	11	8	13	4	1	2	14
11	13	4	14	2	8	1	7

5. Consider the affine geometry of order 7. It has 49 points and 56 lines divided into 8 parallel classes. The two classes with slope 0 and slope ∞ are set aside for coordinate references. Each of the other six classes will yield one of the six mutually orthogonal Latin squares we seek. Each of the 7 lines in a class is a transversal of the Latin square, which we will mark by the constant term of its equation.

6	5	4	3	2	1	0	6	4	2	0	5	3	1
5	4	3	2	1	0	6	5	3	1	6	4	2	0
4	3	2	1	0	6	5	4	2	0	5	3	1	6
3	2	1	0	6	5	4	3	1	6	4	2	0	5
2	1	0	6	5	4	3	2	0	5	3	1	6	4
1	0	6	5	4	3	2	1	6	4	2	0	5	3
0	6	5	4	3	2	1	0	5	3	1	6	4	2

$$
\begin{array}{ccccccc|ccccccc}
6 & 3 & 0 & 4 & 1 & 5 & 2 & 6 & 2 & 5 & 1 & 4 & 0 & 3 \\
5 & 2 & 6 & 3 & 0 & 4 & 1 & 5 & 1 & 4 & 0 & 3 & 6 & 2 \\
4 & 1 & 5 & 2 & 6 & 3 & 0 & 4 & 0 & 3 & 6 & 2 & 5 & 1 \\
3 & 0 & 4 & 1 & 5 & 2 & 6 & 3 & 6 & 2 & 5 & 1 & 4 & 0 \\
2 & 6 & 3 & 0 & 4 & 1 & 5 & 2 & 5 & 1 & 4 & 0 & 3 & 6 \\
1 & 5 & 2 & 6 & 3 & 0 & 4 & 1 & 4 & 0 & 3 & 6 & 2 & 5 \\
0 & 4 & 1 & 5 & 2 & 6 & 3 & 0 & 3 & 6 & 2 & 5 & 1 & 4 \\
\hline
6 & 1 & 3 & 5 & 0 & 2 & 4 & 6 & 0 & 1 & 2 & 3 & 4 & 5 \\
5 & 0 & 2 & 4 & 6 & 1 & 3 & 5 & 6 & 0 & 1 & 2 & 3 & 4 \\
4 & 6 & 1 & 3 & 5 & 0 & 2 & 4 & 5 & 6 & 0 & 1 & 2 & 3 \\
3 & 5 & 0 & 2 & 4 & 6 & 1 & 3 & 4 & 5 & 6 & 0 & 1 & 2 \\
2 & 4 & 6 & 1 & 3 & 5 & 0 & 2 & 3 & 4 & 5 & 6 & 0 & 1 \\
1 & 3 & 5 & 0 & 2 & 4 & 6 & 1 & 2 & 3 & 4 & 5 & 6 & 0 \\
0 & 2 & 4 & 6 & 1 & 3 & 5 & 0 & 1 & 2 & 3 & 4 & 5 & 6 \\
\end{array}
$$

Section 6.3.

1. (a) An affine geometry of order 5 has 25 points and 30 lines in 6 parallel classes. Each line passes through 5 points, and each point lies on exactly 1 line in each parallel class. Every 2 points determine 1 line. Thus we have a (30,25,6,5,1)-design.

   (b) To obtain a projective geometry of order 5 from an affine geometry of order 5, we add an ideal point to each line in the same parallel class, and add an ideal line joining these 6 points. Now each line passes through 6 points and each point lies on 6 lines. Every 2 points still determine 1 line. Thus we have a (31,31,6,6,1)-design.

3. (a) The resulting structure has $v - 1$ blocks on $v - k$ elements. Each of these elements still appears k times and appears together with every other element λ times. Since every two blocks in the original design have λ common elements, the size of each new block is $k - \lambda$. Thus we have a $(v - 1, v - k, k, k - \lambda, \lambda)$-design.

   (b) The resulting (14,8,7,4,3)-design consists of the following blocks:

$$\{8,9,10,11\} \quad \{12,13,14,15\} \quad \{8,9,12,13\} \quad \{10,11,14,15\}$$
$$\{8,9,14,15\} \quad \{10,11,12,13\} \quad \{8,10,12,14\} \quad \{8,11,13,15\}$$
$$\{9,11,12,15\} \quad \{9,10,13,14\} \quad \{9,11,13,14\} \quad \{9,10,12,15\}$$
$$\{8,10,13,15\} \quad \{8,11,12,14\}$$

5. Consider the contestants who solved a particular problem. Each of them solved 5 others, and each of the other 30 problems was solved by exactly one of them. Hence this problem was solved by 6 contestants. By symmetry, each problem was solved by exactly 6 contestants. It follows that there were 31 contestants.

Let there be n problems in the afternoon. Each contestant solved at most 5 of them, since each solved at least one problem in the morning. Suppose it was not true that at least one contestant solved at most 2 problems in the afternoon. Then each solved 3, 4 or 5. Let the numbers of such contestants be x, y and z respectively. Then $x + y + z = 31, 3x + 4y + 5z = 6n$ and $3x + 6y + 10z = \binom{n}{2}$. Multiplying these equations by -15, 7 and -2 respectively and adding, we have $y = -(n - \frac{43}{2})^2 - \frac{11}{4} < 0$, which is a contradiction.

Section 6.4.

1. The non-zero squares in $\mathbf{Z}_{11}$ are $1^2 = 1 = 10^2$, $2^2 = 4 = 9^2$, $3^2 = 9 = 8^2$, $4^2 = 5 = 7^2$ and $5^2 = 3 = 6^2$. Hence $\{1,3,4,5,9\}$ is a $(11,5,2)$-set.

3. To construct a $(37,9,2)$-set, consider the cyclotomic cosets induced by the prime divisor $p = 7$ of $k - \lambda = 9 - 2 = 7$. They are

$$(0), (1, 7, 12, 10, 33, 9, 26, 34, 16), (2, 14, 24, 20, 29, 18, 15, 31, 32),$$

$$(3, 21, 36, 30, 25, 27, 4, 28, 11) \text{ and } (5, 35, 23, 13, 17, 8, 19, 22, 6).$$

Any of the cosets of size 9 is a $(37,9,2)$-set. In particular, the table below shows that $\{5,6,8,13,17,19,22,23,35\}$ is indeed a difference set.

$-$	5	6	8	13	17	19	22	23	35
5	$-$	36	34	29	25	23	20	19	7
6	1	$-$	35	30	26	24	21	20	8
8	3	2	$-$	32	28	26	23	22	10
13	8	7	5	$-$	33	31	28	27	15
17	12	11	9	4	$-$	35	32	31	19
19	14	13	11	6	2	$-$	34	33	21
22	17	16	14	9	5	3	$-$	36	24
23	18	17	15	10	6	4	1	$-$	25
35	30	29	27	22	18	16	13	12	$-$

5. If a $(29,8,2)$-set exists, then the prime divisor $p = 3$ of $k - \lambda = 8 - 2 = 6$ satisfies $p > \lambda$. By the Multiplier Theorem, the difference set is a union of cyclotomic cosets induced by $p = 3$. These are

$$(0) \text{ and } (1, 3, 9, 27, 23, 11, 4, 12, 7, 21, 5, 15, 16, 19,$$

$$28, 26, 20, 2, 6, 18, 25, 17, 22, 8, 24, 14, 13, 10).$$

No union of them has size 8. Hence a $(29,8,2)$-set cannot exist.

Section 6.5.

1. The addition table modulo 5 is shown in the left of the diagram below, and the modified table in the right.

+	0	1	2	3	4
0	0	1	2	3	4
1	1	2	3	4	0
2	2	3	4	0	1
3	3	4	0	1	2
4	4	0	1	2	3

$\oplus$	0	2	4	1	3
0	0	1	2	3	4
2	1	2	3	4	0
4	2	3	4	0	1
1	3	4	0	1	2
3	4	0	1	2	3

The elements are $a_0, a_1, a_2, a_3, a_4, b_0, b_1, b_2, b_3, b_4, c_0, c_1, c_2, c_3, c_4$. The blocks are:

$(a_0, a_1, b_3),\quad (a_0, a_2, b_1),\quad (a_0, a_3, b_4),\quad (a_0, a_4, b_2),\quad (a_1, a_2, b_4),$
$(a_1, a_3, b_2),\quad (a_1, a_4, b_0),\quad (a_2, a_3, b_0),\quad (a_2, a_4, b_3),\quad (a_3, a_4, b_1),$
$(b_0, b_1, c_3),\quad (b_0, b_2, c_1),\quad (b_0, b_3, c_4),\quad (b_0, b_4, c_2),\quad (b_1, b_2, c_4),$
$(b_1, b_3, c_2),\quad (b_1, b_4, c_0),\quad (b_2, b_3, c_0),\quad (b_2, b_4, c_3),\quad (b_3, b_4, c_1),$
$(c_0, c_1, a_3),\quad (c_0, c_2, a_1),\quad (c_0, c_3, a_4),\quad (c_0, c_4, a_2),\quad (c_1, c_2, a_4),$
$(c_1, c_3, a_2),\quad (c_1, c_4, a_0),\quad (c_2, c_3, a_0),\quad (c_2, c_4, a_3),\quad (c_3, c_4, a_1),$
$(a_0, b_0, c_0),\quad (a_1, b_1, c_1),\quad (a_2, b_2, c_2),\quad a_3, b_3, c_3),\quad (a_4, b_4, c_4).$

3. The starter blocks are: $(0,8,12)$, $(3,7,10)$, $(4,5,13)$, $(6,9,11)$, $(1,2,\infty)$. The remaining blocks are generated from the starter blocks by adding 2 at a time modulo 14.

$(0,8,12)$	$(3,7,10)$	$(4,5,13)$	$(6,9,11)$	$(1,2,\infty)$
$(2,10,0)$	$(5,9,12)$	$(6,7,1)$	$(8,11,13)$	$(3,4,\infty)$
$(4,12,2)$	$(7,11,0)$	$(8,9,3)$	$(10,13,1)$	$(5,6,\infty)$
$(6,0,4)$	$(9,13,2)$	$(10,11,5)$	$(12,1,3)$	$(7,8,\infty)$
$(8,2,6)$	$(11,1,4)$	$(12,13,7)$	$(0,3,5)$	$(9,10,\infty)$
$(10,4,8)$	$(13,3,6)$	$(0,1,9)$	$(2,5,7)$	$(11,12,\infty)$
$(12,6,10)$	$(1,5,8)$	$(2,3,11)$	$(4,7,9)$	$(13,0,\infty)$

5. (a) The operation table of $\oplus$ is shown in the diagram below.

$\oplus$	0	2	4	6	1	3	5
0	0	1	2	3	4	5	6
2	1	2	3	4	5	6	0
4	2	3	4	5	6	0	1
6	3	4	5	6	0	1	2
1	4	5	6	0	1	2	3
3	5	6	0	1	2	3	4
5	6	0	1	2	3	4	5

This yields the triples

$$
\begin{array}{lll}
(a_0, a_1, b_4) & (a_0, a_2, b_1) & (a_0, a_3, b_5) \\
(a_0, a_4, b_2) & (a_0, a_5, b_5) & (a_0, a_6, b_3) \\
(a_1, a_2, b_5) & (a_1, a_3, b_2) & (a_1, a_4, b_6) \\
(a_1, a_5, b_3) & (a_1, a_6, b_0) & (a_2, a_3, b_6) \\
(a_2, a_4, b_3) & (a_2, a_5, b_0) & (a_2, a_6, b_4) \\
(a_3, a_4, b_0) & (a_3, a_5, b_4) & a_3, a_6, b_1) \\
(a_4, a_5, b_1) & (a_4, a_6, b_5) & (a_5, a_6, b_2).
\end{array}
$$

(b) The other 7 triples added are (b_1, b_2, b_4), (b_2, b_3, b_5), (b_3, b_4, b_6), (b_4, b_5, b_0), (b_5, b_6, b_1), (b_6, b_0, b_2) and (b_0, b_1, b_3) from the Fano plane. The five starter blocks are (c, a_0, b_0), (b_1, b_2, b_4), (a_1, a_i, b_ℓ), (a_2, a_j, b_m) and (a_4, a_k, b_n), where $\{i, j, k\} = \{\ell, m, n\} = \{3, 5, 6\}$. It is easy to check that $i = 5$, $j = 3$, $k = 6$, $\ell = 3$, $m = 6$ and $n = 5$. The remaining resolution classes are obtained from the first one by adding 1 at a time modulo 7, except that c remains the same. We have a Kirkman system of order 15.

$$
\begin{array}{lllll}
(c, a_0, b_0) & (b_1, b_2, b_4) & (a_1, a_5, b_3) & (a_2, a_3, b_6) & (a_4, a_6, b_5) \\
(c, a_1, b_1) & (b_2, b_3, b_5) & (a_2, a_6, b_4) & (a_3, a_4, b_0) & (a_5, a_0, b_6) \\
(c, a_2, b_2) & (b_3, b_4, b_6) & (a_3, a_0, b_5) & (a_4, a_5, b_1) & (a_6, a_1, b_0) \\
(c, a_3, b_3) & (b_4, b_5, b_0) & (a_4, a_1, b_6) & (a_5, a_6, b_2) & (a_0, a_2, b_1) \\
(c, a_4, b_4) & (b_5, b_6, b_1) & (a_5, a_2, b_0) & (a_6, a_0, b_3) & (a_1, a_3, b_2) \\
(c, a_5, b_5) & (b_6, b_0, b_2) & (a_6, a_3, b_1) & (a_0, a_1, b_4) & (a_2, a_4, b_3) \\
(c, a_6, b_6) & (b_0, b_1, b_3) & (a_0, a_4, a_2) & (a_1, a_2, b_5) & (a_3, a_5, b_4)
\end{array}
$$

Section 6.6.

1. By Example 6.6.2, $f(4, 3k) = 5k$. We have $f(4, 3k + 1) \geq 5k + 1$ by Observation 2. Suppose the building has $5k + 2$ floors. The total number of elevator doors is $4(3k+1) < 3(5k+2)$. Hence some floor is served by at most two elevators, and it cannot be just one. Let these be elevators A and B. Between them, they must serve every floor. Hence there are exactly $2(3k+1) - (5k+2) = k$ floors served by both of them. Denote by A the set of floors served by elevator A, and by B the set of floors served by elevator B. Then $|A - B| = |B - A| = 2k + 1$. Each floor in $A - B$ must be linked to each floor in $B - A$ by elevator C or elevator D. Since $2(3k+1) < 2(4k+2)$, some floor in $(A - B) \cup (B - A)$ is served by exactly one of elevator C and elevator D. We may assume that it is in $A - B$ and served by elevator C. Then every floor in $B - A$ must be served by elevator C. Now some other floor in $A - B$ is served by elevator D but not elevator C. Again, every floor in $B - A$ must be served by elevator D. Hence at most $2k - 2$ floors in $A - B$ can be served by elevator C or elevator D, which is a contradiction.

3. By Example 6.6.3, $f(5, 5k) = 9k$. We have $f(5, 5k + 1) \geq 9k + 1$ by Observation 2. Suppose the building has $9k + 2$ floors. None of them may be served by only one elevator. Let x be the number of floors served by exactly two elevators. Then the total number of elevator doors is $5(5k + 1) \geq 2x + 3(9k + 2 - x)$. Hence $x \geq 2k + 1$. Consider a floor served by exactly two elevators. Call them A and B. Between them, they must serve every floor. Hence there are exactly $2(5k + 1) - (9k + 2) = k$ floors served by both of them. Denote by A the set of floors served by elevator A, and by B the set of floors served by elevator B. Then $|A - B| = |B - A| = 4k + 1$. Consider a floor not in $A \cap B$ which is served by exactly two elevators. By symmetry, we may assume that it is in $A - B$. It must be linked to all floors in $B - A$ by say elevator C. Then elevator C serves all $4k + 1$ floors in $B - A$ and k floors in $A - B$. Now the remaining $3k + 1$ floors in $A - B$ must be linked to all floors in $B - A$ by the remaining two elevators. However, each of them can link to only $2k$ floors in $B - A$, but the set has $4k + 1$ floors.

5. Suppose in a $(9k + 5)$-floor building, the first elevator stops on floors with remainders 0, 1, 3, 5 and 7 when divided by 9, the second on 0, 2, 4, 6 and 8, the third on 1, 5, 6, 7 and 8, the fourth on 2, 3, 4, 5 and 6, and the fifth on 2, 3, 4, 7 and 8. The fifth elevator also stops on two arbitrarily chosen extra floors. Then each elevator stops on $5k + 3$ floors and every two floors are linked directly by at at least one elevator. Hence $f(5, 5k + 3) \geq 9k + 5$. Suppose the building has $9k + 6$ floors. None of them may be served by only one elevator. Let x be the number of floors served by exactly two elevators. Then the total number of elevator doors is $5(5k + 3) \geq 2x + 3(9k + 6 - x)$. Hence $x \geq 2k + 3$. Consider a floor served by exactly two elevators. Call them A and B. Between them, they must serve every floor. Hence there are exactly $2(5k + 3) - (9k + 6) = k$ floors served by both of them. Denote by A the set of floors served by elevator A, and by B the set of floors served by elevator B. Then $|A - B| = |B - A| = 4k + 3$. Consider a floor not in $A \cap B$ which is served by exactly two elevators. By symmetry, we may assume that it is in $A - B$. It must be linked to all floors in $B - A$ by say elevator C. Then elevator C serves all $4k + 3$ floors in $B - A$ and k floors in $A - B$. Now the remaining $3k + 3$ floors in $A - B$ must be linked to all floors in $B - A$ by the remaining two elevators. However, each of them can link to only $2k$ floors in $B - A$, but the set has $4k + 3$ floors.

Section A.0.

1. (a) By the Multiplication Principle, the number of complete dinners is $5 \times 6 \times 3 \times 2 \times 4 = 720$.

    (b) For the other item, the number of choices is $5+6+3+4=18$ by the Addition Principle. Along with one of the 2 main courses, the total number of choices is $2 \times 18 = 36$ by the Multiplication Principle.

3. Let A, B, C and D be the sets of dinners Ethel has with Anita, Betty, Celia and Daisy respectively. By the Addition Rule,

$$
\begin{aligned}
|IA \cup B \cup C \cup D| &= |A| + |B| + |C| + |D| - |A \cap B| - |A \cap C| \\
&\quad - |A \cap D| - |B \cap C| - |B \cap D| - |C \cap D| \\
&\quad + |A \cap B \cap C| + |A \cap B \cap D| + |A \cap C \cap D| \\
&\quad + B \cap C \cap D| - |A \cap B \cap C \cap D \\
&= 4 \times 12 - 6 \times 6 + 4 \times 4 - 1 \times 3 \\
&= 25.
\end{aligned}
$$

5. Take aside one student who solved exactly one problem, one student who solved exactly two problems and one student who solved exactly three problems. They had taken care of six of the 35 problems. The remaining 29 problems were solved by the other seven students. If each of them had solved only four problems, then there would have been only 28 problems solved. Therefore, one student must have solved at least 5 problems.

7. For $n = 1$, the left side is equal to $1^2 = 1$ while the right side is equal to $1^3 = 1$. Assume that the result holds for some $n \geq 1$. Then

$$
\begin{aligned}
&(1 + 2 + \cdots + n + (n+1))^2 \\
&= (1 + 2 + \cdots + n)^2 + 2(1 + 2 + \cdots + n)(n+1) + (n+1)^2 \\
&= 1^3 + 2^3 + \cdots + n^3 + n(n+1)^2 + (n+1)^3 \\
&= 1^3 + 2^3 + \cdots + n^3 + (n+1)^3.
\end{aligned}
$$

By mathematical induction, the result holds for all $n \geq 1$.

9. We shall prove the desired result by mathematical induction. If there is only one planet, there is no other planet to observe it. Suppose the number of planets is an odd number $n \geq 1$. Consider the two planets closest to each other. They must observe each other. Suppose at least one other planet also observes one of them. Then there are only $n - 3$ planets left to observe the other $n - 2$ planets, and one of them will not be observed. If no other planet observes these two, we can remove them. Now we have $n - 2$ planets left, and this is an odd number. By the induction hypothesis, one of these planets will not be observed.

11. We shall use as a critical measure the number of brave spots, which we now define. At the moment it is occupied, a seat between two children of the same gender is a brave spot. If a boy sits between two girls, he uses up a brave spot, and is brave. If he sits between a boy and a girl, he creates a brave spot. If he sits between two boys, he changes one brave spot into two. Hence each boy either uses up a brave spot or creates one. By symmetry, the same can be said about each girl. At the beginning, there are no brave spots, which is also the case at the end. Hence of the twenty children who come later, half of them create brave spots and the other half use them up. It follows that the number of the latter, namely those who are brave, is ten.

Section A.1.

1. (a) Starting from the lone L, there are 2 ways of reaching the second diagonal of As. From any of the As, there are 2 ways of reaching the third diagonal of Rs. Thus the total number of ways of reaching the sixth diagonals of Rs is $2^5 = 32$ by the Multiplication Principle.

   (b) Starting from the L at the bottom left, the number of ways is $\binom{5}{0}$. Starting from the other Ls in order, the numbers are $\binom{5}{1}$, $\binom{5}{2}$, $\binom{5}{3}$, $\binom{5}{4}$ and $\binom{5}{5}$.

   (c) Starting from the lone R, there are 2 ways of reaching it from the fifth diagonal of Es. For each of the Es, there are 2 ways of reaching it from the fourth diagonal of Ds. Thus the total number of ways of reaching the lone R from any of the Ls is $2^5 = 32$ by the Multiplication Principle.

   (d) We have $\binom{5}{0} + \binom{5}{1} + \binom{5}{2} + \binom{5}{3} + \binom{5}{4} + \binom{5}{5} = 2^5$.

3. Let there be m girls and n boys shaking hands with one another. There are mn handshakes between a girl and a boy, $\binom{m}{2}$ handshakes between two girls and $\binom{n}{2}$ handshakes between two boys. Since the total number of handshakes is $\binom{m+n}{2}$, we have $\binom{m+n}{2} - \binom{m}{2} - \binom{n}{2} = mn$.

5. By the Binomial Theorem, $(z + 1)^3 = z^3 + 3z^2 + 3z + 1$. Setting z equal to 1, 2, ..., n and adding, we have

$$
\begin{array}{rclclclcl}
2^3 &=& 1^3 &+& 3 \times 1^2 &+& 3 \times 1 &+& 1 \\
3^3 &=& 2^3 &+& 3 \times 2^2 &+& 3 \times 2 &+& 1 \\
4^3 &=& 3^3 &+& 3 \times 3^2 &+& 3 \times 3 &+& 1 \\
\cdots &=& \cdots &+& \cdots &+& \cdots &+& 1, \\
\hline
(n+1)^3 &=& n^3 &+& 3n^2 &+& 3n &+& 1 \\
\hline
(n+1)^3 &=& 1^3 &+& 3S_2 &+& 3S_1 &+& n
\end{array}
$$

Thia yields $2S_2 = (n+1)^3 - (n+1) - 3S_1 = n(n+1)(n+2) - \frac{3n(n+1)}{2}$. It follows that $S_2 = \frac{n(n+1)(2n+1)}{6}$.

7. We have

$$\binom{100}{0} + \binom{100}{1} + \binom{100}{2} + \binom{100}{3} + \cdots + \binom{100}{100} = 2^{100}$$

and

$$\binom{100}{0} - \binom{100}{1} + \binom{100}{2} - \binom{100}{3} + \cdots + \binom{100}{100} = 0.$$

Let i be a primitive fourth root of unity. Then $1 + i^2 = 0$. Hence $(1 \pm i)^4 = 1 \pm 4i - 6 \mp 4i + 1 = -4$ so that $(1 \pm i)^{100} = (-4)^{25} = -2^{50}$. Setting $x = i$ in the expansion of $(1 + x)^{100}$, we have

$$\binom{100}{0} + \binom{100}{1}i - \binom{100}{2} - \binom{100}{3}i + \cdots + \binom{100}{100} = -2^{50}.$$

Setting $x = -i$ in the expansion of $(1 + x)^{100}$, we have

$$\binom{100}{0} - \binom{100}{1}i - \binom{100}{2} + \binom{100}{3}i + \cdots + \binom{100}{100} = -2^{50}.$$

(a) Add these identities and divide by 4. Then we have the desired result.

(b) Multiply these identities by 1, -1, $-i$ and i respectively, add and divide by 4. Then we have the desired result.

(c) Multiply these identities by 1, 1, -1 and -1 respectively, add and divide by 4. Then we have the desired result.

(d) Multiply these identities by 1, 1, 1 and 1 respectively, add and divide by 4. Then we have the desired result.

Note that the answers to (b) and (d) are the same since $\binom{n}{k} = \binom{n}{n-k}$.

9. We consider three cases.
Case 1. Both elevators are above the person. The first elevator to arrive will be coming from above. This case occurs with probability $\frac{1}{5} \times \frac{1}{5}$, which is equal to $\frac{1}{25}$.
Case 2. Neither elevator is above the person.
The first elevator to arrive will not be coming from above.
Case 3. Exactly one elevator is above the person.
Since either elevator may be above the person, this case occurs with probability $2 \times \frac{1}{5} \times \frac{4}{5} = \frac{8}{25}$. Either elevator may be on the way up or on the way down. Hence the elevator above is within two floors of arrival while the elevator below is within eight floors of arrival. In the latter case, it is beyond two floors of arrival $\frac{3}{4}$ of the time. In the remaining $\frac{1}{4}$ of the time, it has equal probability of arriving before or after the other elevator. Hence the probability in this case is $\frac{8}{25} \times \frac{7}{8} = \frac{7}{25}$.
In summary, the final answer is $\frac{1}{25} + \frac{7}{25} = \frac{8}{25}$.

11. The total number of ways of distributing the 26 missing cards between West and East is $\binom{26}{13} = 10400600$.

   (a) Suppose West holds 3 Spades and 10 non-Spades. This can happen in $\binom{6}{3}\binom{20}{10} = 20 \times 184756 = 3695120$ ways. Suppose West holds 4 Spades. This can happen in $\binom{6}{4}\binom{20}{9} = 15 \times 167960 = 2519400$ ways. There are also the same number of ways for East to hold 4 Spades. Together, there are $2 \times 2519400 = 5038800$ ways. Thus there are only $10400600 - 3695120 - 5038800 = 1834640$ ways for all other distributions. Hence a 4-2 distribution is the most likely.

   (b) Suppose West holds 3 Spades and 10 non-Spades. This can happen in $\binom{5}{3}\binom{21}{10} = 10 \times 352716 = 3527160$ ways. There are also the same number of ways for East to hold 3 Spades. Together, there are $2 \times 3527160 = 7054320$ ways, which exceed half the overall total. Hence a 3-3 distribution is the most likely.

   (c) Suppose West holds 2 Spades and 11 non-Spades. This can hap pen in $\binom{4}{2}\binom{22}{11} = 6 \times 470288 = 2821728$ ways. Suppose West holds 3 Spades. This can happen in $\binom{4}{3}\binom{22}{10} = 4 \times 646646 = 2586584$ ways. There are also the same number of ways for East to hold 3 Spades. Together, there are $2 \times 2586584 = 5173168$ ways. Thus there are only $10400600 - 2821728 - 5173168 = 2640848$ ways for a 4-0 distribution. Hence 3-1 distribution is the most likely.

   (d) Suppose West holds 2 Spades and 11 non-Spades. This can happen in $\binom{3}{2}\binom{23}{11} = 3 \times 1352078 = 4056234$ ways. There are also the same number of ways for East to hold 3 Spades. Together, there are $2 \times 4056234 = 8112468$ ways, which exceed half the overall total. Hence a 2-1 distribution is more likely.

   (e) Suppose West holds 1 Spade and 12 non-Spades. This can happen in $\binom{2}{1}\binom{24}{12} = 2 \times 2704156 = 5408312$ ways, which exceed half the overall total. Hence a 1-1 distribution is more likely.

Section A.2.

1. Note that
$$\binom{-\frac{1}{2}}{n} = \frac{(-\frac{1}{2})(-\frac{1}{2}-1)\cdots(-\frac{1}{2}-n+1)}{n!} = \left(-\frac{1}{2}\right)^n \binom{2n}{n}.$$

Hence $(1 + (-4x))^{-\frac{1}{2}} \sum_{n=0}^{\infty} \binom{-\frac{1}{2}}{n}(-4x)^n = \sum_{n=0}^{\infty} \binom{2n}{n} x^n$. The sequence generated is $\{\binom{2n}{n}\}$.

3. (a) Since the generating function for each child is $x^3 + x^4 + x^5 + \cdots$, the overall generating function is

$$\frac{x^{12}}{(1-x)^4} = \sum_{n=0}^{\infty} \binom{-4}{n} (-1)^n x^{n+12} = \sum_{n=0}^{\infty} \binom{n-9}{3} x^n.$$

Hence there are no ways if the number of apples is less than 12, but if it is at least 12, the number of ways is $\binom{n-9}{3}$.

(b) There are no ways if the number of apples is less than 12, but if it is at least 12, give each child 3 and distribute the remaining apples among them. The number of ways is $\binom{n-12+4-1}{4-1} = \binom{n-9}{3}$.

5. (a) If there are $2n$ mice, Sylvester will eat all the even-numbered ones in the first round. He is then left with n mice. If the k-th mouse is the lucky one when there are n mice, then the k-th odd numbered mouse is the lucky when there are $2n$ mice. The k-th odd number is $2k - 1$, and it follows that $f(2n) = 2f(n) - 1$. If there are $2n + 1$ mice, Sylvester will eat all the even-numbered ones in the first round, with mouse number 1 as dessert. He is again left with n mice. Now the $(k+1)$-st odd-numbered mouse is the lucky one, and it follows that $f(2n + 1) = 2f(n) + 1$.

(b) Using the two formulae in (a), we have

$$\begin{aligned}
f(100) &= 2f(50) - 1 \\
&= 2(2f(25) - 1)) - 1 \\
&= 4(2f(12) + 1) - 3 \\
&= 8(2f(6) - 1) + 1 \\
&= 16(2f(3) - 1) - 7 \\
&= 32(2f(1) + 1) - 23 \\
&= 64f(1) + 9 \\
&= 73.
\end{aligned}$$

7. (a) The characteristic equation is $x^2 - 2x + 1 = 0$, which may be factored as $(x-1)^2 = 0$. Hence the general solution is given by $a_n = C_1 + C2n$. From the initial conditions, we have $C_1 = 1$ and $C_1 + C_2 = 2$. Hence $C_2 = 1$ also, and we have $a_n = n + 1$.

(b) Let $A(x)$ be the generating function for $\{a_n\}$. We have

$$\sum_{n=2}^{\infty} a_n x^n - 2x \sum_{n=2}^{\infty} a_{n-1} x^{n-1} + x^2 \sum_{n=}^{\infty} a_{n-2} x^{n-2} = 0,$$

This yields $(A(x) - 1 - 2x) - 2x(A(x) - 1) + x^2 A(x) = 0$ or

$$A(x) = \frac{1}{(1-x)^2} = \sum_{n=0}^{\infty} \binom{-2}{n} (-x)^n = \sum_{n=0}^{\infty} \binom{n+1}{1} x^n.$$

It follows that $a_n = n + 1$.

9. Let the minimum number of moves required to transfer a tower of n disks be a_n. We have $a_0 = 0$, $a_1 = 1$ and $a_2 = 3$. Suppose $n \geq 3$. To pave the way for the largest disk to move from the first peg to the fourth peg, we must have the smaller disks out of the way. To pave the move of the second largest disk from the first peg to say the second peg, the smaller disks must be assembled on the third peg. This requires a_{n-2} moves. Now we move the second largest disk to the second peg, the largest disk to the fourth peg and the second largest disk on top of the largest disk, a total of 3 moves. Then the smaller disks are transferred to the fourth peg as well, taking another a_{n-2} moves. It follows that the recurrence relation is $a_n = 2a_{n-2} + 3$. The associated homogeneous recurrence relations is $a_n - 2a_{n-2} = 0$. Setting $a_n = x^n$ for some non-zero number x, we have $x^2 - 2 = 0$, so that $x = \pm\sqrt{2}$. Thus the solution is $a_n = K_1(\sqrt{2})^n + K_2(-\sqrt{2})^n$. For a particular solution, we try $a_n = A$. Then $3 = a_n - 2a_{n-2} = A - 2A$ so that $A = -3$. Hence the general solution to the original recurrence relation $a = K_1(\sqrt{2})^n + K_2(-\sqrt{2})^n - 3$. Now $0 = a_0 = K_1 + K_2 - 3$ while $1 = a_1 = K_1\sqrt{2} - K_2\sqrt{2} - 3$. Solving these equations yields $K_1 = \frac{3+2\sqrt{2}}{2}$ and $K_2 = \frac{3-2\sqrt{2}}{2}$. It follows that

$$a_n = \left(\frac{3+2\sqrt{2}}{2}\right)\sqrt{2})^n + \left(\frac{3-2\sqrt{2}}{2}\right)(-\sqrt{2})^n - 3.$$

11. (a) Let $b_n = 2^n$ and $c_n = 3^n$. Then the generating function for $\{b_n\}$ is $B(x) = \frac{1}{1-2x}$, and the generating function for $\{c_n\}$ is $C(x) = \frac{1}{1-3x}$. Now $a_n = b_nc_0 + b_{n-1}c_1 + \cdots + b_0c_n$. It follows that $A(x) = \frac{1}{(1-3x)(1-2x)}$.

    (b) From $\frac{1}{(1-3x)(1-2x)} = \frac{C_1}{1-3x} + \frac{C_2}{1-2x}$, $1 = C_1(1-2x) + C_2(1-3x)$. Setting $x = \frac{1}{3}$, we have $1 = \frac{1}{3}C_1$ or $C_1 = 3$. Setting $x = \frac{1}{2}$, we have $1 = -\frac{1}{2}C_2$ or $C_2 = -2$. Hence $A(x) = 3\sum_{n=0}^{\infty} 3^n x^n - 2\sum_{n=0}^{\infty} 2^n x^n$. It follows that $a_n = 3^{n+1} + 2^{n+1}$.

Section A.3.

1. (a) Tie Johnny and Jackie together and treat them as a single person. Then there are 5!=120 cyclic seating arrangements. This must be multiplied by 2 since Johnny can sit on either side of Jackie after they are untied, for a total of 240 different cyclic seating arrangements.

    (b) Since there are 6!=720 cyclic seating arrangements overall, the Subtraction Principle yields $720 - 240 = 480$ cyclic seating arrangements in which Johnny and Jackie are separated.

3. Clearly there are three letter strings of length 1, namely, M, A and D. For letter strings of length 2, there are two cases. If both letters are the same, we have MM and AA. If not, there are 3 choices for the first letter and 2 for the second. Hence the total number is $2 + 3 \times 2 = 8$. For letter strings of length 3, again there are two cases. If two of the letters are the same, they may be Ms or As. The third letter may be chosen in 2 ways and can be in any of the 3 positions. This yields $2 \times 2 \times 3 = 12$ letter strings. If all three letters are different, the number is $3! = 6$. Hence the total number is $12 + 6 = 18$. For letter strings of length 4, there are also two cases. If both pairs of identical letters are used, the number is $\binom{4}{2} = 6$. If only one pair is used, it can be chosen in 2 ways, and the other two letters may be placed in $4 \times 3 = 12$ ways. Hence the total number is $6 + 2 \times 12 = 30$.

5. We have $e^{2x} - e^{-2x} = \sum_{n=1}^{\infty} 2^n (1 - (-1)^n) \dfrac{x^n}{n!}$. Hence

$$\frac{e^{2x} - e^{-2x}}{4x} = \sum_{n=1}^{\infty} 2^{n-2}(1 + (-1)^n)\frac{x^{n-1}}{n!} = \sum_{n=0}^{\infty} \frac{2^{n-1}(1 + (-1)^n)}{n+1}\frac{x^n}{n!}.$$

Thus the sequence generated is $\left\{ \dfrac{2^{n-1}(1+(-1)^n)}{n+1} \right\}$.

7. (a) The exponential generating function for the digit 0 is $\frac{1}{2}(e^x + e^{-x})$. The exponential generating function for the digit 1 is e^x. Hence the overall exponential generating function is

$$\frac{1}{2}(e^x + e^{-x})e^x = \frac{1}{2}(e^{2x} + 1) = \frac{1}{2}\left(\sum_{n=0}^{\infty} 2^n \frac{x^n}{n!} + 1 \right).$$

It follows that the sequence generated is $a_0 = \frac{1}{2} + \frac{1}{2} = 1$ and $a_n = 2^{n-1}$ for $n \geq 1$.

(b) Let the numbers be a_n. Clearly, $a_0 = 1$ for the empty sequence. For $n \geq 1$, the total number of binary sequences of length n is 2^n. We claim that exactly half of them have an even number of 0s. Each binary sequence is paired with the one which differs from it in only the first term. Clearly, in each pair, exactly one sequence has an even number of 0s. This justifies our claim, so that $a_n = 2^{n-1}$ for $n \geq 1$.

9. The exponential generating function is

$$(1 + x)\left(1 + x + \frac{1}{2}x^2\right)^2 = 1 + 3x + 4x^2 + 3x^3 + \frac{5}{4}x^4 + \frac{1}{4}x^5$$

$$= 1 + 3x + 8\frac{x^2}{2!} + 18\frac{x^3}{3!} + 30\frac{x^4}{4!} + 30\frac{x^5}{5!}.$$

The total number is $1 + 3 + 8 + 18 + 30 + 30 = 90$.

11. The total number of permutations is $100!$. If one of them contains the element 13 in a cycle of length 13, the remaining spots of the cycle may be filled in $99 \times 98 \times \cdots \times 88$ ways, and the remaining elements may be permuted in $87!$ ways. Hence the desired probability is $\frac{99 \times 98 \times \cdots \times 88 \times 87!}{100!} = \frac{1}{100}$.

Section A.4.

1. (a) Let the number of ways be a_n. Consider the boy in seat n. If he reoccupies his own seat, the others can be seated in a_{n-1} ways. Suppose he takes seat $n-1$. Then the only boy who can take seat n is the one who sits in seat $n-1$ initially. Hence they trade places, and the others can be seated in a_{n-2} ways. It follows that $a_n = a_{n-1} + a_{n-2}$. Checking initial values, we have $a_n = F_{n+l}$.

   (b) Consider the girl in seat n. If she reoccupies her own seat, the others can be seated in $a_{n-1} = F_n$ ways where a_n is as in (a). If she trades places with either neighbor, the others can be seated in $a_{n-2} = F_{n-1}$ ways. If she does not reoccupy her own seat and does not trade places with either neighbor, then all the girls must shift one seat clockwise or counterclockwise. The total number of ways is $F_n + 2F_{n-1} + 2$.

3. Suppose there exists a positive integer N with two such expressions. Let F_m be the largest Fibonacci number used in either expression, and we may assume that it appears in only one of them. Let the other expression be $N = F_{i_1} + F_{i_2} + F_{i_3} + \cdots + F_{i_{k-1}} + F_{i_k}$, with $i_2 - i_1 > 1$, $i_3 - i_2 > 1, \ldots, i_k - i_{k-1} > 1$. Then

$$
\begin{aligned}
N \;&<\; F_{i_2-1} + F_{i_2} + F_{i_3} + \cdots + F_{i_{k-1}} + F_{i_k} \\
&=\; F_{i_2+1} + F_{i_3} + \cdots + F_{i_{k-1}} + F_{i_k} \\
&\leq\; F_{i_3-1} + F_{i_3} + \cdots + F_{i_{k-1}} + F_{i_k} \\
&=\; F_{i_3+1} + \cdots + F_{i_{k-1}} + F_{i_k} \\
&\leq\; \cdots \\
&=\; F_{i_{k-1}+1} + F_{i_k} \\
&\leq\; F_{i_k-1} + F_{i_k} \\
&=\; F_{i_{k+1}} \\
&=\; F_m \\
&\leq\; N.
\end{aligned}
$$

This is a contradiction.

5. A term F_n ends in k zeros if $F_n \equiv 0 \pmod{10^n}$. Let a_i be the remainder when F_i is divided by 10^n. We claim that one of a_i is 0 for some $i \leq 10^{2n} + 1$. Consider the 10^{2n} pairs (a_1, a_2), (a_2, a_3), …. Note that the pair $(0,0)$ cannot occur. Thus there are only $10^{2n} - 1$ possible pairs. Hence one pair will repeat. So the sequence $\{a_i\}$ is purely periodic and the period length is at most $10^{2n} - 1$. The first pair to repeat is $(a_1, a_2) = (1, 1)$. If the period is $(1, 2, 3, …, a_p)$, then $a_p = 1 - 1 = 0$. Thus the term 0 will occur in the sequence as the last term of the period.

7. We have an unordered partition of a four-element set into arbitrary subsets. Hence the answer is give by the fourth Bell number, namely 15. The configurations are shown in the diagram below.

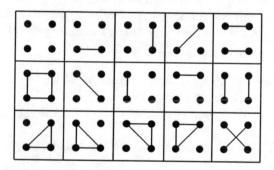

9. We may have three people who know one another, but none knows a fourth person. This shows that four people are not enough. With five people, we may assume that P knows Q. Then no two of R, S and T know each other. If any of these three does not know either P or Q, say R does not know P, then we have R does not know P and S does not know T. Otherwise, we have R knows P and S knows Q.

11. By Exercise A.4.9, with five people P, Q, R, S and T, we may assume that P knows Q while R knows S. If either P or Q knows R or S, the desired conclusion follows. If not, then P does not know R, R does not know Q and Q does not know S. Four people are not enough because we may have three people who know one another, but none knows the fourth person.

Section A.5.

1. The cycle index is $\frac{x_1^{13}+x_1x_2^6+2x_1x_4^3+2x_1^3x_2^5+2x_1^5x_2^4}{8}$. Let $x_1 = x_2 = x_4 = 3$. Then we have $\frac{3^{13}+3^7+2\times3^4+2\times3^8+2\times3^9}{8} = 206145$.

3. (a) There are 6 rotational symmetries of this solid.
   (1) The identity symmetry.
   (2) Two $120°$ symmetries about the axis joining the centers of the opposite bases.
   (3) Three $180°$ symmetries about axes joining the midpoints of the lateral edges to the center of the opposite lateral faces.

   (b) There are 5 cells each of which is a cycle of length 1 under (1). Under (2), three of them form a cycle of length 3 while the other two are cycles of length 1. Under (3), one of them is a cycle of length 1 while the other four form two cycles of length 2. Hence the cycle index is $\frac{x_1^5+2x_1^2x_3+3x_1x_2^2}{6}$.

5. (a) There are 16 rotational symmetries of this solid.
   (1) The identity symmetry.
   (2) A $180°$ symmetry about the axis joining the centers of the bases.
   (3) Two $90°$ symmetry about the axis joining the centers of the bases.
   (4) Four $45°$ symmetry about the axis joining the centers of the bases.
   (5) Four $180°$ symmetries about axes joining the centers of opposite lateral faces.
   (6) Four $180°$ symmetries about axes joining the midpoints of opposite lateral edges.

   (b) There are 10 cells each of which is a cycle of length 1 under (1). Under (2), they form two cycles of length 1 and four cycles of length 2. Under (3), they form two cycles of length 1 and two cycles of length 4. Under (4), they form two cycles of length 1 and one cycle of length 8. Under (5), they form two cycles of length 1 and four cycles of length 2. Under (6), they form five cycles of length 2 . Hence the cycle index is $\frac{x_1^{10}+x_1^2x_2^4+2x_1^2x_4^2+4x_1^2x_8+4x_2^5}{16}$.

7. (a) The planar representation is shown in the diagram below.

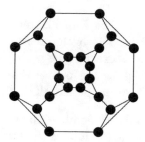

(b) The rotational symmetries of this solid are exactly the same al those of the cube. Hence the cycle index is

$$\frac{x_1^{14} + 3x_1^2x_2^6 + 6x_1^2x_4^3 + 6x_2^7 + 8x_1^2x_3^4}{24}.$$

9. Let the variables r, y, b and g represent the four colors. Then the inventory function is

$$\frac{(r+y+b+g)^6 + 3(r+y+b+g)^2(r^2+y^2+b^2+g^2)^2}{24}$$

$$+\frac{6(r+y+b+g)^2(r^4+y^4+b^4+g^4)}{24}$$

$$+\frac{6(r^2+y^2+b^2+g^2)^3 + 8(r^3+y^3+b^3+g^3)^2}{24}.$$

We are interested in the 4 terms of the form r^3ybg. Each is given by $\binom{6}{3,1,1,1} = 120$. We are also interested in the 6 terms of the form r^2y^2bg. Each is given by $\binom{6}{2,2,1,1} + 3\binom{2}{1}\binom{2}{1} = 192$. It follows that the number of coloring patterns is given by $\frac{4\times120+6\times192}{24} = 68$.

11. A direct count yields the following six distributions.
    (1) All four objects in the same box.
    (2) The three red objects in one box and the blue object in another box.
    (3) Two red objects and the blue object in one box and the other red object in another box.
    (4) Two red objects in one box and the other two objects in another box.
    (5) Two red objects in one box and each of the other two objects in a different box.
    (6) A red object and the blue object in one box and each of the other red objects in a different box.

Section A.6.

1. For any pony Q that A has bitten, there is exactly one lion that has bitten both P and Q by Exercise 6.1.6(c). For any lion B that has bitten P, there is exactly one pony that both A and B have bitten by Postulate 3. Hence the ponies that A has bitten and the lions that have bitten P can be paired off, and their numbers are equal.

3. We first multiply all numbers in the first Latin square by 5, resulting in the Latin square below on the left. Then we add the corresponding numbers of this Latin square to the second Latin square, and obtain the magic square below on the right.

$$
\begin{bmatrix}
0 & 5 & 10 & 15 & 20 \\
5 & 10 & 15 & 20 & 0 \\
10 & 15 & 20 & 0 & 5 \\
15 & 20 & 0 & 5 & 10 \\
20 & 0 & 5 & 10 & 15
\end{bmatrix}
\begin{bmatrix}
0 & 6 & 12 & 18 & 24 \\
8 & 14 & 15 & 21 & 2 \\
11 & 17 & 23 & 4 & 5 \\
19 & 20 & 1 & 7 & 13 \\
22 & 3 & 9 & 10 & 16
\end{bmatrix}
$$

5. The constructions are given in the diagram below.

7. The constructions are given in the diagram below.

A	B	B	F
A	A	B	F
A	B	B	F
A		F	F

B	B	C	D
B	D	C	D
B	B	C	D
C		C	D

C	E	C	A
C	E	E	A
C	E	A	A
C		E	A

A	A	A	A
Q	A	R	R
Q	Q	R	R
Q		Q	R

B	B	R	R
P	B	R	R
B	B	R	P
P		P	P

C	C	C	C
P	P	Q	P
P	Q	Q	C
Q		Q	P

D	D	D	D
P	E	E	E
E	D	E	P
P		P	P

Q	E	Q	F
E	Q	Q	F
E	E	Q	F
E		F	F

D	R	R	F
D	R	R	F
D	R	D	F
D		F	F

9. (a) The numbers in the three categories are (1) 0, 7, 14, 21 and 28; (2) 1, 4, 9, 11, 16 and 29; (3) 3, 12, 13, 17, 27 and 33. The incomplete table below serves the purpose of defining the term quadrant which is used in the solution to (b).

−	0,7,14, 21,28	1,3,4,9,11,12,13, 16,17,27,29,33
0,7,14 21,28	Second Quadrant	First Quadrant Quadrant
1,3,4,9,11,12,13, 16,17,27,29,33	Third Quadrant	Fourth Quadrant

(b) Consider first the numbers of the form $(0, n)$ with $n \neq 0$. Each row of the second quadrant of the difference table contains each of $(0, n)$ exactly once. Hence the total number of appearances of each here is $2s - 1$. These numbers do not appear at all in the first or third quadrant. In the fourth quadrant, they appear as differences of the form $(a_1, b) - (a_2, b)$. There are $2s$ choices for b and $2\binom{s-1}{2}$ choices for a_1 and a_2. The total number of such differences is $2s(s - 1)(s - 2)$. Since each of the $2s - 2$ multiples of $2s + 1$ appears the same number of times, this number is

$$\frac{2s(s - 1)(s - 2)}{2(s - 1)} = s^2 - 2s.$$

It follows that the overall number of appearances of each is given by $2s - 1 + s^2 - 2s = s^2 - 1$. Consider now the numbers of the form $(m, 0)$ with $m \neq 0$. In the first quadrant, they appear as differences of the form $(a, 0) - (a, b)$. For each number, there is only 1 choice for b and $s - 1$ choices for a, for a total of $s - 1$ appearances. The situation is analogous in the third quadrant.

In the fourth quadrant, they appear as differences of the form $(a, b_1) - (a, b_2)$. There are $2s - 2$ choices for a and $\binom{s}{2}$ choices for b_1 and b_2. The total number of such differences is $2s(s-1)^2$. Since each of the $2s$ multiples of $2s-1$ appears the same number of times, this number is

$$\frac{2s(s-1)^2}{2s} = s^2 - 2s + 1.$$

It follows that the overall number of appearances of each is given by $s^2 - 2s + 1 + 2(s-1) = s^2 - 1$.

11. First, we show that $k = 4$ is not enough. We may have the students solving the following eight groups of four problems:(1,2,3,6), (1,2,4,5), (1,3,4,8), (2,3,4,7), (1,5,6,7), (2,5,6,8), (3,5,7,8) and(4,6,7,8). Then each problem is solved by exactly four students, and no two students solve every problem between them. We now prove that $k = 5$ is enough. The average number of problems solved by each student is five. Suppose the largest number of problems solved by one student is five. If one of the other seven students solve all three remaining problems, we have the two students we seek. If not, each solves at most two of those three problems, turning in $7 \times 2 = 14$ solutions. However, the total number of solutions for those three problems is $3 \times 5 = 15$. We have a contradiction. If the largest number of problems solved by one student is greater than five, the argument is similar and much simpler.

Bibliography

- M. Aigner, Combinatorial Theory, Springer-Verlag, New York (1979).

- V. K. Balakrishnan, Combinatorics, Schaum's Outlines, McGraw-Hill (1995).

- A. Benjamin & J. Quinn, Proofs that Really Count, Dolciani Series #27, MAA, Washington DC (2003).

- N. Biggs, Discrete Mathematics, Oxford University Press, (2003).

- K. Bogart, Introductory Combinatorics, Harcourt/Academic Press, San Diego (2000).

- M. Bona, Combinatorics of Permutations, Chapman and Hall/CRC, Boca Raton, (2004).

- M. Bona, Introduction to Enumerative Combinatorics, McGraw-Hill Higher Education, Boston (2007).

- R. A. Brualdi, Introductory Combinatorics, 4th ed. Pearson Prentice Hall, New Jersey (2004).

- P. Cameron, Combinatorics: Topics, Techniques, Algorithms, Cambridge University Press, Cambridge (1994).

- C. C. Chen & K. M. Koh, Principles and Techniques in Combinatorics, World Scientific (1992).

- D. Cohen, Basic Techniques of Combinatorial Theory, Wiley, New York (1978).

- Susanna Epp, Discrete Mathematics with Applications (third edition), Thomson-Brooks/ Cole, Belmont, CA (2004).

- Ronald Graham, Donald Knuth, and Oren Patashnik, Concrete Mathematics: a foundation for computer science, Addison-Wesley, Reading (1989).

- T. Koshy, Fibonacci and Lucas Numbers with Applications, Wiley, New York (2001).

- T. Koshy, Catalan Numbers with Applications, Oxford University Press, (2009).

- C. L. Liu, Introduction to Combinatorial Mathematics, McGraw-Hill (1968).

© The Author(s), under exclusive license to Springer Nature Switzerland AG 2021
S. W. Golomb, A. Liu, *Solomon Golomb's Course on Undergraduate Combinatorics*,
https://doi.org/10.1007/978-3-030-72228-9

- N. Loehr, Bijective Combinatorics, Chapman & Hall/CRC, Boca Raton (2011).

- L. Lovasz, Combinatorial Problems and Exercises, American Mathematical Society (1979).

- J. Matousek & J. Nesetruk, An Invitation to Discrete Mathematics, Oxford University Press (2008).

- D. Mazur, Combinatorics: A Guided Tour, MAA, Washington DC (2010).

- Percy MacMahon, Combinatory Analysis, AMS Chelsea Publishing, Providence (1983).

- I. Niven, Mathematics of Choice, New Mathematical Library #15, MAA Washington DC (1965),

- Marko Petkovsek, Herb Wilf, and Doron Zeilberger, A=B, A K Peters Ltd., Wellesley, MA (1996).

- John Riordan, An Introduction to Combinatorial Analysis, Wiley, New York (1958).

- Fred Roberts and Barry Tesman, Applied Combinatorics, Prentice Hall, Upper Saddle River, NJ (2005).

- H. J. Ryser, Combinatorial Mathematics Mathematical Monograph, #14 MAA, Washington DC (1963).

- Richard Stanley, Enumerative Combinatorics (2 volumes), Cambridge University Press, Cambridge (1997 and 1999).

- Dennis Stanton and D. White, Constructive Combinatorics, Springer-Verlag, New York (1986).

- Alan Tucker, Applied Combinatorics, John Wiley and Sons, New York (2002).

- J. H. van Lint and R. M. Wilson, A Course in Combinatorics, Cambridge University Press, Cambridge (1992).

- N. Vilenkin, Combinatorial Mathematics, Academic Press, New York (1972).

- W. D. Wallis, Introduction to Combinatorial Designs, 2nd ed. Chapman & Hall/CRC, Boca Raton (2007).

- W. D. Wallis & J. C. George, Introductino to Combinatorics, Chapman & Hall/CRC, Boca Raton (2011),

Index

Printed in the United States
by Baker & Taylor Publisher Services